T0331882

Composites Science, Technology, and Engineering

Understand critical principles of composites, such as design of durable structures, choice of fibre, matrix, manufacturing process, and mechanics with this interdisciplinary text. The book features up-to-date coverage of hybrids of fibres and particles and explanation of failure criteria, and includes a comprehensive discussion on choice of fibres, matrices, manufacturing technology, and micromechanics for durable composite structures. It provides the structure and properties of reinforcing fibres, particulates, and matrices together with a discussion of fracture mechanics. This is an essential guide for scientists and engineers wishing to discover the benefits of composite materials for designing strong and durable structures.

Frank R. Jones was Emeritus Professor of Polymers and Composites at the University of Sheffield.

Composites Science, Technology, and Engineering

FRANK R. JONES
University of Sheffield

CAMBRIDGE
UNIVERSITY PRESS

University Printing House, Cambridge CB2 8BS, United Kingdom

One Liberty Plaza, 20th Floor, New York, NY 10006, USA

477 Williamstown Road, Port Melbourne, VIC 3207, Australia

314–321, 3rd Floor, Plot 3, Splendor Forum, Jasola District Centre, New Delhi – 110025, India

103 Penang Road, #05–06/07, Visioncrest Commercial, Singapore 238467

Cambridge University Press is part of the University of Cambridge.

It furthers the University's mission by disseminating knowledge in the pursuit of
education, learning, and research at the highest international levels of excellence.

www.cambridge.org
Information on this title: www.cambridge.org/9781107036123
DOI: 10.1017/9781139565943

First published 2022

A catalogue record for this publication is available from the British Library.

Library of Congress Cataloging-in-Publication Data
Names: Jones, Frank R., 1944– author.
Title: Composites science, technology, and engineering / Frank R. Jones, University of Sheffield.
Description: Cambridge, United Kingdom ; New York, NY : Cambridge University Press, [2022] |
    Includes bibliographical references and index.
Identifiers: LCCN 2021041793 (print) | LCCN 2021041794 (ebook) | ISBN 9781107036123 (hardback) |
    ISBN 9781139565943 (epub)
Subjects: LCSH: Composite materials. | BISAC: TECHNOLOGY & ENGINEERING /
    Materials Science / General
Classification: LCC TA418.9.C6 J587 2022 (print) | LCC TA418.9.C6 (ebook) | DDC 620.1/18–dc23
LC record available at https://lccn.loc.gov/2021041793
LC ebook record available at https://lccn.loc.gov/2021041794

ISBN 978-1-107-03612-3 Hardback

Additional resources for this publication at www.cambridge.org/CompositesScience.

Contents

Preface

Composite materials have matured during my scientific career. When I joined Scott Bader and Co. in 1960 as a laboratory technician, glass fibre or fibreglass was a novel material for contact moulding using hand lay-up processing. Typically, boat hulls were manufactured using this technology. However, with the discovery of high-strength carbon fibre in the early 1960s by William Watt and colleagues at the Royal Aircraft Establishment, carbon-fibre plastics or CFRP stimulated the development of reinforced plastics or composites because the reinforcement enabled high performance and low weight to be achieved together. In parallel, Akio Shindo in Japan had also prepared strong carbon fibres, but the properties were inferior. It is considered that the constraint used in Farnborough during carbonization was critical. Metal matrix (MMC) and ceramic matrix (CMC) composites also benefited from the discovery of fine (small diameter) high-strength fibres that could resist the process temperatures. Boron fibres were first reported by C. P. Talley in 1959. In 1978, S. Yajima reported the synthesis of 'silicon carbide' fibres from a polymeric precursor, which encouraged the development of CMCs. Derek Birchall showed in the early 1980s how fine alumina fibres could be spun from an aqueous sol-gel precursor. These discoveries stimulated interest in metal reinforcement. Prior to these developments, boron fibres and silicon carbide fibres were available from chemical vapour deposition onto tungsten wire, and as such meant that the fibres had diameters of 2–300 μm, which limited composite manufacturing processes and shapes. We should not forget that fibrous reinforcements have been used since biblical times, when straw-reinforced clay was used for bricks.

This book has been written as a student text and concentrates on the needs of individual scientists, technologists, and engineers, providing a wide-ranging interdisciplinary understanding. It is not intended to provide a detailed engineering solution for complex loading arrangements.

The book attempts to provide scientists and engineers with the understanding of the choice of materials and fabrication techniques. To achieve this, a brief historical perspective is given in Chapter 1, and descriptions of available fibres and matrices are given in Chapters 2 and 3. In Chapter 2, the discussion aims to provide the advantages and disadvantages of use of differing fibres. Particulate reinforcements are also reviewed. In Chapter 3, available resins are discussed in the context of their properties and curing chemistry. This is presented in a style intended to inform technologists about how to select the polymer or resin or other matrix materials.

The text explains why control of curing schedules is essential for achieving the required performance. The author recognizes that metal and ceramic matrices are also used in specific applications, but polymeric materials provide the range of potential uses. As a result, the choice of polymeric matrix dominates the discussion.

Chapter 4 reviews the choice of manufacturing methods and explains the technological need to compromise between mechanical properties and performance and shape. We see that fibres provide high stiffness and strength in the direction of their alignment, which limits the shapes that can be achieved. Moulded artefacts will therefore tend to employ a random distribution of fibres and hence have a poorer mechanical performance than a flat lamina with unidirectional continuous fibres.

Chapter 5 describes the micromechanics of failure of composites with simple arrangements ranging from continuous to discontinuous fibres. We describe the models used to understand stress transfer between matrix and fibre, explaining the role of interfacial bond strength and how this influences the fracture behaviour.

Chapter 6 extends the understanding to laminates. The fracture processes of a laminate as a function of ply angle are discussed. The contributing factors such as thermal strains are also included. The relationship between multiple matrix cracking and transverse cracking in laminates is described. The accumulation of damage under complex loads such as impact is also described. The cascade of damage under impact is used to explain the phenomenon of barely visible impact damage (BVID). The book is intended to be used as a textbook, so simple stress transfer models are employed. It is beyond the scope of the text to extend the description to the complex stress situations that exist in loaded laminates. Thus, edge stresses and the statistical models of transverse cracking are not included. Consideration of complex stresses arising in laminated materials under load is adequately covered in a number of other texts.

Chapter 7 describes the fatigue loading of laminated composites and shows how the load–life curves are affected. The basics of life prediction are considered with regard to the fibre–matrix combinations and the lay-up of the laminate. Clearly this is a complex problem, and more detailed discussions are outside the scope of this text.

Chapter 8 describes the environmental factors crucial to the employment of a particular composite system in a specific application. Thus, moisture absorption by matrix resins under a variety conditions is considered. Resins, which are more prone to high moisture absorption, are discussed. The design of a durable glass-reinforced plastic artefact for corrosive environments is also described.

Chapter 9 considers joining, repair, self-healing, and recycling strategies, providing up-to-date potential for the use, repair, and recovery of composite materials.

Chapter 10 describes a number of case histories and future applications. Interestingly, confidence in applications in aerospace has initiated many applications in automotive and civil engineering, which are discussed. The aim of this chapter is to provide the reader with information about future applications.

The text includes a number of 'Discussion Points', which are designed to prompt the reader to re-read in detail the appropriate sections. The 'Solutions' are available to lecturers at www.cambridge.org/CompositesScience.

Frank R. Jones was Emeritus Professor of Polymers and Composites at the University of Sheffield, UK. He studied chemistry and polymer technology while working at Scott Bader and Co., a pioneering company in reinforced plastics. The company encouraged him to undertake post-graduate studies for a PhD. After post-doctoral research and a spell as a teacher in a technical college, he joined the University of Surrey, where as a lecturer he began research into environmental and interfacial aspects of composites. A move to the University of Sheffield enabled the broadening of research into interdisciplinary studies in the fracture of composite materials. He was a pioneer in research linking matrix chemistry and structure with mechanical performance and composites failure using surface analytical techniques. He was a World Fellow of the International Committee of Composite Materials.

1   Introduction

Fibrous reinforcement of materials has been employed over many centuries to increase performance. Many early plastics materials of the late nineteenth and early twentieth centuries relied on 'fibrous' inclusions, while the development of glass fibres for polymer reinforcement in the 1930s introduced the material known as fibreglass. Eventually, with the development of boron fibres for metal reinforcement and the discovery of high-strength carbon fibres in 1964, the term *composites* came into general use. More recently, carbon nanotubes and related materials and graphene have led to the development of nano-composites. The Composites Age has arrived.

1.1   Historical Context

Composite materials have played their part in structural artefacts since biblical days. At that time straw was used to reinforce clay in bricks. This was an early form of natural fibre composite. More recently, the term 'composite' arose from the discovery of high-strength fibres (boron, carbon) in the mid-1960s. Prior to that, the terms 'glass fibre', 'fibreglass', and 'GRP' (glass-reinforced plastic) were used to describe reinforced plastics, which are now referred to as composite materials. While glass fibres have been known for centuries (the first glass-fibre textile was shown in Paris in 1713; Napoleon's coffin was drawn by a horse adorned with glass-fibre cloth in 1840; in Sheffield, UK, the second Mrs W.E.S. Turner's wedding dress and trousseau, made from glass fibre in 1943, are exhibited [1]), it was not until the 1930s that continuous fibres suitable for reinforcing plastics were manufactured, in Newark, Ohio in the USA [2–4]. Suitable resins for supporting the strong fibres were developed in the late 1940s and the 1950s. The most important resins for this application were the unsaturated polyester resins because their low viscosity enabled the fibres to be readily impregnated by hand. Cold curing with the use of catalysts and accelerators is possible. Control over the impregnation time was possible by choosing the concentrations of catalysts and accelerators. The curing regime could be tailored to the manufacturing process, with hot or cold curing options.

One of the earliest applications of glass-fibre reinforced plastics was the hand-lay manufacture of boat hulls [5–7]. Figure 1.1 shows a boat hull produced in this way. With the development of the industry, more advanced structures were possible, including naval ships such as the minesweeper shown in Figure 1.2.

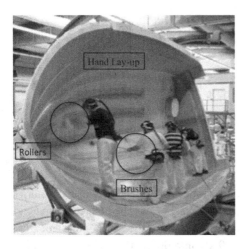

Figure 1.1 A GRP boat during manufacture (courtesy of www.bavariayachts.com).

Figure 1.2 GRP minesweeper: an example of scale-up of early techniques for larger ships.

However, the use of reinforcements in plastics to improve strength and stiffness goes back to the beginnings of the plastics industry. While early plastics were developed in the mid-1800s (celluloid in 1862 was an early mouldable material), the first truly synthetic plastic, Bakelite, discovered by Leo Baekeland in 1909, led to the development of reinforced plastics. Phenolic or phenol formaldehyde plastics could be moulded into complex shapes, but the material needed to be reinforced with a variety of particulate and fibrous fillers. Figure 1.3 shows an early mass-produced moulding for a TV set. While this represents an early use of a discontinuous fibre-reinforced plastic, the foundations of the concept of lamination

Figure 1.3 Phenolic composite (Bakelite) application: the casing of an early mass-produced TV (1953).

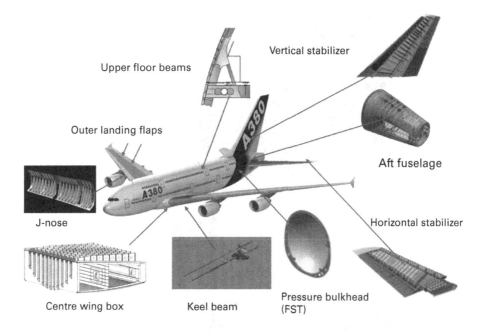

Figure 1.4 Composite applications in the Airbus A380 commercial aircraft [8].

came, in 1904, with the use of phenolic-resin-impregnated paper for electrical insulators. Further, celluloid 'dopes' were laminated and cut to create plastic materials that imitated 'tortoise shell'.

More recently, glass- and carbon-fibre based composites have gained the confidence of structural engineers, who have introduced large volumes into commercial airliners. The Airbus A380 comprises 50% composites by volume in its structure. The fuselage employs the material 'Glare', which is a glass-fibre/aluminium laminate, while the centre wing box was the largest carbon-fibre structural composite in use at the time it was introduced. The applications are summarized in Figure 1.4. In 2013, the Boeing 787 Dreamliner was launched with composites comprising 90% by volume (>50% by weight), of which the fuselage and wings were manufactured from

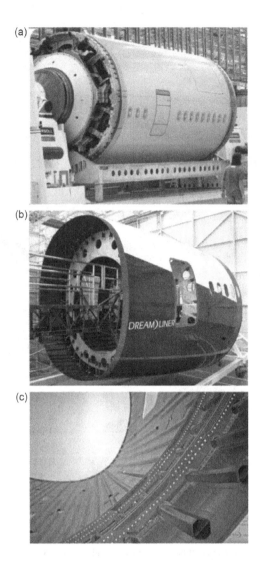

Figure 1.5 Fuselage of the Boeing 787 Dreamliner made from carbon-fibre composite. (a) Constructed on a rotating mandrel using automated tape laying; (b) the resulting monocoque shell has (c) built-in internal longitudinal stiffeners. The highly integrated internal structure requires fewer fasteners than conventional airframes [9].

carbon fibre. Figure 1.5 shows the section of the Dreamliner fuselage manufactured from carbon-fibre composites.

1.2   Definition of a Composite

The range of composite materials has to be considered from the historical development of reinforced plastics and related materials. The definition embraces some natural as well as synthetic materials.

Composites materials are best described as either:

Materials in which the individual components are physically distinct but for which the mechanical properties are synergistically enhanced in combination;

or

A discrete dispersion of one phase in another, to provide a superior performance.

These definitions embrace a number of materials, as follows.

1.2.1   Natural Composites

Natural composites have evolved over time and act as an opportunity to mimic natural materials synthetically. Typically, these are wood, bone, muscle tissue, and bamboo.

1.2.2   Micro-composites

Micro-composites generally utilize reinforcements in one dimension, ranging from 0.01 to 100 μm. Typical examples are: carbon black particles in rubber for tyres and belting; short-fibre-reinforced plastics (SFRP); long-fibre-reinforced plastics, often referred to as 'carbon fibre' (CFRP) or 'glass fibre' (GRP); and straw-reinforced clay.

1.2.3   Macro-composites

In macro-composites the reinforcement is on a larger scale. This concept embraces laminates of dissimilar plies, such as plywood (a laminate of fibrous wooden boards). In conventional composites, fibres are oriented at different angles in the individual plies.

Another example of a macro-composite is steel-reinforced concrete, using reinforcing rods ('rebar') of 1–5 cm diameter (typically 2–3 cm). Currently, steel rebar is being replaced by fibre-reinforced composite rods, which also have a multiscale microstructure.

1.3   Types of Composites

Composites should be considered with respect to their structural properties. Since fibres tend to be strong in the fibre direction, once they are impregnated in a matrix to form a composite, the mechanical properties will be highly anisotropic – that is, there could be large difference in stiffness when the load is applied at different angles. This means there are many configurations of fibre arrangement available to meet the differing requirements of the structure. These might be optimum strength, stiffness, or ease of manufacture. Principally the fibres can be *continuous* or *discontinuous*, leading to a range of composite materials.

1.3.1   Laminates

In order to achieve high performance, a common arrangement of continuous fibres would be in the form of a laminate. Figure 1.6 shows a schematic of a laminate illustrating the principal loading angles. The fibres are unidirectionally aligned in each ply so that they can be arranged at different angles to one another. The 0° angle refers to plies with fibres aligned to the principal stress; 45° and 90° plies are used in the lay-up to achieve quasi-isotropy. A typical lay-up is given schematically in Figure 1.7. It is usual to arrange the plies symmetrically to ensure the artefact is not warped. Alternative lay-ups are given in Figure 1.8. A discussion of the theoretical considerations of this phenomenon is given later. This manufacturing approach leads to relatively simple 2D shapes; one way of improving the complexity of achievable shapes is to employ woven fibres in the plies.

1.3.2   Discontinuous Fibrous Materials

The arrangement of fibres discussed in Section 1.3.1 will lead to simple shapes; however, for the moulding of more complex shapes the fibres will need to flow around

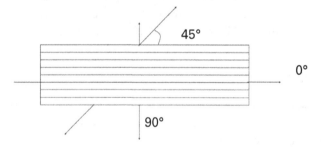

Figure 1.6 Schematic of composite unidirectional laminate structure showing definitions of loading angles.

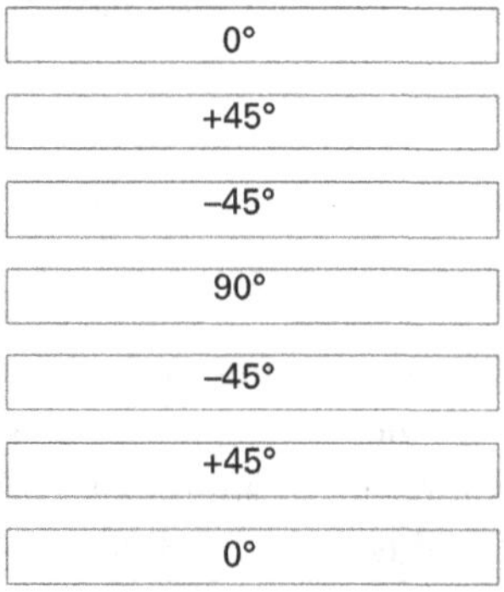

0°
+45°
−45°
90°
−45°
+45°
0°

Figure 1.7 Schematic of an expanded lay-up of unidirectional plies at differing angles in a typical quasi-isotropic laminate, defined as 0°/±45°/90°/±45°/0°, illustrating the principles employed in designing structural composites.

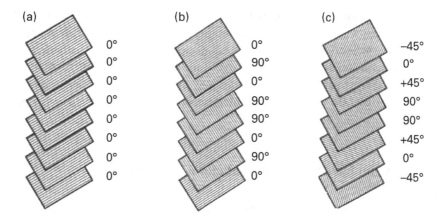

Figure 1.8 Alternative lay-ups illustrating differing laminate structures: (a) unidirectional; (b) cross-ply, $0°/90°/0°/90°_2/0°/90°/0°$; (c) quasi-isotropic, $-45°/0°/+45°/90°_2/+45°/0°/-45°$

the mould with the resin, in the case of a polymer matrix. Clearly the shorter the fibres the more easily the 'plastic' can flow around the mould. However, we shall see later that this is at the expense of the average mechanical properties of the composite material.

It can be recognized that four arrangements of fibres are possible: (1) aligned; (2) random in-plane; (3) random 3D; and (4) a mixture of these fibre configurations. Process technology is often tuned to achieve the optimum arrangement of the fibres.

1.3.3 Particulate Materials

Inorganic particles, generally referred to as fillers, are often added to matrices such as a polymer to reduce cost. This is often at the expense of mechanical performance. Some fillers are reinforcing, and these materials can be considered to be composites. Reinforcing fillers comprise the family of additives, which are either acicular (i.e. elongated in shape) or, if spherical, they will have been surface-treated to optimize the interfacial bonds between the matrix and the 'reinforcement'. The nature of such treatments is discussed in Chapter 2.

1.3.4 Multicomponent Materials

In order to ensure that discontinuous or short fibres are randomly dispersed in a moulding, it is common to include a reinforcing filler in the formulation to provide the correct flow properties. Thus, many reinforced plastics are multicomponent composite materials.

In polymer matrix composites (PMCs), the low modulus of the matrix limits the stiffness of a fibre composite at 90° to the fibres; therefore, using a reinforcing filler in the matrix phase of a laminate has many advantages.

Other multicomponent materials that can be considered as composites as similar mechanics apply are: (1) interpenetrating network polymers, which can be considered to be molecular composites and (2) dispersion-strengthened polymers and metals, where the reinforcing phase is a directionally solidified crystal.

Table 1.1 summarizes the applications and structures of these materials. In order to understand the choice of fibre arrangement we must consider a number of factors in the following discussion.

1.4   Why Do Fibres Dominate Structural Composites?

Fibres are the preferred reinforcement for composite materials. In order to understand why, we can consider the strengths of materials.

1.4.1   Theoretical Strength

According to Orowan [10,11], the theoretical strength (σ_t) of a solid is given by

$$\sigma_t = \left(\frac{E\gamma}{a}\right)^{\frac{1}{2}}, \tag{1.1}$$

$$\sigma_t \approx \frac{E}{10}, \tag{1.2}$$

where

$\gamma =$ surface free energy;
$a =$ interplane spacing in a crystal.

Therefore, the highest strength will be found in dense (i.e. high modulus) crystalline materials with covalent or metallic bonds, such as ceramics and metals.

1.4.2   Actual Strength

Bulk materials contain flaws of length c, often referred to as 'Griffith flaws'. This means the strength of a material will be significantly less than the theoretical strength as given by the Griffith equation [12]. The actual strength (σ_u) will be typically half the theoretical strength:

$$\sigma_u = \sqrt{\frac{E\gamma}{\pi c}} \approx \frac{E}{20}. \tag{1.3}$$

To maximize the strength of a material we need to maximize the modulus and reduce the number and size of flaws present in the material. Fibres achieve this because an oriented molecular structure is obtained in the drawing process. The covalent bonds are aligned to the fibre direction. This is why textile fibres have superior mechanical properties in the fibre direction.

Table 1.1. Typical applications of composites materials with fibre arrangements

Typical application	Reinforcement	Volume (%)	Manufacturing technique	Method
GRP boat hull and similar artefacts	CSM or woven mat (G) • Random in-plane • Continuous fibres	13–30	Hand lay-up	Resin impregnation by hand
GRP boat hull and similar artefacts	Chopped GF • Random in-plane	15–30	Spray-up	Spray application of chopped rovings and resin
Aircraft wing, rotor blade, or fuselage	Continuous fibres (G, C, A) • Aligned • Controlled geometry	60	Autoclave	Forming of stacked prepreg in vacuum bag
Pipes and containers	Continuous GF with possible added particulate fillers • Continuous CF/GF/AF	50	Filament winding	1. Fibres wound onto mandrel with resin 2. Prepreg tape 3. Fibre preform
Pipes	Chopped rovings (G), with particulate fillers (e.g. sand) • Random or aligned in-plane Polyester and related resins Alkali-resistant GF in cement	40–60	Centrifugal casting	Rotating cylindrical mould
I-beam or rods	Continuous fibres and/or CSM and fillers (G) • Aligned	60–80	Pultrusion	Wet resin impregnation in a die

Table 1.1. (*cont.*)

Typical application	Reinforcement	Volume (%)	Manufacturing technique	Method
Medium-sized complex mouldings (e.g. lorry body panels)	Chopped or continuous and fillers (G, A, C) • Random	10–60	Compression moulding	SMC, DMC,
Luggage/suitcases	Continuous HM thermoplastic fibres (HM PE or HM PP)			Self-reinforced composites
Medium-sized thermoplastic mouldings (e.g. car fascias)	Continuous and discontinuous fibres (G) • Random	15–50	Hot stamping of thermoplastics	Fibrous thermoplastic paper
GRP car body panels	Continuous or discontinuous fibres (G, C, A) • Random	25–30	Resin injection or transfer moulding	Woven or CSM preform
Large, complex parts (e.g. auto parts for impact resistance but with low stiffness)	Short-milled glass fibre or fibre preform (G) • Random	5–20	Reinforced reaction injection moulding	Rapid mixing for PU
Numerous small to medium domestic and automotive mouldings	Short glass fibres • Random	5–20	Injection moulding	DMC, BMC, FRTP, FRTS
Compressor blade for jet engines	Directionally crystalline titanium (MMC) Aligned CF in resin	60	Casting	Directional solidification casting Prepreg or tow placement

AF, aramid fibre; BMC, bulk moulding compound; C, carbon; CF carbon fibre; CSM, chopped strand mat; DMC, dough moulding compound; FRTP, fibre-reinforced thermoplastic; FRTS, fibre-reinforced thermoset; G, glass; GF, glass fibre, A, aramid; HM, high modulus; PE, polyethylene; PP, polypropylene; PU, polyurethane.

Figures 1.9 and 1.10 give the moduli and strengths of available reinforcing fibres in comparison to typical plastics and metals (aluminium and steel). We see that fibres have larger properties than bulk materials, especially aluminium, which is a major competitor of composites.

The highest modulus will be achieved when covalent bonds are oriented in the direction of applied stress. Graphite has a structure of stacked planar graphene sheets; since the in-plane bonding consists of efficiently linked π bonded aromatic carbon rings, the in-plane modulus is 1,000 GPa. Through thickness, the modulus is much lower since the only bonding is of a van der Waals nature. Carbon fibres or 'graphite' fibres consist of turbostratic graphite in the fibre direction, where the degree of orientation determines the actual modulus in the fibre direction. They also have a 'low' density and high specific modulus. This leads to the range of carbon fibres and their domination in applications for high stiffness at low weight. Fibres are stronger

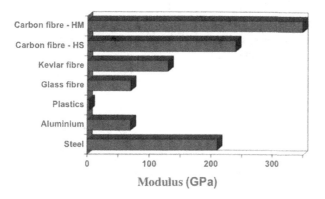

Figure 1.9 Comparison of the typical moduli of reinforcing fibres with plastics, aluminium, and steel.

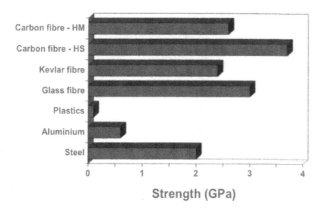

Figure 1.10 Comparison of the typical strengths of reinforcing fibres with plastics, aluminium, and steel.

because of a lower probability of a critical flaw being present compared to a bulk material. Flaws introduced into the surface of a fibre will cause a reduction in strength. Therefore, fibres need to be coated with a protective material, usually polymeric and referred to as a 'size'.

Fibres do not have a single strength, but exhibit a range of strengths. For example, glass fibres will exhibit a distribution of strengths characterized by flaws of different size. As shown in Figure 1.11 [13–17], three levels of strength are observed, which are typical of three characteristic flaw sizes. The strength populations are σ_1, which is the population with the lowest strength; σ_2, that of intermediate strength, and σ_3, that with the highest strength. σ_2 was considered by Bartenev and Izmailova [14,15] to be characteristic of flawless glass, while σ_3 represented glass with a tempered 10 nm layer. Metcalfe and Schmitz [18] considered that the fibres had populations of flaws of differing severity and statistically a strength of 4.5 GPa was attributed to flaws of 0.1 mm, while a strength of 3 GPa could be attributed to flaws spaced at 20 mm. An average strength of 5 GPa was considered to be representative of glass with internal flaws of 10^{-4} mm spacing and an uninterrupted surface layer. The presence of a tempered layer is considered unlikely for commercial glass fibres, which are prepared under ultra-fast cooling conditions.

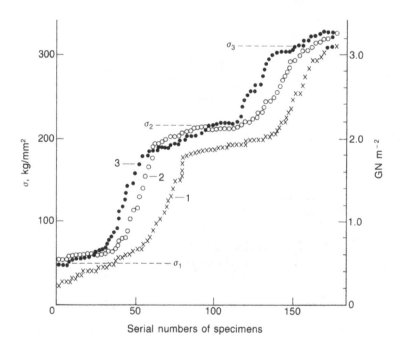

Figure 1.11 Distribution of strengths of filaments within industrial 10 μm diameter glass fibres as a function of test length. (1) 40 mm with average strength of 1.48 GPa (x). (2) 25 mm and 1.76 GPa (O) (3) 25 mm and 1.89 GPa (●), after Bartenev and Izmailova [14,15]. Reproduced from McCrum [16] and Jones [17]. This illustrates populations of differing fibre strengths.

Recent developments in nanotechnology attempt to exploit the concept of theoretical strength by ensuring the alignment of the molecular structure and reducing the density of flaws. Carbon nanotubes (CNTs) and graphene represent an opportunity to achieve reinforcement with a strength that can approach theoretical values. From a practical viewpoint, to realize this strength the CNTs need to be dispersed into a matrix with precision and at sufficiently high volume fraction. Single-walled CNTs are preferred because the graphene tube can be loaded through the interface between the wall and the matrix. Multi-walled CNTs exhibit a telescopic sliding failure and cannot be loaded to fracture. Similar practical difficulties exist with monomolecular graphene sheets. Composites based on these reinforcements therefore utilize the significant benefits in electrical and electronic performance, which arise from the perfection in graphite structure and can be exploited at volume fractions <1%. The exploitation of CNTs and related reinforcements for mechanical performance is in its infancy, since many engineering difficulties need to be solved practically. Recent work has used exfoliated graphite or exfoliated nano-clay minerals to provide plate-like reinforcement on the nano scale. The materials have largely random dispersion, so improvements in strength are not as great as might be anticipated.

1.5 The Concept of Specific Strength and Stiffness

For lightweight structures, the strength of the material is a definitive requirement, so to select the reinforcement we need to consider the densities of the fibres, reinforcing particles, and the matrix.

Many applications require lightweight materials; thus, the strength-to-weight ratio, which is referred to as the specific strength, is a useful design parameter:

$$\text{Specific strength} = \sigma_u/\rho \tag{1.4}$$

and

$$\text{Specific modulus} = E/\rho, \tag{1.5}$$

where ρ is the density.

By plotting specific tensile strength against specific modulus we can identify the materials that offer the highest performance at the lowest weight. Figure 1.12 shows this plot for a range of fibres made from differing classes of materials. The highest reinforcing effect in terms of stiffness (modulus) and strength is provided by polymeric and carbon fibres. The benefits of composite materials are demonstrated by comparison with the specific properties of bulk aluminium, which is a common metallic competitor. The disadvantage of using fibres is the need to combine them with a support matrix, which is most often a polymeric resin. This means the specific properties are lowered by the addition of the matrix.

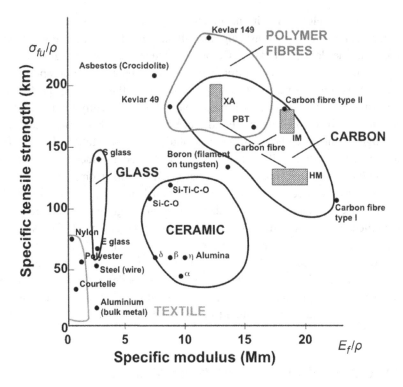

Figure 1.12 Specific strength (σ_{fu}/ρ) versus specific modulus (E_f/ρ) for a series of fibres and bulk aluminium [17].

Figures 1.13 and 1.14 compare the actual and specific moduli and strengths of plastics and composite materials. The benefits of composites over monolithic metals and plastics are demonstrated. It is shown that long, continuous, unidirectional reinforced plastic composites exhibit the highest strength at the lowest weight. The case for composites is apparently less advantageous in terms of modulus. Aluminium has a similar specific modulus to GRP. Thus, designs based on strength are ideally manufactured from composites. Both GRP and CFRP provide strength at low weight. Figure 1.15 compares the relative costs of materials at a thickness required for achieving the same strength as that of BS 4340 steel [17,18]. Figure 1.16 also provides the relative costs for achieving the same rigidity as BS 4340 steel. While the discontinuous fibrous plastics are strong competitors to the metals, the high-performance unidirectional fibre composites tend to be expensive. Therefore, unfortunately the cost of utilizing a high specific modulus material can be prohibitive. However, from a strength perspective the composite design solution can be price-competitive. This means that designing structures for rigidity at low weight may not be the best philosophy, whereas design for high strength at low weight is more efficient.

The overall cost of a manufacturing process may, however, be lower because of reduced processing costs. For example, a moulding can be manufactured in one

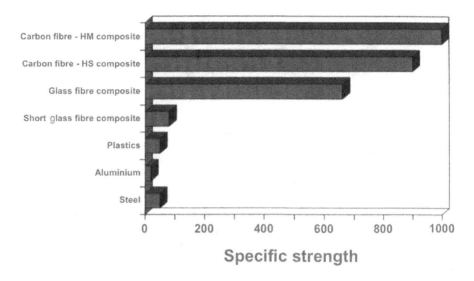

Figure 1.13 Specific strength (in 10^{-1} km) of composite materials in comparison to competitor materials.

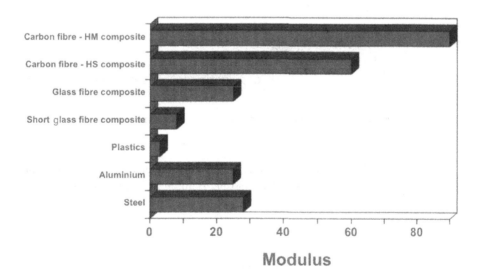

Figure 1.14 Specific modulus (in 10^{-1} Mm) of composite materials in comparison to competitor materials.

step, whereas the equivalent metal design may require assembly of several individually manufactured components. Figure 1.17 illustrates the fact that composite materials can be more expensive than traditional materials, such as metals (in this case aluminium), but the additional costs can be offset by reduced machining,

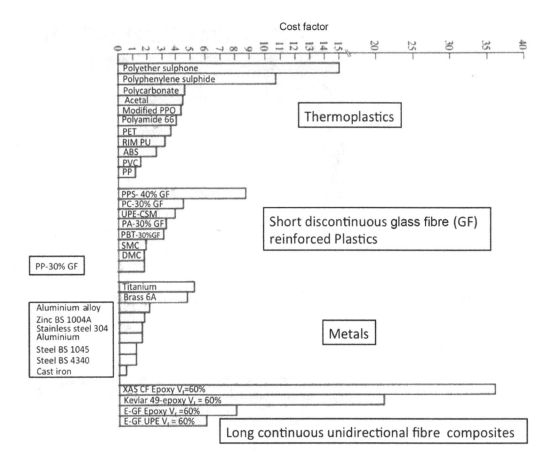

Figure 1.15 Relative costs of materials to achieve the same rigidity as BS 4340 steel [19,20].

assembly, and wastage costs. Metal artefacts can also be manufactured in new processing routes that have arisen from the competition coming from plastics and composites.

1.6   Choice of Material and Process

The selection of reinforcement and matrix for processing into a specific article has a number of interrelated requirements:

1. design and function
2. manufacturing volume
3. mechanical properties
4. complexity of shape
5. costs.

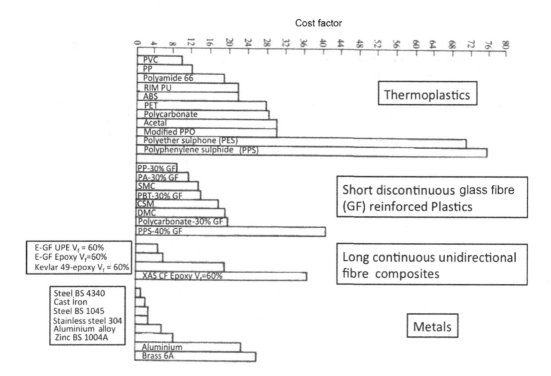

Figure 1.16 Relative costs of materials at a thickness to achieve the same strength as BS 4340 steel [19,20].

Table 1.1 provides a number of products prepared from either continuous or discontinuous glass and carbon fibres and the range of composite moulding materials. The fibres in these materials have a differing arrangement, depending on the application. High performance requires the fibres to be organized so that the principal stress is directed along the continuous fibres, which dictates that a simple shape should be designed. There is clearly a compromise between the required shape and the efficiency of reinforcement of the fibres. A complex-shaped component such as a drill housing (Figure 1.18a) or inlet manifold (Figure 1.18b) would be best manufactured by a plastics injection moulding process, but the fibres will be relatively short and randomly distributed. The performance of the material then dictates the thickness of the walls needed to provide the desired rigidity and weight. A high-performance structure such as a satellite dish requires a relatively high stiffness and a zero thermal expansion coefficient. This can be achieved using high-modulus carbon fibres arranged in a configuration that provides dimensional stability in space where the large thermal excursions in temperature occur.

Carbon fibres are mainly used in structures in which they are arranged in precise continuous geometries to maximize the efficient use of their mechanical performance. Glass fibres, on the other hand, are used in a wider range of applications and a variety of shapes. Table 1.1 shows the arrangement of glass fibres achieved in a number glass-fibre reinforced plastics. There is a great deal of diversity of fibre arrangement and

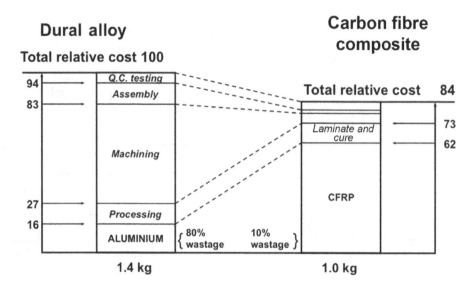

Figure 1.17 Relative costs of the manufacture of a complex component from either aluminium or a carbon-fibre composite, illustrating the different stages and costs and weight savings.

(a)

(b)

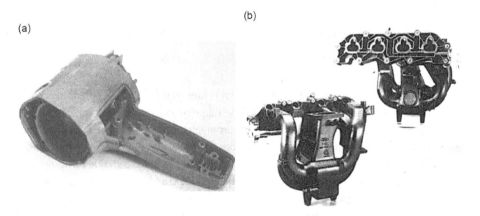

Figure 1.18 Typical complex-shaped mouldings of discontinuous composites: (a) electric drill housing; (b) inlet manifold for automotive engines.

length in this class of materials. To provide a key to the selection of material and process, here we concentrate on glass-fibre artefacts. Carbon and other polymeric fibres are less likely to be incorporated into the matrix as a short or discontinuous reinforcement because the higher cost of reinforcement inhibits their use in less efficient forms. Carbon and high-performance polymer fibres would be preferably used in a continuous form, either as tows or as woven textiles.

Table 1.1 shows that manufactured glass-fibre composites can contain fibres from <5 mm to >20 mm in length, and because of the manufacturing process they are randomly oriented within the moulded artefact. Further discussion of this is given in Chapter 4. The reinforcing efficiency of a discontinuous fibre is significantly lower than that of a continuous fibre (see Section 5.2). The length of a discontinuous fibre needs to be $>10L_c$, where L_c is the critical length for that combination of fibre and matrix. High performance is achieved by ensuring that the fibres are continuous and aligned precisely to the direction of maximum stress.

Apart from fibre alignment, the other aspect that determines the properties of a discontinuous fibre composite is the fibre volume fraction (V_f) that can be achieved with that moulding compound during manufacture. Clearly, fewer fibres can be efficiently dispersed randomly into a matrix. The benefit of short (or discontinuous) fibres is the ease with which they flow around a mould so that more complex shapes can be formed. With good fibre alignment, higher V_f can be obtained in the artefact, but only relatively simple shapes can be moulded. As a result, we can consider the selection of reinforcement and process using mechanical performance as a criterion.

In Chapter 4, a plot of the tensile modulus and strength as a function of V_f is given. The performance of a moulding is determined by the efficiency of reinforcement with fibre length and V_f. Therefore, the choice of reinforcement depends on a number of factors, which means there is a compromise between the complexity of the shape and the mechanical performance required. Complex shapes can be made using techniques such as injection moulding (see Chapter 4), whereas high performance can only be achieved with simple shapes manufactured by relatively slow techniques that ensure that fibre alignment is accurately maintained. This requires the use of prepreg technologies, which employ autoclaves for consolidation (see Chapter 4). The required volume of production and the mechanical performance needed in the product determines the choice of reinforcement. Glass fibres are relatively inexpensive and are therefore used extensively in the manufacture of complex-shaped mouldings such as casings for electric drills (Figure 1.18a). On the other hand, carbon fibres are used for simple-shaped, high-performance artefacts, such as those introduced into commercial and military aircraft (Figures 1.4 and 1.5). Advanced polymer fibres such as aramid (e.g. Kevlar, Twaron) are mostly used in specific applications where high impact strength is required.

1.7　Summary

This chapter has set the scene for the following discussion. The following aspects have been highlighted:

1. the scope of the subject with respect to the nature of materials with regard to the range of reinforcements;
2. the reasons for the choice of fibres as reinforcements for composite materials;
3. the arrangement of the fibres, which is determined by mechanical performance required and the economics of processing; and
4. the concept of specific strength and specific stiffness.

1.8   Discussion Points

1. What is the difference between the construction of the fuselage of the Airbus A380 and Boeing 787 airliners?
2. Why is the theoretical strength of a material not realized? Which *form* of material provides an opportunity to obtain values that approach those for the highest available values?
3. Which parameters best provide a selection of material from a mechanical performance perspective?
4. What aspects of a manufacturing process for composite artefacts need to be considered when choosing a material and manufacturing technology?

References

1. BBC. The glass fibre wedding dress. http://news.bbc.co.uk/local/sheffield.
2. F. R. Jones and N. T. Huff, Structure and properties of glass fibres. In *Handbook of tensile properties of textile and technical fibres*, ed. A. R. Bunsell (Cambridge: Woodhead, 2009), pp. 529–573.
3. F. R. Jones, Glass fibres. In *High-performance fibres*, ed. J. W. S. Hearle (Cambridge: Woodhead, 2001), pp. 191–238.
4. Owens Corning Heritage, The evolution of excellence: Owens Corning. Owens Corning Publication 15-GL-23185. Ward M. Canaday Center for Special Collections (1998). www.ocpreferred.com/acquainted/about/history/1930.asp.
5. Crystic, *Crystic handbook of composites* (Wollaston: Scott Bader, 2005). www.scottbader.com/uploads/files/3381_crystic-handbook-dec-05.pdf
6. B. Parkyn, Historical background. In *Glass reinforced plastics*, ed. B. Parkyn (London: Iliffe-Butterworth, 1970), pp. 3–7.
7. T. M. Wilks, Boat hulls. In *Glass reinforced plastics*, ed. B. Parkyn (London: Iliffe-Butterworth, 1970), pp. 30–41.
8. J. Pora, Composite materials in the Airbus A380: from history to future. In *ICCM 13 proceedings* (Beijing, 2001), paper 1695.
9. V. Giurgiutiu, *Structural health monitoring of aerospace composites* (London: Academic Press, 2016), pp. 1–23.
10. E. Orowan, Fracture and strength of solids. *Rep. Prog. Phys.* **12** (1949), 185–232.
11. A. Kelly, N. H. MacMillan, *Strong solids*, 3rd ed. (Oxford: Clarendon, 1990).
12. A. A. Griffith, The phenomena of rupture and flow in solids. *Phil. Trans. Roy. Soc, Lond. A* **221** (1920), 163–198.
13. G. M. Bartenev, *The structure and mechanical properties of inorganic glasses* (Groningen: Wolters-Noordhoff, 1970).
14. G. M. Bartenev, L. K. Izmailova, The structure and properties of glass fibres. *Dan SSR* **146** (1962), 1136–1138.
15. G. M. Bartenev, L. K. Izmailova, Nature of the high strength and structure of glass fibres. *Sov. Phys. Solid State* **6** (1964), 920.
16. N. G. McCrum, *Review of the science of fibre reinforced plastics* (London: HMSO, 1971).

17. F. R. Jones, Fibre reinforced plastic composites. In *Aluminium alloys: contemporary research and applications*, ed. A. K. Vasudevan, and R. D. Doherty (New York: Academic Press, 1989), pp. 605–649.

18. A. G. Metcalfe, G. K. Schmitz, Mechanism of stress corrosion in E-glass filaments. *Glass Technol.* **13** (1972), 5–16.

19. NEDO, *Replacement of metals with plastics, FG 4/1462* (London: NEDO, 1985).

20. F. R. Jones, A guide to selection. In *Handbook of polymer-fibre composites*, ed. F. R. Jones (Harlow: Longman, 1994), pp. 133–138.

2 Fibres and Particulate Reinforcements

This chapter describes the synthesis of the principal fibres and provides the range of acicular reinforcing particles, nanofibres, nanotubes, and nanosheets. The properties of the most common fibres – carbon, glass, ceramics, and natural and advanced polymers – are considered. The differing grades and their structural property relationships are also discussed. Surface treatments for adhesion and compatibility are described.

2.1 Introduction

The term *composite* now embraces a range of materials based on impregnated fibres with high performance and 'modified' plastic, ceramic, or metal matrices. The polymer matrices may also be polymer blends and contain so-called fillers, which provide some enhancement to the mechanical properties of the matrix. Fillers and reinforcing fillers can be mainly spherical particles or acicular particles, which are non-spherical with an elongated shape. But these 'reinforced' plastics can be moulded rapidly into complex-shaped artefacts. Generally, the term 'filler' refers to an agent intended to reduce the cost of the material. However, we are mainly concerned with so-called reinforcing fillers that have been surface-treated to provide adhesion to the polymer or, in the case of metal and ceramic matrices, to prevent powerful embrittling reactions at the interface that occur at the processing temperatures.

Table 2.1 shows the range of fibres available for use in high-performance composites. Of those listed, carbon, glass, and aramid fibres provide the major reinforcements for use in engineering applications. As discussed in Chapters 1 and 4, we can see that glass fibres tend to be used where cost is a critical design factor. These composites are often referred to as 'industrial' composites. Carbon fibres provide better mechanical properties at a cost penalty, and are used in 'high performance' or 'advanced' composites. Advanced polymer fibres such as the aramids introduce specialist properties such as absorption of impact energy. Other reinforcements, such as carbon nanotubes, introduce functional properties such as electrical conductivity. We discuss the major classes of fibres in detail where this understanding is essential to the choice of reinforcement for good design.

Table 2.1. Typical properties of reinforcing fibres and bulk materials [1]

Material	Grade	Relative density	Young's modulus (GPa)	Tensile strength (GPa)	Failure strain (%)	Fibre diameter (μm)
Fibres						
Glass	E	2.55	72	1.5–3.0	1.8–3.2	10–16
	S	2.5	87	3.5	4.0	12
	S2	2.49	86	4.0	5.4	10
Carbon	Mesophase Pitch	2.02	380–800	2.0–2.4	0.5	10
Carbon	PAN					
	High strength	1.8	220–240	3.0–3.3	1.3–1.4	7
	High performance	1.8	220–240	3.3–3.6	1.4–1.5	7
	High strain	1.8	220–240	3.7	1.5–1.7	7
	Intermediate modulus	1.9	280–300	2.9–3.2	1.0	5
	High modulus	2.0	330–440	2.3–2.6	0.7	7
Silicon carbide	SiCO continuous	2.5	200	3.0	1.5	10–15
	SiC whisker	3.2	480	7.0	–	1–50
	SiTiCO continuous	2.35	200	2.8	1.4	8–10
Boron		2.6	410	3.4	0.8	100
Alumina						
α-alumina	FP	3.9	380	1.7	0.4	20
β-alumina	Saffil	3.3	300	2.0	0.5	3
δ/θ-alumina	Safimax SD	3.3	250	2.0	–	3.1
η-alumina	Safimax LD	2.0	200	2.0	–	3.2
Aramid	Poly p-phenylene paraphthalamide					
	High modulus	1.47	180	3.45	1.90	12
	Intermediate modulus	1.46	128	2.65	2.4	12
	Low modulus	1.44	60	2.65	4.0	12

Table 2.1. (*cont.*)

Material	Grade	Relative density	Young's modulus (GPa)	Tensile strength (GPa)	Failure strain (%)	Fibre diameter (µm)
Aramid staple fibre	Poly m-phenylene isophthalamide	1.4	17.3	0.7	22.0	12
PBT	Poly p-phenylene benzothiazole	1.5	250	2.4	1.5	20
Polyamide 66		1.44	5.0	0.9	13.5	≈10
Polyethylene	Theoretical	–	>200	–	–	–
	Solution spun	1.0	100–120	1.0–3.0	–	–
	Drawn (Tenfor)	≈1	60	1.3	5	–
Bulk materials						
Steel		7.8	210	0.34–2.1	–	–
Aluminium alloys		2.7	70	0.14–0.62	–	–
Inorganic glass		2.6	60	-	–	–
Resins	Phenolic	1.4	7	–	≈0.5	–
	Epoxy	1.2	2–3.5	0.05–0.09	1.5–6.0	–
	Unsaturated polyester	1.4	2–3.0	0.04–0.085	1–25	–
Thermoplastics	Nylon 66	1.4	2.0	0.07	60	–
	High-density polyethylene	0.96	1.3	–	–	–
	Low-density polyethylene	0.91	0.25	–	–	–

2.2   Carbon Fibres

Carbon fibres (CFs) are manufactured from three different precursors: polyacrylonitrile (PAN) homo- and copolymers, rayon, and pitch. Rayon has been used for many years to make carbon filaments, but the discovery of the technology to convert PAN into high-strength CF [2–4] led to the growth of composites applications. Thus, industrial glass-fibre reinforced plastics (GRP) or 'glass fibre' were included in the range of composite materials, which now includes high-performance glass, aramid, and CF composites, as well as specialist products such as self-reinforced polyethylene and polypropylene. Pitch-based CF, which have a different microstructure to PAN-based fibres, have also found some niche applications.

2.2.1   Carbon Fibres from PAN Precursors

Most high-strength CFs are manufactured from PAN precursors. The starting point is a polymer fibre, which has been wet-spun from a polymer solution, or 'dope', into a fibre, which can then be drawn to provide the molecular orientation required for stabilization and graphitization. PAN is only soluble in powerful polar solvents such as dimethyl formamide (DMF), which is only removed from the spun fibre with difficulty. Historically the 'Courtelle' process for polymerization of acrylonitrile to PAN was conducted in an aqueous salt (sodium thiocyanate) solution (NaSCN), where the polymer remained in solution and could be wet-spun directly into a fibre. The PAN fibre from this technique used a copolymer containing 6 mol% methyl acrylate and 1 mol% itaconic acid. The former reduced the glass transition temperature (T_g) from 100 °C to 90 °C. This means that the copolymer fibre can be drawn in steam (at 100 °C) to the draw ratio required to ensure the carbonization achieves a strong CF. The homopolymer PAN fibre could only be drawn at higher temperatures in, for example, a glycerol bath.

The formation of CFs is described in Figure 2.1. Figure 2.2 describes the effect of the heat treatment temperature on the strength and modulus of CF achieved. Also included is the benefit associated with use of filtered dope for spinning the PAN fibres. Special acrylic fibre (SAF) differs from textile grade in three respects: the degree of

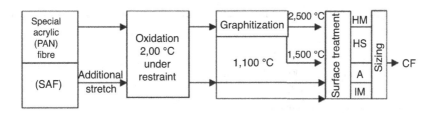

Figure 2.1 Schematic of the stages involved in the manufacture of CF from PAN precursors. A, Type A (7 μm diameter); HM, high modulus (7 μm); HS, high strength (7 μm); IM = intermediate modulus (5 μm) [8].

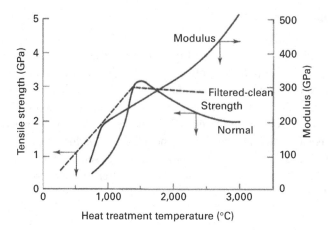

Figure 2.2 The effect of heat treatment temperature and clean-room conditions on the strength and modulus of CFs from PAN precursor [2,5–7].

draw (spin–stretch factor), the careful filtration of the polymer dope, and spinning in a clean room. In this way the number and size of strength-reducing defects in the CF are limited. This was demonstrated by Reynolds, Sharp, and Moreton [5–7], who showed that high strength could be achieved by carbonization at 1,100 °C as opposed to 1,500 °C. Carbon fibres produced at 1,500 °C were originally known as high strength, whereas those produced at 1,200 °C are referred to as Type A. High modulus (HM) CF is produced by graphitization at higher temperatures, approaching 2,500 °C. Intermediate modulus (IM) CF is formed at a similar temperature to Type A from a PAN yarn that has received an additional drawing to a higher spin–stretch factor. Watt [4] showed that the modulus of the CF was directly proportional to the spin–stretch factor of the yarn. Figure 2.3 illustrates the importance of the spin–draw factor of the PAN precursor for the tensile modulus of the CF. For HM CF, a precursor with a large spin–draw factor is required.

2.2.1.1   The Chemistry of CF Synthesis from PAN

The spin–stretch factor applied to the PAN precursor fibre is an important factor in the conversion of the polymer fibre into a high-strength CF. An understanding of the chemical changes occurring explains this observation. The process involves five stages:

1. oxidation (200–220 °C)
2. carbonization (400–600 °C)
3 graphitization (600–2,500 °C)
4. fibre surface treatment
5. sizing or coating.

The **oxidation stage** is undertaken in an oxygen atmosphere. It is critical since the molecular orientation induced in the precursor in the drawing stage is stabilized. This

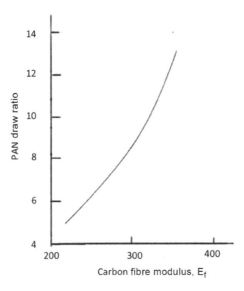

Figure 2.3 The effect of draw ratio applied to PAN precursor on the tensile modulus of CF [4].

determines the size and number of nuclei, which initiate the growth of the graphite crystals in the carbonization and graphitization stages.

The PAN polymer is atactic. That means the molecular structure contains random short sequences of syndiotactic and isotactic nitrile (CN) side-groups on the chain (Figure 2.4). An extended syndiotactic polymer chain has the CN groups on alternative sides, whereas on an isotactic chain they are on the same side. This explains why the degree of draw of the PAN fibre determines the modulus of CF because the first stage involves self-polymerization of the CN groups to create sequences of ladder polymer. Polymer molecules are normally arranged in random coils so only the extended isotactic sequences, in the drawn polymer, are in a reaction volume for thermal polymerization of the pendant CN groups to occur.

To avoid the relaxation of the extended molecular arrangement, the oxidation stage is carried out with the fibre under *restraint*. The thermal polymerization of the pendant CN groups is initiated by the comonomer, itaconic acid, as shown in Figure 2.5.

According to Watt [4], the thermal polymerization of the CN groups precedes the oxidation stage. The ladder polymer created has a structure of short sequences of 'puckered' rings. Therefore, the oxygen is considered to catalyse the conversion of the puckered ladder sequence into a planar aromatic structure. Some oxygen-containing groups are also introduced into some of the rings, which aid the condensation of these aromatic ring structures in the carbonization stage. Since the PAN precursor is atactic, with short sequences of isotactic pendant CN groups, the cyclized sequences of the ladder polymer will be limited to 5–15 groups. Thus, the nuclei for the subsequent carbonization are of similar size, with relatively low numbers of aromatic rings. It can also be envisaged that the involvement of syndiotactic CN groups will cause the molecular chains in the so-called 'oxidized PAN' to take on a curvature rather than the planar ribbon-like form.

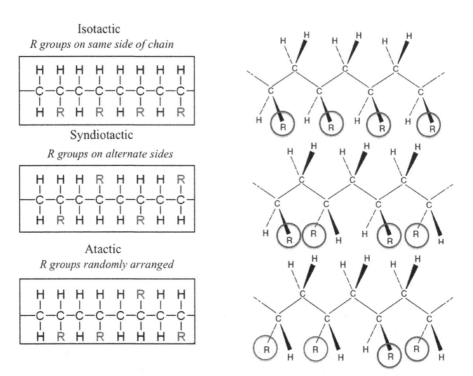

Figure 2.4 Tacticity in a polymer; isotactic is where the pendant side-groups are on one side of the fully extended chain; syndiotactic is where the side-groups alternate; atactic is where a random arrangement leads to short sequences of each. For PAN $-(CH_2\text{-}CHCN)-$, $R = CN$.

The **carbonization stage** is carried out in an inert atmosphere over the temperature range 400–600 °C. In this stage, hydrogen and water are evolved in the condensation of the aromatic ladder sequences.

The **graphitization stage** is also carried out in an inert atmosphere at temperatures of 600–1,300 °C. Further condensation occurs with the loss of nitrogen (N_2), hydrogen cyanide (HCN), and oxygen-containing molecules. The fibres obtained after heat treatment to 1,200 or 1,300 °C are usually referred to as Type A, and have high tensile strength.

Further heat treatment to temperatures as high as 2,500 °C increases the dimensions of the graphitic domains and crystals because the C atoms become mobile at around 1,100°C. The increase in the degree of graphitization leads to a higher modulus. Thus, HM fibres are usually subjected to temperatures near 2,500 °C. Unfortunately, the fibre strength tends to decrease when the graphitization temperature exceeds 1,200 °C. Johnson [11] showed that the critical flaw size for failure of CF is always larger than the measured crystallite dimensions. Thus, the strength-reducing flaws arise from the catalysed graphitization around inclusions, which leads to sufficient continuity within crystals with highly mis-oriented angles (to the fibre axis) to form a flaw of critical size. Typical inclusions are poorly dissolved polymer precursor nodules and other

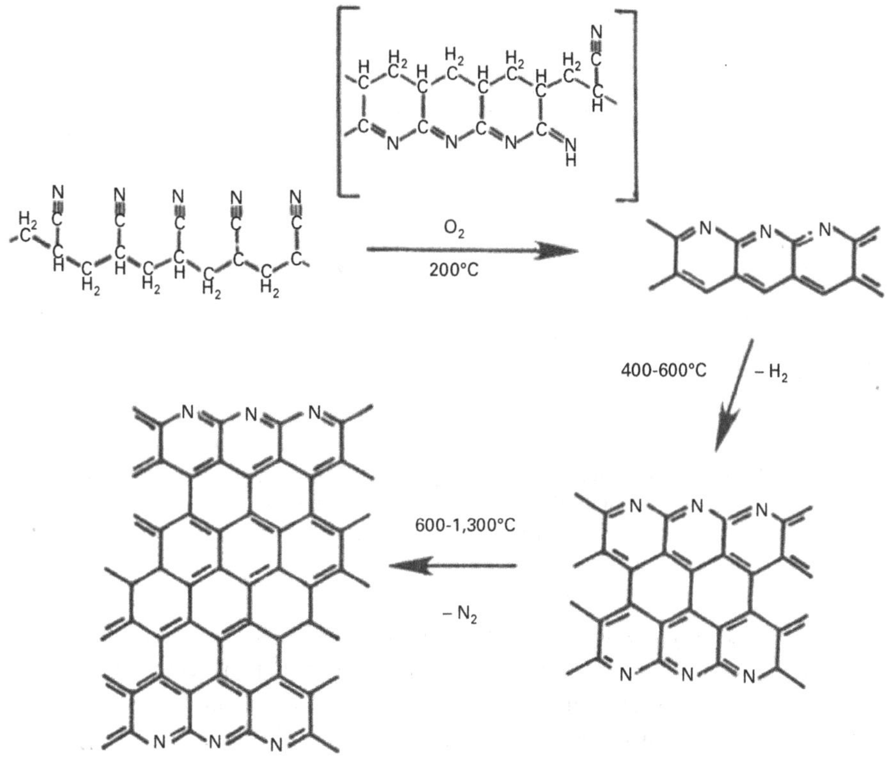

Figure 2.5 Structure of ladder polymers and carbonization structures [9,10].

foreign particulate matter. Hence, the essential use of SAF for CF production. The SAF is spun from filtered PAN dopes under clean-room conditions to reduce the probability of inclusions, which can promote the formation of continuous off-axis graphitized regions in the carbonized fibre with sufficient continuity for crack growth into a flaw of critical size. This is the explanation for the reduced strength of CFs not prepared from SAF precursor. Hence the difference between CFs synthesized from industrial and SAF precursors.

Fibre surface treatment is an essential stage in the production of CF. The graphitic structure of CF differs for high-modulus or high-strength versions. In the former the orientation of the graphene sheets to the fibre axis is more perfect than in the high-strength fibre. The theoretical modulus of graphite in the plane is 1,000 GPa, so the perfect graphite orientation is ~20% (high strength) to 40% (high modulus), so that the edges of the graphene planes exit the surface to differing degrees. The surface is complicated by the apparent core–skin structure whereby the degree of graphite perfection is higher in the skin. Thus, the surface of the fibres oxidize to differing degrees on exiting the graphitization oven. Control of the atmosphere in the process is therefore important.

The fibre surface is insufficiently polar for wetting by the matrix resins; therefore, the surface of the fibre needs to be oxidized to ensure a strong interface between fibre

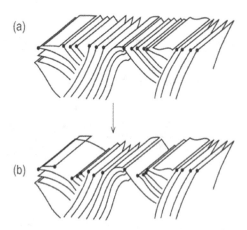

Figure 2.6 Axial cross-sections of Type A CFs showing the mechanisms of electrolytic oxidation; etching, activation, and functionalization of the edges of the graphite basal planes, which reach the fibre surface with the formation of micropores. (a) Before oxidation; (b) after oxidation [1,8,12,13].

and matrix is formed in the cured composite. Several surface treatment techniques have been described [3], but generally electrolytic oxidation is employed.

It is easier to treat HS fibres because the oxidation occurs at the edges of the graphene sheets, where they reach the surface (Figure 2.6). Also, the thin, more highly ordered skin on HS fibres can be removed to expose the graphene sheet edges where oxidation occurs. HM fibres have a much more ordered structure, with the graphene planes more highly oriented to the fibre axis. In this case, oxidation occurs at the crystalline tilt boundaries, as shown in Figure 2.7.

Once the fibres are surface-treated they are washed, dried, and coated with a protective size. This is the **fibre sizing and coating** stage. A size is a polymer, which is usually deposited from an aqueous emulsion, and is used to prevent the introduction of strength-reducing flaws during handling. It should also enable the impregnation of the fibres with resin during processing. Since many matrices for composites are based on epoxy resins (see Chapter 3), sizing polymers have similar epoxy-like chemical structures.

Figure 2.8 shows a schematic of the electrolytic surface treatment process. The important point is that the CF makes up the anode, and carbon electrodes are used as cathodes for applying a voltage. The electrolyte is generally either aqueous NH_4HCO_3 or NH_4SO_4. Morgan [3] has reviewed the literature and other aqueous electrolytes have been examined. The degree of oxidation is quantified by the current density flowing through the cell, which is typically 10–100 C m^{-2}.

In mild alkaline conditions the following half-reactions are involved. Nascent oxygen (O) forms at the anode, facilitating oxidation:

$$H_2O \rightarrow HO^* + H^+ + e$$
$$2HO^* \rightarrow H_2O + O^* \tag{2.1}$$
$$2O^* \rightarrow O_2.$$

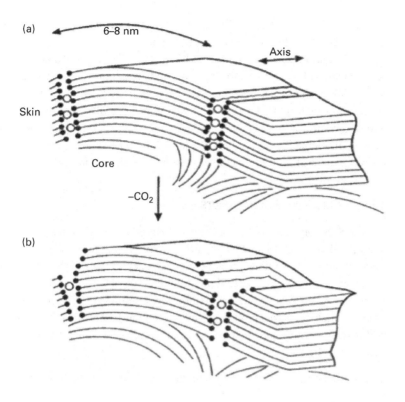

Figure 2.7 Electrolytic oxidation of HM CFs showing the formation of micropores, functionalized and adsorbent sites, by etching at crystal boundaries. • is an active site or functional group, o is an adsorbate molecule such as water. (a) Before oxidation; (b) after oxidation [12,13].

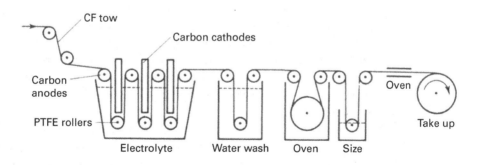

Figure 2.8 Schematic of the electrolytic oxidation surface treatment process for CF [13].

The reaction of the CF surface with O leads to the introduction of a number of functional groups: $-COOH$, $=C=O$, $-OH$, and their reaction products such as ester, anhydride, lactone, and amide ($-COONH$), which arises by reaction with the NH_3 in the electrolyte. As with other carbon surfaces, active sites will be introduced into the fibre surface during oxidation. These sites occur in micropores and will chemisorb water and also contribute to formation of the interface in a composite, as shown in Figure 2.7. The presence of nitrogen (N) in the fibre surface analysis has led to speculation that this arises from exposure and oxidation of residual CN from the precursor, but a significant fraction seems to arise from reaction of the active surface with the electrolyte [14]. Other functional groups are also reported. The complex nature of the surface-treated fibres has led to attempts to control the chemistry at the surface with alternative techniques. These include plasma treatment [15] in selected gases or the deposition of a functional plasma polymer coating [16].

2.2.2   Carbon Fibres from Pitch Precursors

While PAN copolymer and homopolymer precursors are commonly used for manufacturing high-strength CFs, other precursors have been used. Early CF filaments were prepared from cellulose fibres, but these did not achieve high strength. Pitch precursors, however, could be carbonized into high modulus CF [17,18]. Unfortunately, the strength was relatively low because of their low failure strain. Despite the low cost of petroleum pitch (a waste product from oil refining), the process required several complex stages, which added significantly to the price of the CFs. The following five stages are required:

1. transformation of isotropic pitch into anisotropic or mesophase pitch
2. melt-spinning
3. oxidative stabilization
4. carbonization at temperatures of $1,500-2,800\,°C$
5. surface treatment and sizing.

Formation of a mesophase pitch is the most important stage because it introduces the nuclei for the growth of graphitic regions in the fibres. Thus, the isotropic pitch needs to be heat-soaked for long periods for it to be converted into a liquid crystalline morphology. The process is encouraged by bubbling through an inert gas such as nitrogen to eliminate small molecular volatiles, which originate from the aliphatic pendant groups on the aromatic elements. This occurs with an increase in the molecular weight of the planar polyaromatic moieties, which crystallize to form mesophase regions.

Melt-spinning of the pitch into fibres is possible once the optimum mesophase content is achieved. The fibres are melt-spun with a diameter of 100–500 µm and then drawn down to 10–30 µm. The presence of the crystalline polyaromatic regions embedded in the amorphous 'matrix' means that the spinning involves the organization of these disc-like regions. Therefore, the transverse morphology of CFs obtained from these precursors is strongly determined by shear forces developed in the die. As a result, the radial textures of the graphitic regions can be classified as either random, radial, or circumferential or onion-skin. Figure 2.9 illustrates the structures of pitch-based CFs. In reality, a wide variety of hybrid structures will result. The viscosity of the mesophase pitch will determine the shear forces in the die and hence the structure

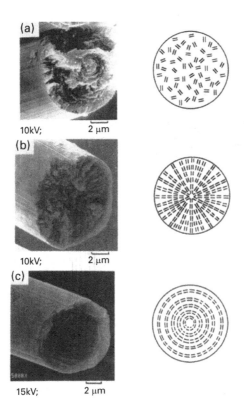

Figure 2.9 Typical cross-section textures of CFs spun from pitch precursors, illustrating the crystalline arrangements achieved in drawn liquid crystal fibres: (a) random; (b) radial; (c) circumferential. This is also applicable to other liquid crystal spun fibres such as aramid [19].

and mechanical properties of the CFs. Since pitch has a variable structure, it is clear that the effectiveness of the transformation stage will determine the final CF properties.

Oxidative stabilization of the melt-spun fibres is required before the fibres can be carbonized. Morgan [3] reports that temperatures in the range of 275–350 °C are needed for gas-phase oxidation. The as-spun fibre is extremely brittle, with low strength, so oxidation is used to introduce crosslinks into the molecular structure to render it infusible and handleable in the graphitization stage. Oxygen is chemically incorporated into the molecular structure as crosslinks and stabilizes the orientational order of the crystalline polyaromatic domains. This order is retained in the CF after carbonization.

Carbonization occurs at temperatures of 1,500–2,800 °C. On heating to this temperature, oxygen incorporated during stabilization is evolved as molecular fragments with the remaining pendant aliphatic hydrocarbons resulting from the continued polymerization of the polyaromatic elements. As shown in Figure 2.10, above 1,500°C the planar polyaromatic layers grow and become extensive, but the graphitic structure is distorted by lattice defects and special dislocations parallel to the layer

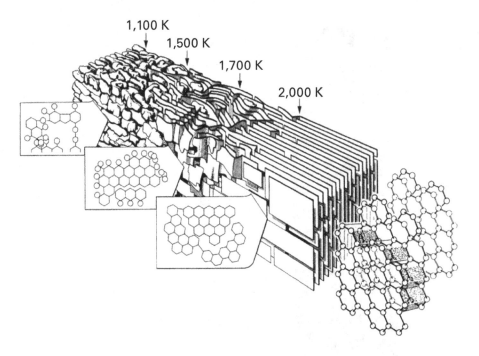

1,100 K

1,500 K

1,700 K

2,000 K

Figure 2.10 Development structure occurring in mesophase CF during graphitization at different temperatures [17,20].

planes and is turbostratic. Above 2,000°C, the C atoms are sufficiently mobile for true graphitization to take place, producing extensive regions of graphitic structure. At the higher temperatures the extent of graphitization increases and fibres with modulus of 894 GPa, approaching the theoretical value for in-plane graphite of 1,060 GPa, can be achieved.

Surface treatment and sizing: the literature and information on surface treatment is limited. Since the graphitic structure is more extensive, the oxidation chemistry needs to be more vigorous. Furthermore, the transverse strength of the fibres is lower than that of PAN-based CF, so interfacial failure could occur within the fibre surface even with a mild surface treatment. A weak interfacial bond could be beneficial.

2.2.3   Interface and Interphase in CF Composites

The **interface** refers to the distinct region between the fibre surface and the matrix, which is usually a cured resin, where there is an instantaneous change in properties. In the early development of PAN-based composites the electrolytic oxidation was used to change the micromechanics of failure at the interface from complete debonding to partial debonding at a fibre-break. With a strong interfacial bond the fibre-break propagates across the interface into the matrix. The latter composites fail in a brittle

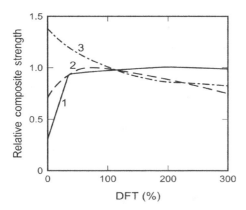

Figure 2.11 The effect of the degree of fibre surface treatment (DFT) on the mechanical performance of Type II CF composites. DFT is defined as 100% from the optimal performance. (1) Interlaminar shear strength of a unidirectional composite; (2) impact strength; (3) notched tensile strength of a $(0°/\pm45°/0°)_S$ laminate [13,21].

manner. Therefore, the degree of fibre surface treatment was optimized to maximize the impact strength, where partial debonding of the interface dominates the micro-mechanics. This is illustrated schematically in Figure 2.11 [2,21].

Currently, all CFs will be coated with a polymer size after fibre surface treatment. The surface oxidation treatment creates a highly 'active' carbon surface so that the polymer coatings are strongly adsorbed. Consequently, during composite manufacture, the sizing will dissolve to differing degrees into the matrix while remaining bonded to the fibre surface, and a 'bonded interphase' will form. With matrix-compatible sizings this is a **graded interphase**; with low compatibility the interphase can be referred to as **distinct**. Figure 2.12 illustrates the differing interphase structures, which were confirmed from a comparison of the interfacial characteristics of CF composites from the thermoplastic matrix, polyether sulphone (PES), with that from an epoxy resin matrix [22].

2.2.4 Mechanical Properties of Carbon Fibres

The range of manufacturers of CFs and the available grades is extensive. Within the group of PAN-based fibres the modulus varies from 230 to >500 GPa. Rather than provide a comprehensive list of the mechanical properties of the differing grades, we will provide data from two major manufacturers. A comprehensive set of data is given elsewhere by Morgan [3]. Table 2.2 gives the properties of the range of fibres available from Hexcel, while Table 2.3 details the range of fibres available from Toray. The author apologizes to other manufactures for the selective choice of data. The choice illustrates that the maximum modulus obtained from PAN precursors is 588 GPa (Table 2.3), which has been achieved by additional drawing of the precursor with a consequence of a diameter of <5 μm and reduced strength.

Table 2.2. The mechanical properties of CFs manufactured from PAN precursors by Hexcel (tradename HexTow) [24]

Designation	Diameter (μm)	Density (g cm^{-3})	Modulus (GPa)	Strength (GPa)	Failure strain (%)	Tow filament no. (k)
AS4	7.1	1.79	231	4.62	1.8	3,6,12
	7.1	1.79	231	4.41	1.7	
AS4C	6.9	1.78	231	4.65	1.8	3
		1.78	231	4.45	1.7	6
		1.78	231	4.48	1.8	12
AS4D	6.7	1.79	241	4.83	1.8	12
AS7	6.9	1.79	248	4.9	1.7	12
IM2A	5.2	1.78	276	5.31	1.7	12
IM2C	5.2	1.78	296	5.72	1.8	12
IM6	5.2	1.76	279	5.72	1.9	12
IM7	5.2	1.78	276	5.52	1.9	6
	5.2	1.78	276	5.65	1.9	12
IM8	5.2	1.78	310	6.07	1.8	12
IM9	4.4	1.80	304	6.14	1.9	12
IM10	4.4	1.79	310	6.96	2.0	12
HM63	4.9	1.83	441	4.68	1.0	12

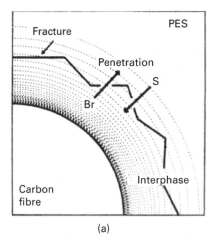 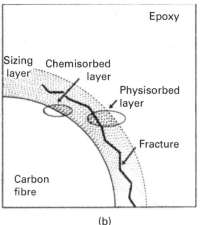

(a)   (b)

Figure 2.12 The concept of the interphase structure in the preparation of CF composites. (a) *Graded interphase* from brominated epoxy resin sized fibre in thermoplastic PES matrix and (b) *distinct interphase* from brominated epoxy resin sized fibre in a thermoset epoxy resin matrix. The locations of the fracture crack are also indicated [2,22,23].

Similarly, a selection of the properties of pitch-based CFs is presented in Table 2.4. Of note is the higher moduli of pitch-based CF, but at the expense of the failure strain. Since the micromechanics of polymer matrix composites are dominated by matrix failure, the failure strain of the fibre needs to be as high as possible. Pitch-based CF finds application in low-failure-strain matrices such as ceramics.

Table 2.3. The mechanical properties of CFs manufactured from PAN precursors by Toray (tradename Torayca) [25]

Designation	Diameter (μm)	Density (g cm^{-3})	Modulus (GPa)	Strength (GPa)	Failure strain (%)	Tow filament no. (k)
T300	7.0	1.76	230	3.53	1.5	1,3,6,12
T300J	7.0	1.78	230	4.21	1.8	3,6,12
T400H	7.0	1.80	250	4.41	1.8	3,6
T600S	7.0	1.79	230	4.31	1.9	24
T700S	7.0	1.80	230	4.90	2.1	12,24
T700G	7.0	1.80	240	4.90	2.1	12,24
T800H	5.0	1.81	294	5.49	1.9	6,12
T1000G	5.0	1.80	294	6.37	2.2	12
M35J	6.0	1.75	343	4.70	1.4	6,12
M40J	5.0	1.77	377	4.41	1.2	6,12
M46J	5.0	1.84	436	4.21	1.0	6,12
M50J	5.0	1.88	475	4.12	0.8	6
M55J	5.0	1.91	540	4.02	0.8	6
M60J	4.7	1.94	588	3.82	0.7	3,6
M30S	6.5	1.73	294	5.49	1.9	18
M40	6.5	1.81	392	2.74	0.7	1,3,6,12

Table 2.4. The mechanical properties of CFs manufactured from pitch precursors by Nippon Graphite Fiber Corp. (tradename Granoc) [26]

Designation	Diameter (μm)	Density (g cm^{-3})	Modulus (GPa)	Strength (GPa)	Failure strain (%)	Tow filament no. (k)
YSH-70A	7	2.14	720	3.63	0.5	1,3,6
YSH-60A	7	2.12	630	3.9	0.6	1,3,6,12
YSH-50A	7	2.10	520	3.9	0.7	1,3,6
YS-90A	7	2.18	880	3.53	0.3	3,6
YS-80A	7	2.17	785	3.63	0.5	3,6
XN-90–60S	10	2.19	860	3.43	0.4	6
XN-80-A2S	10	2.17	780	3.43	0.5	12
XN-80–60S	10	2.17	780	3.43	0.5	6
XN-60-A2S	10	2.12	620	3.43	0.6	12
XN-60–60S	10	2.12	620	3.43	0.6	6
XN-15–30S	10	1.85	155	2.4	1.5	3
XN-10–30S	10	1.7	110	1.7	1.6	3
XN-05–30S	10	1.65	64	1.1	2.0	3

2.3   Glass Fibres

A **glass** in this context is essentially a soda–lime–silica glass similar to the glass used for bottles and flat glass. It has a low tensile strength and chemical durability, especially in acidic media. **AF glasses,** which are used for fibrous thermal insulators and sound absorbers, have a similar composition with the inclusion of 2–10% B_2O_3 to

Table 2.5. Compositions (in wt%) of a range of glasses used for fibres [27,28]

Constituent	E	ECR	Advantex	C	A	S-2	R	Cemfil	AR1	AR2	D
SiO_2	55.2	58.4	59–62	65	71.8	65.0	60	71	60.7	61.0	75.5
Al_2O_3	14.8	11.0	12–15	4	1.0	25.0	25	1	–	0.5	0.5
B_2O_3	7.3	0.09	–	5	–	–	–	–	–	–	20.0
ZrO_2	–	–	–	–	–	–	–	16	21.5	13.0	–
MgO	3.3	2.2	1–4	3	3.8	10.0	6	–	–	0.05	0.5
CaO	18.7	22.0	20–24	14	8.8	–	9	–	–	5.0	0.5
ZnO	–	3.0	–	–	–	–	–	–	–	–	–
TiO_2	–	2.1	<1	–	–	–	–	–	–	5.5	–
Na_2O	0.3	–	0.1–2	8.5	13.6	–	–	11	14.5	–	} 3.0
K_2O	0.2	0.9	<2	–	0.6	–	–	–	2.0	14.0	
Li_2O	–	–	–	–	–	–	–	–	1.3	–	–
Fe_2O_3	0.3	0.26	<0.5	0.3	0.5	Trace	–	Trace	Trace	–	–
F_2	0.3	–	<0.5	–	–	0	–	–	–	–	–

AR, alkali-resistant.

improve the spinning characteristics. Historically, **basalt glasses** have been used in these applications (often called mineral wool). The method of producing these wool glasses is quite different. However, recently some basalt compositions have also been used in the production of continuous roving for use in reinforcement applications. The composition of basalt glass varies significantly with geographical region and has a tendency to form crystals. This has limited their use in composites. Jones and Huff [27,28] have discussed the manufacture and properties of glass fibres in detail. Table 2.5 provides a list of the glass fibres available for use in composites. These have different applications, ranging from conventional E-glass to the low-dielectric D-glass for application in fast-response electrical circuit boards.

2.3.1　E-Glass

Table 2.6 illustrates the composition of a range of E-glass fibres [27–29]. E-glass is primarily used for fibre reinforcements in polymers in both rubbers and plastics. The term E-glass originates from its early use as reinforcement in electrical applications such as printed circuit boards because of its very high electrical resistivity at room temperature. The main oxides used in these glasses are silica, calcia, and alumina. They also contain boron (B). However, E-type glasses without B or fluorine (F) are becoming widespread, despite the higher melt viscosities, because of lower batch costs and process emissions. Moreover, they have even higher chemical durability to acidic solutions than the B-containing E-glass and can be considered a type of ECR (chemically resistant) glass. The B-free E-glass Advantex was introduced by Owens Corning Company in 1998 for environmental manufacturing reasons and because it has improved chemical resistance to acidic aqueous environments. Many plants for E-glass are being converted to Advantex production.

Table 2.6. Development of the composition (wt%) of E-glass fibres

Constituent date	E-glass original 1940	Improved E-glass 1943	621 glass 1951	MgO-free glass	816 glass 1979	F-free glass 1987	B- and F-free glass 1987	Advantex 1998	Low η_D glass 1989
SiO_2	60.0	54.3	54.0	54.3	58.0	55.3	59.0	60.0	55.8
Al_2O_3	9.0	14.0	14.0	15.1	11.0	13.9	12.1	13.5	14.8
B_2O_3	–	10.0	10.0	7.4	–	6.8	–	–	5.2
MgO	4.0	4.5	–	0.1	2.6	1.8	3.4	3.0	–
CaO	27.0	17.5	22.0	22.1	22.5	21.4	22.6	22.5	21.0
ZnO	–	–	–	–	2.6	–	–	–	–
TiO_2	–	–	–	–	2.4	0.2	1.5	–	–
$Na_2O/\,K_2O$	–	1.0	1.0	0.1	1.0	0.4	0.9	1.0	1.4
Fe_2O_3	–	Trace	Trace	0.2	0.1	0.2	0.2	–	n.d.
F_2	–	0.5	0.5	0.6	0.01	–	–	–	0.5

n.d., not detected.

It is important to understand the development of E-glass compositions as shown in Table 2.6 so that the whole literature on glass fibres can be put into the context of the growth of the application of composite materials.

2.3.2 AR Glass

AR (alkali-resistant) glass fibres were developed for use in alkaline environments as reinforcement of cementitious products such as concrete. It is a speciality glass, but the polymer size is essential to ensuring long-term durability. These glasses are all zirconia-based. Cemfil was an early commercial fibre for reinforcement of cementitious products. The polymer coating is essential for good durability in these highly alkaline materials (pH = 13.5) [30].

2.3.3 High-Strength Glass (R- and S-Glasses)

High-strength glasses have higher concentrations of SiO_2 than the other types of fibreglass (Table 2.5). As a result, these glasses have higher melting temperatures than conventional E-glasses. In Europe this glass is referred to as R glass, but the highest tensile strength glass available in relatively large tonnages is called S-2 glass (AGY Holding Corp.). This is a magnesia–alumina–silica glass with a tensile strength ≈50% higher than that of standard E-glasses (Table 2.7). It is relatively expensive because it is melted in small volumes at higher temperatures. Recently, a magnesia–alumina–silicate glass with significantly higher tensile strength than E-glass has been developed which can be processed in a modified conventional E-glass melter. A relatively high throughput is therefore achievable, providing the potential for greatly increasing the use of glass fibre as a reinforcement material in high-performance applications.

Table 2.7. Typical physical and mechanical properties of glass fibres

Property	E	ECR	Advantex	C	A	S-2	R	Cemfil	AR1	AR2	D
σ_{fu} (GPa)	3.7	3.4	3.8	3.4	3.1	4.7	4.5	2.9	3.24	2.5	2.4
E_f (GPa)	76.0	73.0	–	–	72.0	85.0	85.0	–	73.0	80	52
ρ_f (g cm^{-3})	2.53	2.6	2.62	2.49	2.46	2.48	2.55	–	2.74	2.74	2.1
η_D	1.55	1.58	1.56	–	1.541	1.523	–	–	1.562	1.561	1.465
^{a} Fiberizing temp. (°C)	1,200	–	1,250	–	1,280	–	–	1,470	1,290	–	–
Dielectric constant at 25 °C and 10^6 Hz	6.6	6.9	–	6.9	6.2	5.3	6.2	5.21^b	8.1	–	3.85

σ_{fu}, single filament strength at 25 °C; E_f, single filament modulus; ρ_f, density, η_D, refractive index.
^{a} Temperature at which the viscosity of the glass is 10^3 poise; ^{b} dielectric constant at 25 °C at 10^{10} Hz.

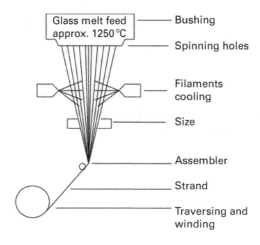

Figure 2.13 Schematic of glass-fibre spinning.

2.3.4   D-Glass

For the specific requirements of fast-response electronic circuit boards, D-glass with a low dielectric constant is available, as shown in Table 2.7. A typical composition for D-glass is given in Table 2.5.

2.3.5   Fibre Manufacture

The manufacture of glass fibres is described in detail by Mohr and Rowe [31] and Loewenstein [29] and only a brief description is given here.

2.3.5.1   Continuous Filament Process

Figure 2.13 shows a schematic of the spinning of continuous glass fibres. They are formed at bushings (or spinnerets) with up to several thousand small tubes (or tips) in the bottom plate, which is made from a platinum–rhodium alloy. One fibre forms at each tip. The temperature of the glass exiting the tip is in the range of 1,150–1,300 °C, depending upon its composition. For direct rovings, the higher number of tips dictates the need for a more rigid plate and the use of reinforced bushings.

The molten glass flows out of the bushing under gravity as fibres of 1 mm diameter, where it is drawn under tension to the final diameter. The 'pull speed' and glass viscosity (i.e. temperature) are used to control the desired diameter. The fiberizing temperature is defined as the temperature at which the viscosity is 1 poise. The 'pull speed' is determined by the rate of winding of the fibres around a rotating tube (collet) placed 1–4 m below the bushing. In some processes the continuous filaments are chopped in-line into short lengths at a large wheel in contact with a smaller rotating bladed wheel. These blades cut (or more precisely break) the continuous filaments into bundles of the appropriate length.

Clean air and water in a fine mist are used to cool the individual fibres just below the bushing at a quench rate of 500,000–1,000,000 K s^{-1}. As a result, the larger-diameter fibres (of the same glass composition) will be cooled more slowly and tend to have a higher density than those with a small diameter.

2.3.5.2 The Marble Process

In the early days of continuous fibres, the glass was made into marbles, which were remelted in specially designed bushings and drawn into continuous filaments. Originally, the marble process produced higher-quality glass of good chemical homogeneity and low bubble populations. Now comparable glass quality can be obtained with large continuous melt furnaces, so the marble process is being phased out. However, for smaller specialist fibre drawing, the marble process is still used. One advantage of remelting on a smaller scale is that the fibres exiting the bushing can be quenched rapidly without the need for water spray. More details are given elsewhere [29,31,32].

2.3.6 Strength of Glass Fibres

Table 2.7 gives the typical properties of a range of glass fibres. The modulus of E-glass is very much a function of the chemical forces operating within the amorphous network. As the number and strength of the chemical bonds in the 3D network decrease and/or become weaker, the modulus of the glass will decrease. Therefore, the introduction of network modifiers such as alkali or alkaline earth oxides will decrease the modulus. Glass formulations used for continuous filaments have a Young's modulus of 70–80 GPa. The tensile strength will vary with composition, and typically E-glass has a value of ≈3.5 GPa. Pure silica glass filaments collected in the same way can exhibit values of up to ≈7 GPa. Alkali network modifiers reduce the tensile strength of the fibres to around 2.5–3 GPa. Borate glasses, phosphate glasses, and lead silicate glasses typically have strengths of ≈1–2 GPa.

Although tensile strength is a function of composition, it is not a fundamental property of the material because it depends strongly on the presence of defects and flaws within the structure. The nature of the flaw(s), which determines the strength of a glass, is still uncertain, but molecular dynamics simulations are providing insights into the mechanism of brittle fracture [33].

2.3.6.1   Griffith Theory of Strength

Griffith [34] validated his theory of strength using glass fibres and demonstrated that strength is a strong function of the presence of flaws or defects. Equation (2.2) defines the critical stress (σ_k) before fracture, where γ is the surface free energy of a solid, E is the modulus, and c is the crack length. If the microcrack exists at the surface (i.e. a flaw), its half-depth can be substituted for the crack length, c, in eq. (2.2) to obtain eq. (2.3) for the maximum technical strength (σ_m) of the glass:

$$\sigma_k = 2\left(\frac{\gamma E}{c\pi}\right)^{\frac{1}{2}}, \tag{2.2}$$

$$\sigma_m = \left(\frac{2\gamma E}{\pi l}\right)^{\frac{1}{2}}, \tag{2.3}$$

where ℓ is the depth of a surface crack which is half that of an inner flaw.

Thus, strength is a function of the surface free energy created when the crack propagates. The surface free energy of glass changes from 122 erg cm^{-2} to 290 erg cm^{-2} in the presence of water, which may explain static fatigue. Thus, in the presence of flaws, the theoretical strength of $\approx E/10$ is reduced further. Thus, a glass with a Young's modulus of 70 GPa would have a theoretical strength of 7 GPa, but its practical strength could be as low as 0.07 GPa [35] for bulk glass. Converting bulk glass into fibres reduces the probability of the presence of a strength-reducing flaw, bringing the actual strength closer to the theoretical value. To reduce the introduction of surface flaws, the fibres are coated with a polymer to protect it from damage and attack by moisture.

2.3.6.2   Determining the Strength of Glass Fibres

Identifying the strength of fibres is not simple, as the value will be an average of a population. Glass fibres were used to test these ideas, but equally these could apply to other fibres. The distribution of strengths in a population of glass fibres has been discussed by Metcalfe and Schmitz [36] and Bartenev [37].

Figure 1.11 [10,38] illustrates the observation that glass fibres may exhibit three strength levels [37,38]. Metcalfe and Schmitz [36] used a statistical analysis to conclude that the fibres had a distribution of flaws of different severities causing the differing populations of strength. The fibres of average strength of 3 GPa could be attributed to severe surface flaws of 20 mm spacing. The population of average strength of 4.5 GPa could be attributed to flaws of 0.1 mm spacing. The population of average strength of 5 GPa could be attributed to internal defects of 10^{-4} mm spacing, characteristic of a defect-free filament with an uninterrupted surface layer. Bartenev [37] also reported three strength populations in alkaline-free aluminium borosilicate glass fibres of diameter 10 μm at gauge lengths of 25 and 40 mm. σ_1 is the lowest strength, σ_2 the intermediate strength, and σ_3 the maximum strength. The population of highest strength, at 3 GPa, was considered to result from the presence of a tempered surface layer of 10 nm thickness, because after treatment with hydrogen fluoride to remove a 10 nm surface layer, the strength decreased to the σ_2 level of 2.0 GPa. The tempered layer was considered to either exhibit a compressive thermal stress or to have a more uniform molecular structure. The role of a tempered surface layer is not generally accepted because the temperature gradient across a 10 µm fibre is calculated to be less than 5 K, which is considered be insufficient to introduce a tempered layer. Therefore, the drawing of fibres under clean conditions with a very low concentration of surface flaws is probably more important. This can also explain the diameter-dependence of strength [39–41]. Low tensile strength can be attributed to significant flaws, such as partially melted batch material.

2.3.6.3 Weibull Statistics of Strength

It is clear from Figure 1.11 that the distribution of strengths within one continuous fibre or within a bundle of glass fibres needs a statistical method. The Weibull statistical method is commonly used to describe the distribution of strengths [42–44]. This can be used to enable the strengths of fibres at different lengths to be calculated. First, the strength is strongly dependent on the gauge length of the fibre under test, and this can lead to uncertainty about the correct value of strength to be used in any predictive analyses of mechanical properties. Weibull statistics are an example of a 'weakest link' model. The probability of failure curve is obtained by ranking the strengths of the individual glass filaments in the population in order. Figure 2.14 is a schematic plot of the probability of failure ($P(\sigma)$) against the strength of the individual filaments, σ_{fu}, of given length L. Since this is a generic analysis we use σ_u in the following analysis.

The probability curve can be described by eq. (2.4):

$$P(\sigma) = 1 - \exp\left[-L\left(\frac{\sigma_u}{\sigma_o}\right)^m\right], \tag{2.4}$$

where m = Weibull modulus or shape parameter, L = fibre length, σ_u = fibre strength, and σ_0 = characteristic strength or scale parameter. The values of σ_0 and m are

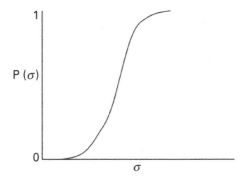

Figure 2.14 Probability of failure ($P(\sigma)$) against σ_u.

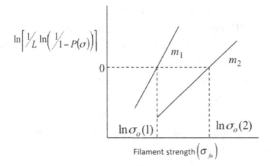

Figure 2.15 Weibull plots of two fibre-strength populations illustrating the Weibull moduli m_1 and m_2 (slope of the lines) and definition of the characteristic strengths, $\sigma_0(1)$ and $\sigma_0(2)$. m_2 is a description of a broader distribution of strengths relative to that given by m_1. A high value of m demonstrates a narrow distribution of strengths.

determined from a plot of eq. (2.5), which is obtained by taking logs twice. Taking logs of eq. (2.4) gives:

$$\ln\left\{\frac{1}{L}\ln\left[\frac{1}{1-P(\sigma)}\right]\right\} = m\ln\sigma_u - m\ln\sigma_o. \tag{2.5}$$

Figure 2.15 shows a plot of eq. (2.5) of the strength of two populations (1 and 2), illustrating how the Weibull modulus and characteristic strength (or scale parameter) are defined. A narrow distribution of strength is identified by a high value for m (say, >20) and a broad distribution would exhibit a low value of m ($\approx$2–5).

For fibres with a unique strength, equal to σ_o, the value of m would be very large, approaching infinity. Typically glass fibres have a broad distribution of strengths and m has a low value of $\approx$5.

The distribution in strength of glass fibres is not always precisely captured, and a bimodal distribution of strength has been proposed. This can provide a better description of the distribution of strengths. (Note that Figure 1.11 could imply that a trimodal distribution function might be preferred.) Equation (2.6) is the equation for a bimodal Weibull distribution:

$$P_f = 1 - \left\{q\exp\left[-L\left(\frac{\sigma_{u1}}{\sigma_{01}}\right)^{m_1}\right] - (1-q)\exp\left[-L\left(\frac{\sigma_{u2}}{\sigma_{02}}\right)^{m_2}\right]\right\}, \tag{2.6}$$

where q is a mixing parameter, the fraction of fractures in population (1) is defined by the Weibull modulus, m_1, and characteristic strength, σ_{01}. Population (2) is defined by the parameters m_2 and σ_{o2}.

2.3.6.4   Static Fatigue of Glass Fibres

Glass fibres suffer from time-dependent fracture under a *constant* load, which is referred to as static fatigue [45,46]. This differs from conventional use of the term fatigue, where a dynamic load is implied. Figure 2.16 shows the time-dependence of fibre strength as a function of time at a constant temperature in distilled water. In high

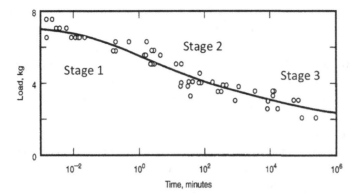

Figure 2.16 Static fatigue of E-glass strands in distilled water. A load of 2 kg provides a strain of 0.5% [45,46].

vacuum, the time-dependence of strength is absent so static fatigue is really a stress corrosion phenomenon in the presence of water. In alkali glasses the sodium ion, Na^+ (and potassium, K^+), acts as a catalyst for the hydrolysis of the silica network as shown in eq. (2.7), where the OH^- is the active species.

$$\begin{aligned}
&\text{Si—O}^-\text{Na}^+ + \text{H}_2\text{O} \longrightarrow \text{Si—OH} + \text{Na}^+ + \text{HO}^- \\
&\text{Si—O—Si} + \text{OH}^- \longrightarrow \text{Si—O}^- + \text{HO—Si} \qquad (2.7) \\
&\text{Si—O}^- + \text{H}_2\text{O} \longrightarrow \text{Si—OH} + \text{OH}^-
\end{aligned}$$

Three stages are observed in static fatigue, depending on the load applied. In **stage 1**, at high loads, fracture is dominated by the rate of crack growth, which is a function of the rate of diffusion of sodium ions to the surface. In **stage 2**, the stress corrosion region, a synergism exists between the applied load and the environment because the rate of crack growth is equal to the rate of corrosion. Thus, the crack always remains sharp and propagates into the weakened material. **Stage 3** is stress-assisted corrosion, where the effect of stress on the failure time is much less significant, because the rate of hydrolysis of the silica network is higher than the rate of crack growth. From the theory of Charles [47], the crack tip is blunted by corrosion so that the stress concentration at the crack tip is reduced. Therefore, according to eq. (2.8), a higher load is required for the crack to propagate at the same rate:

$$\sigma_{max} = 2\sigma_a \left(\frac{x}{\rho}\right)^{\frac{1}{2}}, \qquad (2.8)$$

where x is the crack length and ρ is the radius of the crack tip. σ_{max} is the maximum stress (or stress concentration) at the crack tip under an applied stress of σ_a.

Ghosh et al. [48] studied the subcritical crack growth in E-glass and obtained the static fatigue limit. The threshold stress intensity factor (K_{th}) was found to be 0.15 ± 0.04 MN m$^{-3/2}$, while the critical stress intensity factor (K_{1c}) was 0.93 ± 0.03 MN m$^{-3/2}$ under monotonic mode I loading. To predict the time-to-failure of a filament by calculating the crack growth rate, the stress corrosion exponent of K_{th} is required. This can be estimated from the strain rate dependence on fibre strength.

The static fatigue of glass fibres will determine the maximum life of a structural composite material. In a well-manufactured composite the resin matrix will extend the life of the structure significantly because diffusion of moisture through the resin is rate-determining. A good and durable interfacial bond is essential to avoid rapid capillary transport.

2.3.6.5   Environmental Stress Corrosion of Glass Fibres

In low-pH environments brittle fracture of E-glass fibres occurs (Figure 2.17). This is referred to as environmental stress corrosion (ESC). In this case, the rate of corrosion

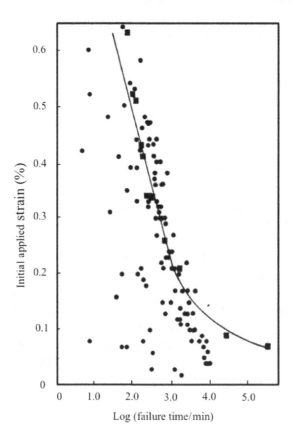

Figure 2.17 Stress corrosion failure times of single E-glass filaments (●) in comparison to equivalent 0° epoxy composites (■) in 0.5 M aqueous sulphuric acid at room temperature [49].

is increased so that the load at which a brittle crack propagates at the same rate is reduced relative to static fatigue.

In acidic environments the corrosion is highly pH dependent because the glass modifiers are soluble. As a result, the fracture of the glass could occur at strains as low as 0.1%. The network modifiers involved are Ca^{2+}, Al^{3+}, Mg^{2+}, Na^+, and K^+. Thus, glasses with compositions containing low concentrations of aluminium, alkaline earths, sodium, and potassium will have higher resistance to ESC. ECR glass (chemical-resistant E-glass) is an example of this. High-strength glasses, such as S-2, also show good resistance to stress corrosion cracking. Owens Corning has recently replaced E-glass with Advantex for glass-fibre production, which is a boron- and fluorine-free E-glass composition (Tables 2.5 and 2.6).

2.3.6.6   Corrosion of E-Glass Fibres

Figure 2.18 shows how the retained strength of unloaded E-glass fibres is affected by the pH of the environment [50]. E-glass fibres have the least durability in an acidic pH of 0.5. At high pH, in alkaline environments, the rate of corrosion is relatively low and stress-assisted corrosion, rather than ESC, occurs. It is strongly dependent on the pH of the environment. At a pH of 7–10 the solubility of the network modifiers decreases, while above pH 10 the O–Si–O bonds in the silica network are hydrolysable, becoming soluble as various silicate salts. The ionization of silicic acid provides the mechanism. This leads to an increase in the rate of corrosion of the glass in highly alkaline environments. But reaction with network modifiers leads to the precipitation

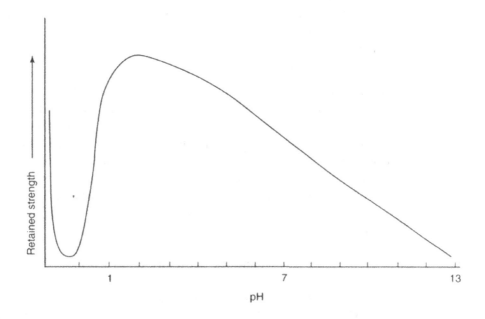

Figure 2.18 Retained strength of unloaded E-glass fibres after immersion in aqueous environments of differing pH [28,50].

of aluminosilicate salts at the crack tip, inhibiting brittle crack growth so ESC at high pH is not observed. However, an analogous fracture of the matrix induced by the crystallization pressure has been observed [51]. A similar failure mechanism also operates at low applied strains in acidic environments whereby a degraded interface leads to ESC fracture above the liquid in half-immersed specimens when 'alum' salts crystallized at the junction between the wet and dry halves of the specimen [52].

2.3.7   Sizings and Binders

To facilitate their commercial production, it is essential to coat glass fibres with a polymeric size at the bushing. Thomason [53] has reviewed the patent literature in this area to give an insight into the approach used commercially. An aqueous-based emulsion is generally applied by contact with a rubber or graphite roller.

The sizing protects the fibre surface, maintaining fibre strength during spinning and haul-off and during handling, weaving of textiles, and manufacture of preforms and composite materials. In addition, the sizing is chosen to ensure compatibility with the selected matrix and the composite manufacturing process. For example, where slow wet-out is needed for the maintenance of strand integrity, hard-sized fibres (with reduced sizing solubility) are used. For preforms or fibre mats, a secondary binder is used to hold the fibres together during manufacture.

The finish, applied to a glass fibre, typically consists of:

1. an adhesion promoter, which is often a silane coupling agent;
2. a protective polymeric film former;
3. lubricants of different composition to aid the flow of the fibres through machinery without damage;
4. surfactants used in the emulsification of the polymeric film former; and
5. an optional polymeric binder.

A typical emulsion applied to the glass fibre will have a solids content of approximately 10%, of which 0.3–0.6% will be the silane coupling agent. In addition, 0–0.3% lubricant, 0–0.5% surfactant, and 0–0.3% optional antistatic agent may also be used.

The term *sizing* can refer to the film former or the compounded finish, with or without the adhesion promoter or silane coupling agent. The generic term *finish* universally refers to the deposited solids on the glass fibre and will include any optional binder for fibre mats and textiles. The film former imparts good handleability and controls the wet-out of the resin during manufacture of composite materials [54,55].

2.3.7.1   Silane Coupling Agents and the Structure of Hydrolysed Silanes on Glass Surfaces

Adhesion promoters are included in the sizing emulsion to provide the glass with compatibility and chemical reactivity to the chosen matrix. The silane needs to displace adsorbed water from the glass surface to create a hydrophobic surface with the correct thermodynamic characteristics for complete wetting by the matrix resin.

Table 2.8. Typical functional silanes for different matrices

Active functional group		Structure	Abbreviation	Matrix application
Amino	$-NH_2$	$NH_2 - (CH_2)_3 - Si(OC_2H_5)_3$	γ-APS	Epoxy general
Epoxy	$\overset{O}{\overset{\diagup \diagdown}{CH_2-CH-}}$	$\overset{O}{\overset{\diagup \diagdown}{CH_2-CH}}-CH_2-O-(CH_2)_3-Si(OCH_3)_3$	γ-GPS	Epoxy
Vinyl	$CH_2 = CH-$	$CH_2 = CH - Si(OCH_3)_3$	VTS	Unsaturated polyester and related resins
Vinyl	$CH_2 = C(CH_3)-$	$\overset{CH_3}{\underset{\mid}{CH_2 = C}} - COOCH_2CH_2CH_2Si(OCH_3)_3$	γ-MPS	Unsaturated resins

This aids the development of a strong interfacial bond between the fibre and the matrix.

Table 2.8 lists some typical silane coupling agents [32,56] for promoting adhesion between polymers and the glass fibres. Most are trialkoxy organo silanes with the generic structure

$$\begin{array}{c} R' \\ | \\ RO - Si - OR \\ | \\ OR, \end{array}$$

where R' is a polymer-compatible or reactable organic group. R is either ethyl or methyl.

The alkoxy (OR) groups in the silane molecule are hydrolysed by the water in the emulsion into hydroxyl groups. The rate of hydrolysis depends on the pH of the emulsion, which is chosen to be slightly acidic (pH = 4). The OH and OR can condense into polymers of hydrolysed silane. The rates of hydrolysis and polycondensation ensure that a mixture of oligomeric siloxane polymers and monomers are deposited onto the glass-fibre surface. The deposit on glass fibre is a complex cross-linked polymer containing oligomers of varying degrees of polymerization (Figure 2.19).

The polycondensation of the alkoxy groups is an equilibrium polymerization whose floor concentration is of the order of 0.1%, depending on the structure of R'. At typical silane concentrations of 0.5%, oligomeric polysiloxanes will be deposited. Since the floor concentrations of the various hydrolysed silanes differ, their concentration in the sizing emulsion can be used to optimize the performance of a composite.

An E-glass-fibre surface can be considered to be silica-rich, with hydroxyl groups introduced by reaction with the cooling water. These silanol groups on the surface react with those formed from hydrolysis of the alkoxy silane. There is competition for

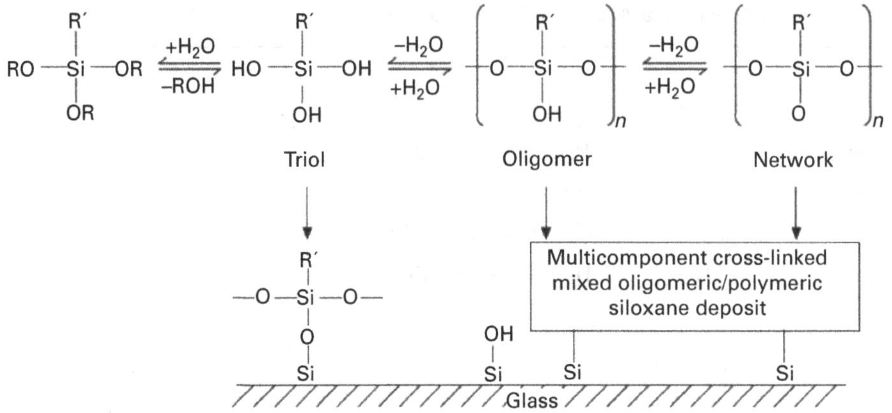

Figure 2.19 The complex molecular structure of the silane deposit on glass fibres.

condensation between the monomeric and oligomeric triols and the glass surface. On drying, the concentration of the silanol groups in the sizing increases, promoting polycondensation [57,58]. Therefore, the deposit on the surface of E-glass fibres will have typically around 100 molecular layers, of which 90% can be extracted into the matrix on fabrication of the composite. The remaining crosslinked component can then accept the penetration of the matrix resin to form an interpenetrating network (IPN).

The presence or absence of boron in the glass composition influences the concentration of silanols on the fibre surface, so the sizing compositions needed to be re-optimized for Advantex [59,60].

2.3.7.2   Interface and Interphase in Glass-Fibre Composites

As discussed above, the coating on a glass fibre is complex, consisting of a bonded crosslinked, functional polysiloxane containing dissolved oligomeric silanols together with the sizing polymer and other process aids. Therefore, during processing the resin will penetrate the hydrolysed silane polymer. For this to happen the film former and silanol oligomers will diffuse into the resin. On curing of the resin, an interphase region will form at the fibre–matrix interface, which consists of a bonded semi-interpenetrating network between the silane and the matrix, together with the monomeric and oligomeric silanols and surfactants, which have diffused over a longer distance. The interphase region, therefore, has a dimension determined not just by the thickness of the silanol deposit, but also on the diffusional length scale of the other additives, which includes the film former. Figure 2.20 illustrates schematically the structure of the interphase that forms in a composite [23]. A distinct interface is unlikely because it is reported [58] that certain elements in the glass surface may be extracted during sizing and incorporated into the 'polymeric hydrolysed silane' deposit. The silane can penetrate the subsurface of the denuded glass so that the interphase can be considered to extend into the fibre surface.

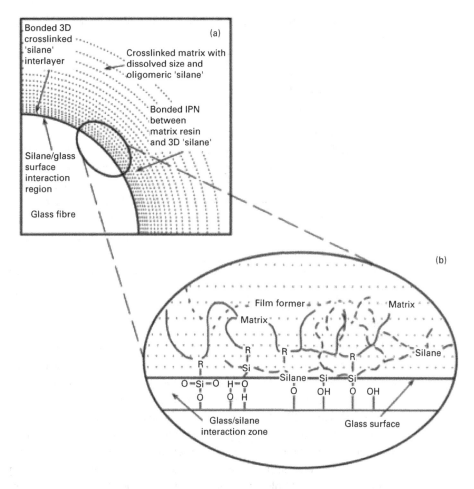

Figure 2.20 Schematic example of the structure of the interphase that forms in glass-fibre composites. (a) A graded interphase formed from the diffusion of the polymeric film former and silane oligomers into the matrix resin. (b) A schematic of the bonded semi-interpenetrating network at the interface [23].

2.3.7.3   Selection of Silane

The silane is usually chosen from the 'chemical bonding' model of adhesion. The assumption is that R′ reacts with the matrix and that RO reacts with the fibre surface. Hence, they are referred to as coupling agents. Table 2.8 gives typical structures of amino, glycidyl, and vinyl silanes. The former two react directly with epoxy groups in epoxy resins, whereas the latter has a vinyl group that reacts with unsaturated groups in unsaturated polyester and vinyl ester resins.

It is shown in Table 2.8 that R can usually vary between C_2H_5- and CH_3- because the rates of hydrolysis differ and provide a variable for affecting the structure of the deposit. There is a variety of commercially available silane coupling agents that can be used to optimize the interfacial response of the fibres and hence the mechanical properties of the composite.

2.3.7.4   Adhesion Mechanisms

The following hypothetical adhesion mechanisms of silanized glass fibre in composites have been proposed [56]:

1. chemical bonding or coupling – molecular bridge formation;
2. deformable layer theory – a ductile tough layer between fibre and matrix;
3. restrained layer theory – highly crosslinked interphase of intermediate modulus;
4. wettability theory – silanes improve wetting of the fibre surface by the polymer;
5. acid–base theory – silane modifies the acidity of the fibre surface;
6. weak boundary layer theory – coupling agents eliminate weak boundary layers;
7. button down theory – the formation of 'silane' islands on the fibre surface and a buttoning-type mechanical mechanism;
8. reversible hydrolytic bonding theory – hydrolysis of –Si–O–Si– interfacial chemical bonds during water contact with reformation on drying to explain recovery in performance;
9. bonded interpenetrating network theory – interphase is a bonded IPN.

As discussed above, an interphase forms in a glass-fibre composite through the diffusion of matrix resin into a partially crosslinked 'silane' layer which is chemically bonded to the fibre surface. Other additives and sizing polymers diffuse into the matrix. On resin curing, copolymerization with the functional silane group (R') occurs to create a bonded IPN. Therefore, the interfacial micromechanics involve mechanism 9. As a result, mechanisms 2 and 3 could also apply to the deformation of the interphase during failure. In this case mechanism 8 does not need to be invoked to explain the recovery in interfacial performance after moisture conditioning followed by drying. Reversible plasticization by water modifies the deformation behaviour of the interphase.

In special cases, such as electrical artefacts and components, a starch or similar size is used for the fibre spinning and is removed after preform manufacture so that direct silanization is possible. In these applications monomolecular silane layers are anticipated so that the chemical bonding/coupling (mechanism 1) may apply.

2.3.7.5   Selection of Sizings and Binders

Sizings [55] are applied directly at the bushing, whereas the binders are used in secondary processes to bind the fibres into a variety of textiles. They are used for the following:

1. handleability of fibres during processing;
2. controlled wet-out with matrix resin during fabrication;
3. protection of fibres from mechanical damage

Typical sizings are based on aqueous emulsions of (1) polyvinyl acetate (general); (2) epoxy resin; and (3) polyurethane. They are selected based on the following:

compatibility with the matrix; process requirements (e.g. textile, moulding technique); and environmental durability of the laminate or moulding.

An additional polymeric binder is required for the manufacture of a random discontinuous or continuous fibre mat or cloth. Typically, a polyvinyl acetate emulsion (PVAc) is used for general-purpose reinforcement, but it has limited compatibility with many matrices. Composites may therefore not be durable in aqueous environments. Powder-bound mats offer higher performance and durability. A variety of thermoplastic polymers are available in powder form for binding the fibres in textiles and preforms.

2.3.7.6   Glass-Fibre Form

Glass fibres are available in a range of forms for use as reinforcement for the manufacture of composite materials. Figure 2.21 shows typical forms of commercial reinforcements. These forms of glass fibre are typically available for subsequent use in the fabrication routes indicated in Figure 2.22.

Continuous fibres are collected immediately at the collet or in-line chopped for incorporation into moulding compounds. In conventional plant each bushing has 200–204 spinnerets or tips. Filaments from $\approx$10 bushings are gathered into a strand and wound onto the collet. The product has a structure determined by assembling

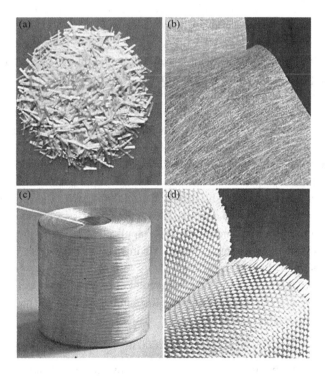

Figure 2.21 Available forms of glass fibres for composite manufacture: (a) chopped strands; (b) chopped strand mat; (c) woven rovings; (d) continuous rovings [32].

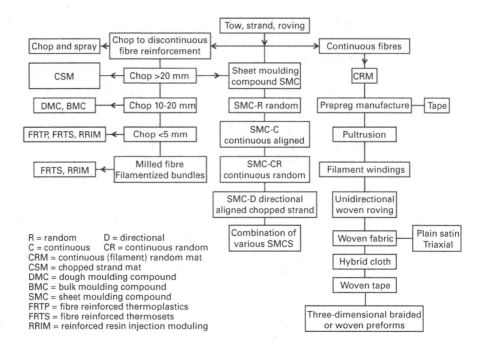

Figure 2.22 The range of glass-fibre forms and moulding compounds used for manufacturing composites. Continuous rovings and fabrics are directly impregnated during pultrusion and 'wet' filament winding and chop-and-spray fabrication routes [32].

tows with ≈200 filaments into **strands** containing ≈2,000 filaments. Individual manufacturing processes require reinforcements with higher total linear density or tex (eq. 2.9) so that multiple strands are assembled in a separate operation into **rovings**. The process introduces a slight twist, which keeps the strands in a stable structure. The reinforcement can be unwound from the centre, as illustrated in Figure 2.21.

The rigidity of the bushing limits the number of filaments that can be drawn. Recent improvements in the properties of bushing materials have enabled the number of filaments, which can be spun, to be increased to 4,000–6,000. These filaments are collected at the bushing and used directly. They are referred to as **direct rovings** and are intended primarily for advanced composite applications such as prepreg, where precise fibre alignment is essential. **Direct rovings** do not have the twist that is introduced in assembled rovings (Figure 2.21).

The rovings are normally designated by the total linear density or tex (T), which is defined by the average filament diameter and the number of filaments per strand. Tex is the number of grams per kilometre:

$$1 \text{ Tex} = g/1,000 \text{ m}. \tag{2.9}$$

In the non-SI system the linear density was defined by the older denier unit:

$$1 \text{ denier} = g/9,000 \text{ m}. \tag{2.10}$$

In the American system an alphabetic letter is followed by the count, which is the number of hundred yards per pound. The relationship between this is given by:

$$\text{Count} = 4{,}961/T. \tag{2.11}$$

The fibre diameter d_f (in μm) is given by:

$$d_f = 15.8(T/0.4961\ N)^2, \tag{2.12}$$

where N is the number of filaments/strand.

Typically in the metric system of designation, EC14 160 refers to E-glass with filaments of nominal diameter 14 μm and a tex 160. The American designation is K 31. The number of filaments per strand is 400. The tow is the smallest unitary element of a strand, which represents the number of filaments drawn from a single bushing. Rovings are reassembled from a number of strand-cakes, typically 30 strands, to give a tex of 2,300.

Continuous rovings can be **woven** into a reinforcement fabric or textile in a range of patterns, depending on the arrangement of the fibres required for the fabricated composite part. Continuous fibres can also be formed into a **continuous random mat** or used directly in several processes such as filament winding or pultrusion. Figure 2.22 lists the available configurations of glass fibres.

The strands can also be chopped either in-line or off-line into relatively long discontinuous fibre strands and assembled into an in-plane random arrangement bound together with a polymeric binder; this is a **chopped strand mat (CSM)**. They can also be chopped into different lengths for incorporation into moulding compounds such as dough moulding compound (DMC) or sheet moulding compound (SMC).

Glass fibres are used in a variety of products, some of which are pre-mixed with the polymeric matrix. Figure 2.22 shows the range of available reinforcements and pre-compounded moulding materials as a function of fibre length.

2.4 Advanced Polymer Fibres

The major reinforcements for composite materials are currently glass and carbon fibres. There are many polymer fibres that are used for textiles. The drawing of a polymer during spinning aligns the molecular structure to the fibre axis. The enhancement in modulus and strength over the bulk polymer explains the benefits for textile usage.

We are interested in HM polymer fibres, which can be employed to reinforce a composite. The most popular are the aramids: Kevlar, Nomex, Twaron, and Technora. High modulus polyethylene and xylon (poly(p-phenylene benzobisoxazole, or PBO) are also significant players. Because of the nature of the spinning technology the crystals are highly ordered to the fibre direction, which means that the properties are highly anisotropic with relatively low radial performance. Therefore, the applications of these fibres tend to differ from those for the PAN-based carbon and glass fibres.

2.4.1   Aramid Fibres [61–63]

The term *aramid* arises from a contraction of **ar**omatic poly**amide**. The structures of aromatic polymers are shown in Figure 2.23.

The meta-aramid (Figure 2.23b) with the tradename Nomex (DuPont) and has been commercially available in discontinuous form since the 1960s, whereas the para-aramid (Figure 2.23a) with the tradename Kevlar (DuPont) has been available in continuous form since 1971. The para-aramid with the tradename Twaron (Akzo now Teijin) has been available in continuous form since the 1980s. A meta-aramid, Teijin-Conex, is also available from Teijin. Technora is a copolymer of the meta- and para-aramids (Figure 2.23c) available from Teijin. Since Technora is a copolymer it can be spun from the isotropic polymerization solution, whereas the homopolymers Kevlar and Twaron are rigid rod polymers, which are insoluble in conventional solvents. Therefore, they are spun from anisotropic liquid crystal solutions. The latter can be spun into highly oriented crystalline fibres with high modulus, as described below.

The para-aramid is dissolved in 100% sulphuric acid. It needs to be water-free otherwise hydrolysis of the polymer can occur at the required temperature of 80–90 °C. At a concentration of 10%, 'rigid segments' of the polymer become ordered into a nematic or liquid crystal solution and the viscosity falls. The 'solubility' limit is 20%; above this concentration, insoluble polymers can be transferred into the spun fibre, causing defects that reduce the fibre strength. Therefore, these fibres are spun from nematic solutions at concentrations ≤20%. Dry jet wet-spinning into water, as coagulant, is used because the orientation of the crystalline domains occurs in the spinneret with further extension of the soluble segments in the dry jet region. The ordering of the liquid crystalline domains in the spinneret is analogous to spinning of

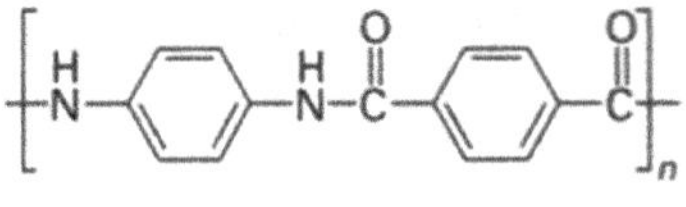

(a)

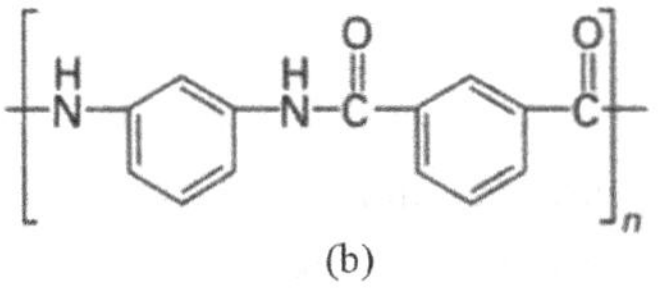

(b)

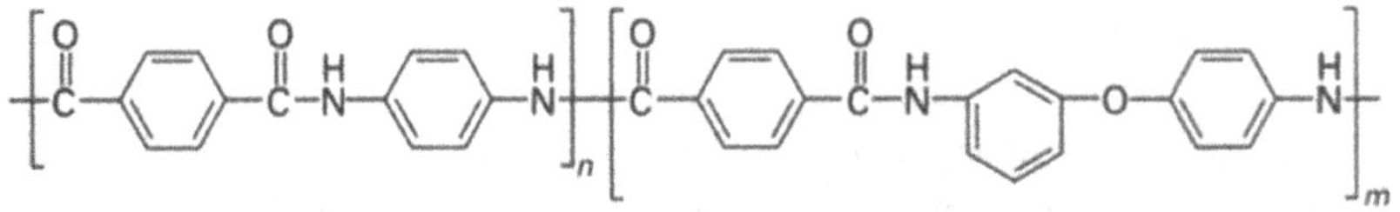

(c)

Figure 2.23 Aramid structures: (a) poly(p-phenylene terephthalamide); (b) poly(m-phenylene isophthalamide); (c) copolymer of p-phenylene/3,4-oxydiphenylene ether terephthalamide.

mesophase pitch for CF production, as shown in Figure 2.24. The schematics of the fibre cross-sections, in Figure 2.25, show how the microstructure reflects the induced orientation of the crystallites. Therefore, the cross-section of the fibres can show either random, radial, or tangential orientation of the crystallites. The radial microstructure can only be achieved using the dry jet technique.

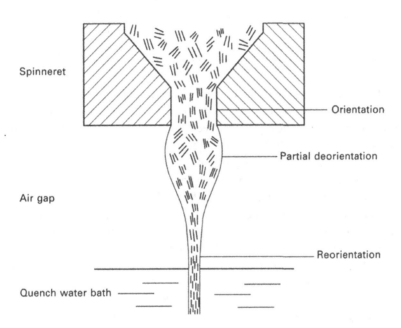

Figure 2.24 Schematic of dry jet spinning illustrating the alignment of crystallites [63].

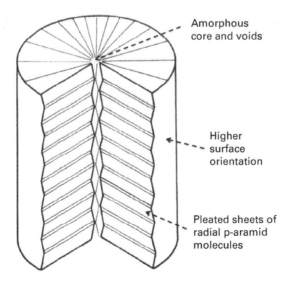

Figure 2.25 Schematic of the pleated structure of a para-aramid fibre [64].

Table 2.9. Typical mechanical properties of aramid fibres

Property	Kevlar 29	Kevlar 49	Kevlar 149	Twaron	Nomex	Teijin-Conex	Technora
Modulus (GPa)	60	128	180	60–120	17	–	78
Strength (GPa)	2.65	2.65	3.45	2.4–3.6	0.7	0.62–0.69	3.4
Failure strain (%)	4.0	2.4	1.9	2.2–4.4	22.0	35–45	4.6
Density (g cm^{-3})	1.44	1.45	1.49	1.44–1.45	1.4	1.38	1.39
Tenacity (N/tex)	2.03	2.08	1.68	1.65–2.5	0.485	0.45–0.5	2.5

The axial orientation of the crystalline regions in the as-spun fibres can be significantly improved by a brief (a few seconds) heat treatment under tension at 500 °C. The orientation angle of the crystallites decreases from 12–15° to ≤9°. This increases the modulus of the fibres from ≈60 to >150 GPa.

The para-aramid fibres are generally accepted as one of a pleated microstructure in which the molecules are arranged axially in radial crystalline sheets. Figure 2.25 shows the pleated structure, which reflects the shape of the para-aramid molecule within the crystallites. The centre of the fibre tends to be dominated by the presence of elongated voids of dimension 5–10 × 25 nm, whereas the skin exhibits a much less perfect arrangement to accommodate the radial stacking of the crystallites. The longitudinal arrangement of the molecular sheets ensures that the chemical bonds can carry their maximum load, which explains the high modulus that can be achieved with these fibres. However, only weak intermolecular forces operate between the molecular sheets, resulting in low transverse modulus and strength. Table 2.9 shows that the degree of longitudinal organization determines the modulus of the fibre. Aramid fibre composites provide impact energy absorption because of the large number of micromechanisms of fracture available through transverse and shear failure of the fibres. Composite applications usually reflect these attributes.

2.4.1.1 Surface Treatment and Sizing

Adhesion of the fibre to the polymer matrix can be improved by plasma treatment, but the transverse fibre strength is low and treatments are apparently not very effective. This is because the locus of failure will simply move from interfacial to within the fibre, where the bonding between individual crystalline sheets is weak. Any treatments are usually bonded coatings, which improve the integrity, moisture resistance, and processability of the fibres.

2.4.2 Other High-Performance Polymer Fibres

A complete review of HM fibres is beyond the scope of this book and full details are given elsewhere [63,65]. Therefore, we will only identify fibres with potential for use in composite applications and indicate their advantages.

Table 2.10. Tensile properties of advanced polymer fibres [65].

Fibre	Modulus (GPa)	Strength (GPa)	Tenacity (N/tex)	Failure strain (%)	Density (g cm^{-3})	Compressive strength (GPa)
PBT	325	4.1		1.5	1.58	0.26–0.41
PBO Zylon AS	180	5.8	3.7	3.5	1.54	
PBO Zylon HM	270	5.8	3.7	2.5	1.56	0.2
PBO Zylon HM+	360					
M5	330	4.0	2.3	1.5	1.7	1.0–1.7
PE theoretical	240	32	33			
PE HM	75–120	3.6	2.6–3.7	2.9–3.5	0.97	0.1

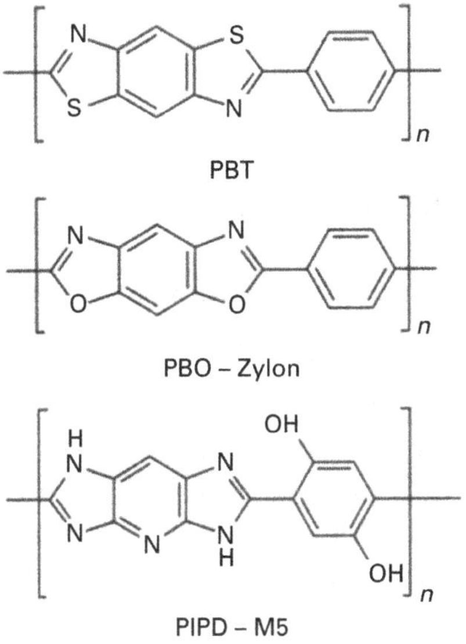

PBT

PBO – Zylon

PIPD – M5

Figure 2.26. The chemical structures of ladder polymers used in continuous fibres.

2.4.2.1 Poly(P-Phenylene Benzobisoxazole) and Related Fibres

Poly(p-phenylene benzobisoxazole) and the related fibre poly(p-phenylene benzo-bisthiazole) (PBT) (Figure 2.26) were originally developed at the Wright Patterson Air Force Laboratory, USA. The aim was to exploit the stiffness and heat resistance of fused aromatic heterocyclic polymer structures. PBO has been commercialized as Zylon by Toyobo in Japan [66]. The spinning technique is analogous to that for the aramids, whereby dry jet wet-spinning technology is used. A subsequent heat treatment in nitrogen at 500–700 °C for a few seconds is used to achieve the optimum modulus. The molecular structure within the crystallites is arranged axially along the fibre and perfected by the heat treatment. A non-aqueous coagulation technique has been developed by Toyobo to produce the Zylon HM+ fibre of higher modulus (360 GPa). The properties of Zylon are given in Table 2.10

These are examples of the use of liquid crystalline aromatic heterocyclic polymers as high-performance fibres. Good mechanical performance is achieved by aligning these rigid rod polymers into liquid crystals prior to ordering them further by dry jet spinning. In this way the strong chemical bonding of the polymer skeleton is axially aligned to the fibre axis so that it can take the applied mechanical load. However, in contrast, in the transverse direction of the fibre only the weak intermolecular bonds can carry a load so that the properties are significantly lower. Poly(2,6-diimidazo 4,5-b:4',5'-pyridinylene-(2,5-dihydroxy) phenylene) (PIPD) or M5 is a development by Akzo Nobel intended to improve the off-axis properties of a HM polymer fibre [67,68]. This was achieved by introducing hydrogen bonding between the molecules through pendant hydroxyl groups. The high compressive strength of up to 1.7 GPa reflects the shear modulus of 7 GPa. The M5 fibre is not currently commercially available. Many other high-performance polymer fibres are available, but do not find application in structural composites. Readers are referred to the work of Eichhorn et al. [65,69].

2.5   High Modulus Polyethylene (HMPE) Fibres [70,71]

There are a number of potential composites applications that employ polyethylene and polypropylene fibres. For completeness we will briefly review these and provide their advantages and disadvantages.

Polymer molecules normally exist in the state of a random coil, where intermolecular forces dominate. This state arises from rotation about the individual skeletal bonds that make up a polymer or macromolecule. A linear polymer could have a typical average molecular weight of $\approx 100,000$ g mol^{-1} with $\approx 10,000$ skeletal bonds. As a result, the dimensions of the random coil, defined by the end-to-end distance of the chain, r (given by the average root mean square end-to-end distance, $< r^2 >^{1/2}$) are significantly smaller than the extended chain length. This is given by

$$C_\infty = \frac{< r^2 >_o^{1/2}}{nl^2},$$
(2.13)

where C_∞ is the characteristic ratio and $< r^2 >_o^{1/2}$ is the unperturbed average root mean square end-to-end distance of the molecular chain with n links of bond length l.

Bulk plastics exhibit a maximum modulus in the glassy state of 4 GPa, where the intermolecular bonds are responsible for resistance to deformation. An elastomer has a modulus of <1 MPa because the intermolecular forces are absent and rotational uncoiling occurs. In order to utilize the rigidity of the skeletal bonds, the molecules need to be in an extended conformation. An approximate estimate from eq. (2.13) shows that a draw ratio of ≈ 100 is needed to achieve an extended chain conformation. This can be achieved by drawing in the solid state or by enabling the crystallization of the molecular segments in an extended conformation, which can be aligned with the fibre axis. Mesophase pitch and liquid crystal solutions of high-performance polymers are examples of the latter.

In a fully extended state the polyethylene chain has a theoretical modulus of 180 GPa, obtained from a summation of the resistance to deformation of the skeletal bonds and bond angles (valence force field (VFF) estimate) [71]. More sophisticated models that include intramolecular interactions give values of 257 GPa and 340 GPa. X-ray measurements on single crystals provide a value of 240 GPa. The latter is generally accepted as the maximum achievable.

Three methods are employed commercially to produce high-performance polyethylene (HPPE) fibres. Ward and coworkers [72] pioneered the development of HPPE fibres using **melt-spinning with subsequent hot drawing**. Tenfor (Snia) and Certran (Hoechst Celanese) fibres have moduli in the range of 40–60 GPa. The limitation of the process is the molecular weight of the polymer that can be processed. In particular, the draw ratio was highly sensitive to the average molecular weight (M_w) of the polymer. M_w is a measure of the distribution in chain lengths of the polymer. Ultrahigh molecular weight (UHMW) polymers have a very narrow distribution in chain length and were examined by Lemstra et al. [73] as potential precursors for HPPE fibres. UHMW polyethylene has a molecular weight of 3–5 million g mol^{-1} and cannot be melt-spun. However, the melt-spun fibres have found applications in self-reinforced composites. In this application the surface of the fibres melts in the compaction process to form a matrix. A classic application is in ultra-light high-strength structures.

Dyneema (DSM) and Spectra (Honeywell) fibres are manufactured using **gel-spinning**, in which UHMW polyethylene is dissolved in a solvent, which gels on cooling through the crystallization of segments with extended conformations. These liquid crystal gels can be super-drawn to extension ratios of 50–100 to ensure a high modulus is achieved. The modulus achieved by gel-spinning is 100–120 GPa. The spinning process is discussed in detail by van Dingenen [71].

Solid-state extrusion (SSE) is also used for the manufacture of HPPE fibres. Zacharides et al. [74] reported the methodology for the solid-state drawing of UHMW polyethylene. In this technique it is critical that the precursor polymer has few chain entanglements, and this is achieved using low-temperature polymerization and ensuring that the polymer product has not been exposed to melting temperatures at which entanglements would be introduced. The powder is converted into a HPPE fibre in a multistage process: (1) powder compression and compaction; (2) rolling; and (3) ultra-drawing to a draw ratio of 12–13 between hot plates at a temperature below the melting point. The total draw ratio of the process is 85–100. The highly drawn tape fibrillates to provide a range of products from tapes and yarns. The SSE process produces Tensylon fibres (Integrated Textile Systems) with a tensile modulus of ≈120 GPa and tensile strength of ≈1.9 GPa and elongation to failure of 2.6%.

2.5.1   Surface Treatment

For inclusion in composite materials polyethylene fibres need to be surface-treated to increase their wettability by potential matrices. For these applications surface oxidation using corona discharge and/or plasma processing have been employed. Similar techniques are used to provide polyethylene and related materials with print capability [75]. However, these fibres readily fibrillate so that improving the adhesion of the surface to other polymer matrices does not provide a significant increase in interfacial shear strength because the locus of failure moves from the interface to the fibre subsurface.

2.5.2   Other Polymer Fibres: Polypropylene Fibres [76]

The theoretical modulus of polypropylene, which is estimated to be 49 GPa, is lower than that for polyethylene because of the helical conformation the molecule adopts within a crystal. The lower modulus reflects the rotation of the skeletal bonds, which occurs before the skeletal bonds can take up the load. These fibres find applications in hot-compacted composites where the conventional co-extrusion melt-spinning with hot drawing enables the fibre 'surface material' to be designed to have a lower melt temperature and hence a wider processing window for hot compaction compared to polyethylene [77]. Polypropylene also has a higher use-temperature.

Polypropylene fibres have many applications as strings and ropes. Fibrillation occurs readily and this has been exploited in the manufacture of reinforced cementitious composites. In the material described by Hannant [78,79], polypropylene string is transversely strained to create a random network that could be readily infiltrated by ordinary Portland cement paste (Figure 2.27).

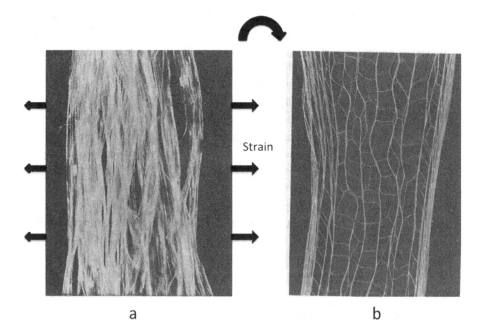

Strain

a   b

Figure 2.27 Fibrillated drawn polypropylene fibres provides a random reinforcing network under a transverse strain for reinforcing cementitious matrices. Mechanical fibre–matrix bonding arises from the fibre entanglements within the network [78].

The composite was intended as a replacement for asbestos cement. This is illustrated in Figure 2.27. The technical problem of transverse loading of the polypropylene string during infiltration limited the commercial exploitation of the Netcem process. However, the research spurred the development of a variety of polypropylene reinforcements for cement, some of which are discontinuous fibres [80].

2.6 Natural Fibres

2.6.1 Bio-cellulose Fibres

A range of natural fibres are available, many of which have been used in textiles and related products for many years. Recently their application as reinforcements for composite materials has grown, largely for cost and ecological reasons. In particular, the automotive industry has shown much interest because of their recycling potential. However, the fibres need to be extracted from the source and treated for compatibility with organic polymer matrices. The fibres are generally discontinuous in nature and can be used as strands for reinforcements or fully defibrillated for dispersion in the matrix.

The cost implications of these treatments determine the use of these natural fibres. The other aspect to be considered is the variability in mechanical properties from a grown resource. Table 2.11 provides a list of potential fibres and their chemical compositions. They are all cellulosic in character, so the surface treatment to make them wettable by potential matrix polymers is an issue, and the reader is advised to research this because current knowledge is rapidly changing. Table 2.12 gives the

Table 2.11. Structure of bio-cellulose natural fibres [81]

Bio-fibre	Cellulose (wt%)	Hemicellulose (wt %)	Lignin (wt%)	Wax (wt%)	Pectin (wt%)
Jute	61–71.5	13.6–20–4	12–13	0.5	0.2
Flax	71	18.6–20.6	2.2	1.5–1.7	2.3
Hemp	68–74.4	15–22.4	3.7–10	0.8	0.9
Kenaf	72	20.3–21.5	9–19	–	3–5
Ramie	68.6–76.2	13.1–16.7	0.6–0.7	0.3	1.9
Sisal	60–78	10–14.2	8–12	2	10
Abaca	56–63	20–25	7–9	3	1
Pineapple leaf	70–82	–	5–12.7	–	–
Date palm leaf	46	28	20	–	–
Curaua	73.6	9.9	7.5	–	–
Coir	32–43	0.15–0.25	40–45	–	3–4
Oil palm	42.7–65	17.1–33.5	13.2–25.3	–	–
Wheat straw	38–48.8	15–35.4	12–23	–	8
Cornhusk	42.3	41	12.6	–	–
Rice straw	64	23	8–19	8	–
Sugar cane bagasse	55.2	16.8	25.3	–	–
Bamboo	26–43	30	21–31	–	–

Table 2.12. The mechanical properties of bio-cellulose fibres in comparison to E-glass [81]

Fibre	Young's modulus (GPa)	Tensile strength (MPa)	Failure strain (%)	Diameter (μm)	Density (g cm^{-3})
Jute	13–26.5	393–800	1.5–1.8	25–250	1.3–1.5
Flax	27–39	345–1,500	2.7–3.2	40–600	1.5
Hemp	38–70	550–900	1.6–4	25–250	1.5
Kenaf	40–53	350–930	1.6	260–400	1.5–1.6
Ramie	24.5–128	400–938	1.3–3.8	34–49	1.5–1.6
Sisal	9.4–22	468–640	2–7	50–200	1.5
Abaca	12	400	3–10	–	1.5
Pineapple leaf	34.5–82.5	400–1,627	1.6	20–80	1.4
Date palm leaf	9	233	2–19	100–1,000	0.9–1.2
Curaua	11.8	500–1,150	3.7–4.3	500–1,150	1.4
Coir	4–6	175	15–51.4	10–460	1.2–1.5
Oil palm	0.5–9	50–400	4–18	150–500	0.7–1.6
Sugar cane bagasse	17	290	–	–	1.25
Bamboo	11–17	140–230	–	–	0.6–1.1
Wheat straw	13	273	2.7	94	1.6
Rice straw	26	449	2.2	–	–
Cornhusk	9.1	351	15.3	20	–
E-glass	70–73	2,000–3,500	2.5–3.7	15–25	2.5

typical mechanical properties of these bio-fibres; in comparison to glass fibres, their lower density is an advantage for use in lightweight structures.

2.6.2   Silk Fibres

Silk fibres [82–84] represent a more appropriate option for high-performance composites since they are continuous. Silk has a protein chemical structure and fibres are spun by the *Bombyx mori* moth caterpillar or the *Argiope aurentia* spider, which produces a dragline of higher modulus. Proteins are essentially polyamides, so Table 2.13 compares the mechanical properties of different silks to aliphatic (nylon 6) and aromatic (Kevlar 29) polyamide fibres. Silk has a low density (1.31 g cm^{-3}), so the specific properties favourably compete with synthetic reinforcing fibres. Given that $\sigma_{fu} = 0.45$–1.97 GPa, specific strength of 34–150 km and specific modulus of 10–26 Mm compare favourably with the specific properties of HM CFs. The surface of silk fibres may need treating to achieve optimum bonding with matrices.

2.7   Ceramic Fibres

The fibres described above do not resist the temperatures employed in the manufacture of metal matrix composites (MMCs) and ceramic matrix composites (CMCs). Low-temperature impregnation routes such as sol-gel or pitch processing still require an elevated temperature to consolidate the matrix. Therefore, these composites need

Table 2.13. Typical mechanical properties of silk and wool fibres in comparison to synthetic polyamide fibres [82]

Fibre	σ_u (GPa)	d_f (µm)	E_l (GPa)	E_t (GPa)	G_l (GPa)	E_l/E_t	E_l/G_l
Argiope aurentia spider silk	0.9	3.57	34.0	–	–	–	–
Nephila clavipe spider silk	12.15	4.2	12.71	0.579	2.38	21.95	5.34
Bombyx mori cocoon silk (silkworm)	0.6*	12.9*	9.90*	–	3.81*	–	4.93*
Merino sheep wool	0.2	25.5	3.50	0.93	1.31	3.76	2.67
Nylon filament	0.95	16.2	2.71	1.01	0.52	2.68	5.21
Kevlar 29	2.6	13.8	79.80	2.59	2.17	30.81	36.77

* Silkworm silk has a triangular cross-section and d_f is an average of bottom and height measurements. σ_u, E_l, E_t, G_l, d_f are the strength, longitudinal, transverse, shear moduli, and diameter, respectively.

Table 2.14. Mechanical properties and composition of SiC and related fibres

Fibre	Manufacturer	Composition (%)	Diameter (µm)	Density (g cm^{-3})	Modulus (GPa)	Tensile strength (GPa)
Hi Nicalon S	Nippon	68.9Si/30.9C/0.2O	12	3.1	420	2.6
Hi Nicalon	Carbon	63.7Si/ 35.8C/0.5O	14	2.74	270	2.8
Nicalon NL 200/201		56.5Si/31.2C/12.3O	14	2.55	220	3.0
Tyranno SA 3	Ube	67.8Si/34.2C/0.3O/<2.0Al	7.5/10	3.1	380	2.8
Tyranno ZMI		56.1Si/34.2C/8.7C/1.0Zr	11	2.48	200	3.4
Tyranno Lox-M		55.4Si/32.4C/10.2O/2.0Ti	11	2.48	187	3.3
Tyranno S		50.4Si/29.7C/17.9O/2.0Ti	8.5/11	2.35	170	3.3
Sylramic-iBN	ATK COI	SiC/BN	10	3.00	400	3.0
Sylramic		96.0SiC/3.0TiB$_2$/ 1.0B$_4$C/0.3O	10	2.95	310	2.7
SCS-Ultra	Speciality	SiC on C core	140	3.0	415	5.87
SCS-9A	Materials	SiC on CF core	78	2.8	307	3.45
SCS-6		SiC on CF core	140	3.0	380	3.45
Sigma	Tisics	SiC on W wire core	100/140	3.4	400	4.0

high-performance reinforcements that can resist the processing temperature and environment.

Early ceramic fibres (1960s) were manufactured by chemical vapour deposition (CVD) onto a support fibre [85]. Boron and silicon carbide (SiC) fibres had a tungsten wire (or CF) core so their diameters were >100 µm. The properties of these CVD fibres are given in Table 2.14, which describes the range of fibres available. Clearly these large-diameter fibres cannot be woven into applicable reinforcements, but they are being considered as a reinforcement for titanium. Mono-crystalline **SiC whiskers** were also developed in the 1980s. These were grown at high temperature from a mixture of rice husks and a carbon source such as petroleum pitch. With a typical diameter of 1 µm, they were difficult to handle and presented potential health concerns.

2.7.1   Non-oxide Small-Diameter Fibres

The above SiC fibres have been largely superseded by the development of polymer precursors that can be melt-spun and converted into 'SiC' fibres with diameters of 10–15 µm [86]. The fundamental studies of Yajima in the 1970s led directly to the commercialization of **Nicalon** fibres by Nippon Carbon in 1982. The first generation of Nicalon fibres were synthesized in the following stages (Figure 2.28):

1. Synthesis of polysilane by condensation of a dichlorosilane using sodium metal.
2. Rearrangement of the polysilane into a polycarbosilane: the polysilane (Figure 2.28a) with Si–Si bonds was converted into a polycarbosilane (Figure 2.28b) with C–Si bonds in the skeleton in an autoclave at 470 °C.
3. Melt-spinning into a fibre: the original polycarbosilane could only be spun at 350 °C. Further work showed that the reaction product of the polysilane with borax, namely polyborodiphenylsiloxane, promoted the transformation to polycarbosilane at ≈300 °C without the need for an autoclave. Fibre spinning also became possible at 250 °C.
4. Stabilization of the polycarbosilane fibre for pyrolysis: the polycarbosilane fibre was crosslinked (cured) in air at 110 °C. Oxygen bridges between the polycarbosilane molecules were created.
5. Conversion of the crosslinked polycarbosilane fibre into an SiC fibre: this occurred at 1,000–1,300 °C in an inert atmosphere, but SiCO was formed with an oxygen content of 15–20%. The presence of O and inclusions of free carbon in the fibre structure was responsible for their limited thermal stability. In particular, at

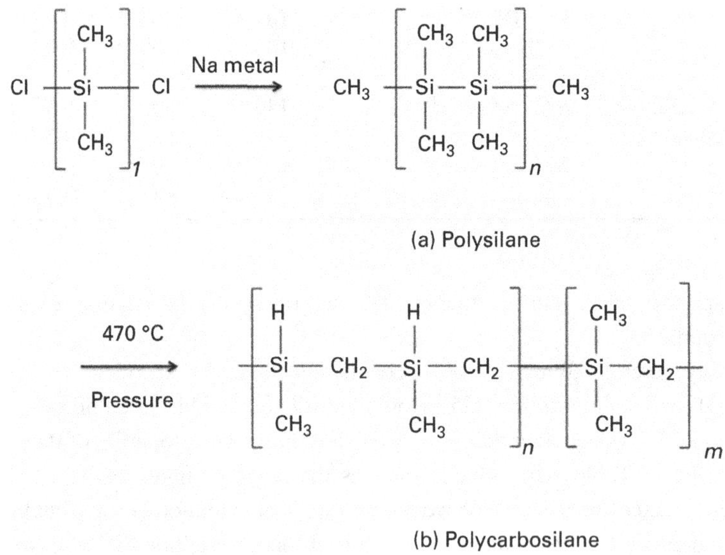

(a) Polysilane

(b) Polycarbosilane

Figure 2.28 Synthesis of the polycarbosilane precursor for the first generation of SiC fibres according to Yajima [86].

>1,100 °C β-SiC crystal growth occurred, reducing the fibre strength. The fibres also had a passivating SiO_2 layer, which promoted strong bonding with potential matrices and led to loss of composite 'toughness'.

The mechanical properties and composition of these fibres is given in Table 2.14.

2.7.1.2 Hi Nicalon Fibres

These fibres have improved thermal stability resulting from a reduction in the O content to $\leq 1\%$. This was achieved by curing the spun polycarbosilane fibres in an inert atmosphere using an electron beam or γ-radiation. Thermal stability is increased to 1,400–1,500 °C. A further benefit of radiation curing is that the pyrolysis temperature could be increased to 1,300 °C, which reduces the free carbon and the SiCO contents. **Hi Nicalon S** fibres have a near unity Si:C ratio arising from the use of H_2 atmospheres in the pyrolysis process.

2.7.1.3 Tyranno Fibres

These fibres, from Ube Industries, have a composition of SiTiCO and are manufactured similarly to Nicalon, except titanium tetraisopropoxide is included in production of the polycarbosilane. As with Nicalon, radiation curing reduces the O content from $\approx 20\%$ to 5% (**Tyranno LOX-M**).

2.7.1.4 Tyranno SA Fibres

These fibres have a composition of SiAlCO as a result of a modified polycarbosilane, which incorporated aluminium (III) acetyl acetonate. The ceramic fibre, which has been sintered at 1,800 °C, has an Al content of <2% and O content of $\approx 0.5\%$. A tensile modulus of >300 GPa and a strength >2.5 GPa can be achieved.

2.7.1.5 Sylramic Fibres

These fibres are a near-stoichiometric SiC fibre originally developed by Dow Corning in a related technique, and now available from ATK COI Ceramics, who use an *in-situ* boron nitride (BN) sintering agent that provides improved creep resistance at higher temperatures.

A number of attempts to synthesize small-diameter SiN fibres have been reported in the literature, but the presence of free Si has limited the development of the mechanical properties for high-temperature performance. The author is unaware of commercial availability.

2.7.2 Surface Finish of Small-Diameter Fibres

Nicalon and related fibres need to be handled in a variety of textile processes and are therefore sized with an appropriate finish for the impregnation technique and matrix. Typical coatings are epoxy, PVAc, polyvinyl alcohol (PVAl), polyimide (PI), and carbon or graphite. The latter is used to prevent the formation of strong interfaces during high-temperature processing as a result of reactions of an SiO_2-rich surface with ceramic matrices. More recently, BN coatings [87] have been shown to provide the correct interfacial response for optimum toughness in SiC–SiC composites.

2.7.3   Oxide Fibres

Early attempts employed melt-spinning of aluminosilicates with alumina concentrations of 45–60%. These glass-like fibres are manufactured in discontinuous form and find applications as high-temperature insulation and asbestos replacements. The form of the material is not really suitable for composite manufacture, but the demand for reinforcement of metals and ceramics led to the development of continuous high-performance ceramic fibres [88,89]. Essentially, alumina provided the potential mechanical and thermal performance and high-content materials were explored.

An early approach employed an aqueous slurry of alumina (Al_2O_3) particles, whose viscosity was controlled with dissolved polymers, to extrude a fibre that could be calcined and sintered to obtain a 20 μm α-Al_2O_3 fibre (99.9%), which was marketed by DuPont as **Fiber FP**. The modulus was 380 GPa, but a maximum strength of 1.5 GPa limited applications. **Fiber-PD166** included 20% tetragonal zirconia (ZrO_2) with 80% Al_2O_3 to achieve a higher strength of 2.2 GPa, resulting from a finer grain size. These fibres are no longer commercially available. Sol-gel processing has been employed in the manufacture of discontinuous fibres (e.g. **Saffil** and **Saffimax** (Unifrax)) and a range of continuous fibres (e.g. **Nextel** (3M)).

Saffil fibres have a diameter of 3–3.5 μm and are in the form of a fibre mat, but have found applications in MMC by liquid metal infiltration. Typical application is the reinforcement of piston crowns in automotive engines [89,90]. Nextel fibres are available with a diameter of 10–12 μm in a continuous form.

In the sol-gel process aqueous solutions of soluble aluminium salts or alkoxides are prepared which increase in viscosity for spinning as the salts or alkoxides hydrolyse and polymerize and eventually gel into an Al_2O_3 precursor. Silica is also added to control the crystalline structure. The composition of Nextel fibres is given in Table 2.15.

Altex (Sumitomo) is a 17 μm 85% Al_2O_3/15% SiO_2 fibre spun from a non-aqueous solution of the polyaluminoxane/siloxane precursor (from aluminium and silicon alkoxides) in glacial acetic acid [89,91]. These fibres are reported to have a modulus of 210 GPa and strength of 1.8 GPa. The reader is referred to more detailed reviews of ceramic fibres and their performance [88,89].

Almax fibres (Mitsui) are also available, but details of their manufacture are unknown. However, the reported properties are attractive.

2.8   Particulate Fillers and Reinforcements

There is a range of mineral particles that are used as fillers for plastics. Most of these are used to reduce material costs. The surfaces need to be modified for compatibility with polymers. For silica and silicate minerals, silane coupling agents are used to provide the

Table 2.15. Mechanical properties of oxide ceramic fibres

Fibre	Manufacturer	Composition (%)	Process	Diameter (μm)	Density (g cm^{-3})	Modulus (GPa)	Strength (GPa)
Saffil	Saffil	96–97 Al$_2$O$_3$/3–4 SiO$_2$	Sol-gel staple	3–3.5	3.3–3.4	300	1.5–2.0
Saffimax	Saffil	96–97 Al$_2$O$_3$/3–4 SiO$_2$	Sol-gel semi-continuous	3–3.5	3.3–3.4	300	2.0
Saffil/Nextel 720	3M	85 Al$_2$O$_3$/15 SiO$_2$	Sol-gel	10–12	3.4	260	2.1
Nextel 610		>99 Al$_2$O$_3$	Sol-gel	10–12	3.9	380	3.1
Nextel 550		73 Al$_2$O$_3$/27 SiO$_2$	Sol-gel	10–12	3.03	193	2.0
Nextel 440		70 Al$_2$O$_3$/28 SiO$_2$/2 B$_2$O$_3$	Sol-gel	10–12	3.05	190	2.0
Nextel 312		62.5 Al$_2$O$_3$/24.5 SiO$_2$/13 B$_2$O$_3$	Sol-gel	10–12	2.7	150	1.7
Altex	Sumitomo	85 Al$_2$O$_3$/15 SiO$_2$	Polyaluminoxane/siloxane	10–15	3.3	210	1.8
ALF	Nitivy	72 Al$_2$O$_3$/28 SiO$_2$	Sol-gel	7	2.9	170	2.0
Almax-B	Mitsui	60–80 Al$_2$O$_3$/40–20 SiO$_2$	Unknown	7–10	2.9	–	–
Almax	Mitsui	>99.5 Al$_2$O$_3$	Unknown	10	3.6	323	1.8

Table 2.16. Principal synthetic acicular reinforcing minerals

Properties	Graphene/graphene oxide	Carbon nanotubes	Aragonite (Maruo)	Anhydrite (US Gypsum)	Phosphate (Monsanto)	Dawsonite (Alcoa)	Brucite (Kyowa)	Magnesium oxysulphate (Ube)
Structure properties	C	C (graphite)	CaCO$_3$	CaSO$_4$	CaNa(PO$_3$)$_3$	NaAl(OH)$_2$CO$_2$	Mg(OH)$_2$	MgSO$_4$·5MgO·8H$_2$O
Aspect ratio (l/d)	Plate	>10^6	1	30	20	30	40	50–100
Diameter (μm)	N/A	*SW: 0.001 †MW: 0.005–0.05	1	2	1–5	0.5	<1	<1
Density (g cm^{-3})	1.3–2.0		2.9	3.0	2.9	2.4	2.4	2.3

* Single-wall; † multi-wall.

mineral surface with matrix-matching chemistry. Since the interfacial bond is improved, these 'fillers' are usually referred to as 'reinforcing fillers'. In the case of chalk fillers, stearic acid is used to provide the surface with hydrocarbon (CH_3 $(CH_2)_{16}$ $COO-$) chains, which can dissolve in polyethylene and polypropylene matrices.

To provide composite-like properties, the minerals should be fibre-like (i.e. needles) so there has been much interest in acicular crystals with high aspect ratio (length/diameter). Natural products such as asbestos, which are hydrated silicates, are examples of acicular particles. A natural low-toxicity acicular mineral is wollastonite (calcium metasilicate), but the maximum aspect ratio is 20. Synthetic minerals have become available as described in Table 2.16 [92,93].

Carbon nanotubes (CNTs) have largely superseded these acicular minerals because of (1) the higher modulus of CNTs (1,000 GPa) and (2) the higher aspect ratio. The major difficulty with CNTs is the random arrangement arising from the high aspect ratio. This is partially overcome by synthesizing CNT forests on catalysed substrates [94].

In composites technology fillers or particulate reinforcements are mainly used to provide different functionalities. Aluminium trihydrate is used to provide fire retardancy; CNTs or carbon black are used for thermal or electrical conductivity (carbon black is a reinforcement for elastomers such as natural rubber); and reinforcing fillers are used as matrix modifiers to reduce expansion coefficients and laminate thermal stresses.

2.9 Summary

Fibres and other fibrous materials available as reinforcements in composite materials are reviewed. The major players such as glass and carbon are described in detail, and sufficient details of the less important fibrous materials are given to enable further research.

2.10 Discussion Points

1. Which aspects of the structure of polyacrylonitrile (PAN) fibre control the achievable modulus of a CF?
2. What processing factor for the PAN precursor fibre is employed for the manufacture of intermediate modulus CF?
3. What processing factor for the carbonization of PAN fibre is responsible for HM CF?
4. Which factors determine the strength of Type A carbon fibre and which parameters are used to maximize the strength?
5. Which surface property of CF is modified to optimize the fracture toughness of a composite? How is this achieved?
6. Briefly discuss the manufacture of glass fibres.

7. Why are glass fibres sized during manufacture?
8. What component is included in the sizing to provide interfacial durability?
9. In which composite applications are aramid fibres preferred and why?
10. Which fibres can be employed for reinforcing ceramics?

References

1. F. R. Jones, ed. *Handbook of polymer–fibre composites* (Harlow: Longman, 1994).
2. F. R. Jones, Serendipity in carbon fibres: interfaces and interphases in composites. In *The structural integrity of carbon fiber composites*, ed. P. Beaumont, C. Soutis, A. Hodzic (Cham: Springer, 2017), pp. 71–97.
3. P. Morgan, *Carbon fibres and their composites* (Boca Raton, FL: Taylor & Francis, 2005).
4. W. Watt, Chemistry and physics of the conversion of polyacrylonitrile fibres into high modulus carbon fibres. In *Strong fibres*, ed. W. Watt, B.V. Perov (Amsterdam: Elsevier, 1985), pp. 327–388.
5. W. N. Reynolds and R. Moreton, Some factors affecting the strengths of carbon fibres. *Phil. Trans. R. Soc. London* A294 (1980), 451.
6. W. N. Reynolds and J. V. Sharp, Crystal shear limit to carbon fibre strength. *Carbon* **12** (1974), 103.
7. R. Moreton, The tensile strengths of PAN based carbon fibres. In *Strong fibres*, ed. W. Watt and B.V. Perov (Amsterdam: Elsevier, 1985), pp. 445–474.
8. F. R. Jones, A review of interphase formation and design in fibre-reinforced composites. *J. Adhes. Sci. Technol.* **24** (2010), 171–202.
9. P. J. Goodhew, A. J. Clarke, and J. E. Bailey, A review of the fabrication and properties of carbon fibres. *Mater. Sci. Eng.* **17** (1975), 3–30.
10. F. R. Jones, Reinforced plastics composites. In *Modern aluminium alloys*, ed. R. Doherty and A. Vasudevan (New York: Academic Press, 1989), pp. 605–469.
11. S. C. Bennett, D. J. Johnson, and W. Johnson, Strength–structure relationships in PAN-based carbon fibres. *J. Mater. Sci.* **18** (1983), 3337–3347.
12. P. Denison, F. R. Jones, and J. F. Watts, *Surf. Interface Anal.* **9** (1986), 43–435.
13. F. R. Jones, Fibre-matrix adhesion-carbon fibres. In *Handbook of adhesion*, ed. D. E. Packham, 2nd ed. (Chichester: Wiley, 2004), pp. 177–181.
14. M. R. Alexander and F. R. Jones, Effect of electrolytic oxidation upon the surface chemistry of type A carbon fibres: III. Chemical state, source and location of surface nitrogen. *Carbon* **34** (1996), 1093–1102.
15. C. Jones and E. Sammann, The effect of low power plasmas on carbon fibre surfaces. *Carbon* **28** (1990), 509–519.
16. A. P. Kettle, A. J. Beck, L. O'Toole, F. R. Jones, and R. D. Short, Plasma polymerisation for molecular engineering of carbon-fibre surfaces for optimised composites. *Compos. Sci. Technol.* **57** (1997), 1023–1032.
17. B. Rand and M. Turpin, Carbon fibres from pitch. In F. R. Jones, ed. *Handbook of polymer–fibre composites* (Harlow: Longman, 1994), pp. 34–38.
18. B. A. Newcomb and H. G. Chae, The properties of carbon fibers. In *Handbook of properties of textile and technical fibres*, ed. A. R. Bunsell, 2nd ed. (Duxford: Woodhead-Elsevier, 2018), pp. 841–863.

19. T. Suzuki and H. Umehara, Pitch-based carbon fiber microstructure and texture and compatibility with aluminum coated using chemical vapor deposition. *Carbon* **37** (1999), 47–59.

20. H. Marsh and R. Menendez. In *Introduction to carbon Science*, ed. H. Marsh (London: Butterworth, 1989), p. 37.

21. D.V. Dunford, J. Harvey, J. Hutchings, and C. H. Judge, *The effect of surface treatment of type 2 carbon fibre on CFRP properties, RAE Technical Report TR81096* (London: HMSO, 1981).

22. S. Yumitori, D. Wang, and F. R. Jones, The role of sizing resins in carbon fibre-reinforced polyethersulfone (PES). *Composites* **25** (1994), 698–705.

23. F. R. Jones, Interphase in fiber-reinforced composites. In *Wiley encyclopedia of composites*, ed. L. Nicolais and A. Borzacchiello, 2nd ed. (New York: Wiley, 2012).

24. Hexcel Corporation. Home page. www.hexcel.com.

25. Toray Industries Inc. Home page. www.toray.com.

26. Nippon Graphite Fiber Corporation. Home page. www.ngfworld.com.

27. F. R. Jones and N. T. Huff, The structure and properties of glass fibres. In *Handbook of textile fibre structure*, vol. 2, ed. S. J. Eichhorn, J. W. S. Hearle, M. Jaffe, and T. Kikutani (Oxford: Woodhead, 2009), pp. 307–352.

28. F. R. Jones and N. T. Huff, The structure and properties of glass fibers. In *Handbook of properties of textile and technical fibres*, ed. A. R. Bunsell, 2nd ed. (Duxford: Woodhead-Elsevier, 2018), pp. 577–803.

29. K. Loewenstein, *The manufacturing technology of continuous glass fibres*, 3rd ed. (Amsterdam: Elsevier, 1993).

30. A. J. Majumdar, Br. Pat. GB 1,243,972/ GB 1,243,973 (1971).

31. J. G. Mohr and W. P. Rowe, *Fibreglass* (New York: van Nostrand, 1978).

32. F. R. Jones, Glass fibres. In *High-performance fibres*, ed. J. W. Hearle (Cambridge: Woodhead, 2001), pp. 191–238.

33. J. H. Simmons, What is so exciting about non-linear viscous flow in glass, molecular dynamics simulations of brittle fracture and semiconductor-glass quantum composites. *J. Non-Cryst. Sol.* **239** (1998), 1–15.

34. A. A. Griffith, The phenomena of rupture and flow in solids. *Phil. Trans. Roy. Soc. Lond. A* **221** (1920), 163.

35. A. Kelly and N. H. MacMillan, *Strong solids*, 3rd ed. (Oxford: Clarendon, 1990).

36. A. G. Metcalfe and G. K. Schmitz, Mechanism of stress corrosion in E-glass filaments. *Glass Tech.* **13** (1972), 5.

37. G. M. Bartenev, *The structure and mechanical properties of inorganic glasses* (Groningen: Walters-Noordhoff, 1970).

38. N.G. McCrum, *Review of the science of fibre reinforced plastics* (London: HMSO, 1971).

39. W. F. Thomas, An investigation of the factors likely to affect the strength and properties of glass fibres. *Phys. Chem. Glass* **1** (1960), 4–18.

40. V. E. Khazanov, Y. I. Kolesov and N. N. Trofimov, Glass fibres. In *Fibre science and technology*, ed. V. I. Kostikov (London: Chapman & Hall, 1995), pp. 15–230.

41. W. H. Otto, Relationship of tensile strength of glass fibers to diameter. *J. American Ceramic Soc.* **38** (1955), 122–124.

42. Weibull W. *A statistical theory of the strength of materials* (Stockholm: Royal Technical University, 1939).

43. Weibull W. A statistical distribution function of wide applicability. *J. Appl. Mech.* **18** (1951), 293–297.

44. J. L. Thomason, On the application of Weibull analysis to experimentally determined single fibre strength distributions. *Comp. Sci. Tech.* **77** (2013), 74–80.

45. J. Aveston, A. Kelly, J. M. Sillwood, Long-term strength of glass reinforced plastics in wet environments. In *Advances in composite materials*, ed. A. R. Bunsell (Paris: Pergamon, 1980), pp. 556–568.

46. F. R. Jones, The effects of aggressive environments on fatigue in composites. In *Fatigue of composite materials*, ed. B. Harris (Cambridge: Woodhead, 2003), pp. 117–146.

47. R. J. Charles, Static fatigue of glass I and II. *J. Appl. Phys.* **29** (1958), 1549.

48. S. B. Ghosh, F. R. Jones, and R. J. Hand, A novel indentation based method to determine the threshold stress intensity factor for subcritical crack growth in Glass. *Eur. J. Glass Sci. Technol.* **A 51** (2010), 156–160.

49. F. R. Jones, J. W. Rock, and J. E. Bailey, The environment stress corrosion cracking of glass fibre reinforced laminates and single E-glass filaments. *J. Mater. Sci.* **18** (1983), 1059–1071.

50. D. R. Cockram, Strength of E glass in solutions of different pH. *Glass Tech.* **22** (1981), 211–214.

51. P. A. Sheard, Transverse and environmental cracking of glass fibre reinforced plastics, PhD thesis, Surrey University, 1986.

52. F. R. Jones, J. W. Rock, and A. R. Wheatley, Stress corrosion cracking and its implications for the long-term durability of E-glass fibre composites. *Composites* **14** (1983), 262.

53. J. L. Thomason, *Glass fibre sizings: a review of the scientific literature* (Glasgow: James L. Thomason, 2012).

54. J. L. Thomason, *Glass fibre sizing: a review of size formulation patents* (Glasgow: James L. Thomason, 2015).

55. J. L. Thomason and L. J. Adzima, Sizing-up the interphase: an insider's guide to the science of sizing. *Composites A* **32** (2001), 313–321.

56. E. Plueddemann, *Silane coupling agents*, 2nd ed. (New York: Plenum 1991).

57. X. M. Liu, J. L. Thomason, and F. R. Jones, XPS and AFM study of interaction of organosilanes and sizing with E-glass fibre surface. *J. Adhesion* **84** (2008), 322.

58. D. Wang, F. R. Jones, and P. Denison, A surface analytical study of the interaction between γ-amino propyltriethoxysilane and E-glass surface, Part 1: time-of-flight secondary ion mass spectrometry (ToF SIMS). *J. Mater. Sci.* **27** (1992), 36–48.

59. C. G. Pantano, R. A. Fry, and K. T. Mueller, Effect of boron oxide on surface hydroxyl coverage of aluminoborosilicate glass fibres: a ^{19}F solid state NMR study. *Phys. Chem. Glasses* **44** (2003), 64–68.

60. X. M. Liu, J. L. Thomason, and F. R. Jones, The concentration of hydroxyl groups glass surfaces and their effect on the structure on E-glass surfaces. In *Silanes and other coupling agents*, vol. 5, ed. K. L. Mittal (Boston, MA: Brill Academic Publishers, 2009), pp. 25–38.

61. S. Rebouillat, Aramids. In *High-performance fibres*, ed. J. W. Hearle (Cambridge: Woodhead, 2001), pp. 23– 61.

62. S. van der Zwaag, The structure and properties of aramid fibres. In *Handbook of textile fibre structure*, vol.1, ed. S. J. Eichhorn, J. W. S. Hearle, M. Jaffe, and T. Kikutani (Oxford: Woodhead, 2009), pp. 394–412.

63. A. Pegoretti and M. Traina, Liquid crystalline organic fibers and their mechanical behaviour. In *Handbook of properties of textile and technical fibres*, ed. A. R. Bunsell (Duxford: Woodhead-Elsevier, 2018), pp. 354–436.

64. M. G. Dobb, D. J. Johnson, and B. P. Saville, Supramolecular structure of a high modulus polyaromatic fibre. *J. Polymer Sci., Polym. Physics* **15** (1977), 2201–2011.

65. S. J. Eichhorn, J. W. S. Hearle, M. Jaffe, and T. Kikutani, eds. *Handbook of textile fibre structure*, vol. 1 (Oxford: Woodhead, 2009).

66. T. Kitagawa, The structure of high-modulus, high tenacity poly-p-phenylenebenzobisoxazole fibres. In *Handbook of textile fibre structure*, vol. 1 ed. S. J. Eichhorn, J. W. S. Hearle, M. Jaffe, and T. Kikutani (Oxford: Woodhead, 2009), pp. 429–454.

67. D. Beers, R. J. Young, C. L. So, et al. Other high modulus-high tenacity (HM-HT) fibres from linear polymers. In *High-performance fibres*, ed. J. W. Hearle (Cambridge: Woodhead, 2001), pp. 93–155.

68. J. W. S. Hearle, The structure of high-modulus, high tenacity PIPD 'M5' fibre. In *Handbook of textile fibre structure*, vol.1, ed. S. J. Eichhorn, J. W. S. Hearle, M. Jaffe, and T. Kikutani (Oxford: Woodhead, 2009), pp. 455–459.

69. S. J. Eichhorn, J. W. S. Hearle, M. Jaffe, and T. Kikutani, eds., *Handbook of textile fibre structure*, vol. 2 (Oxford: Woodhead, 2009).

70. I. M. Ward and P. J. Lemstra, Production and properties of high-modulus and high strength polyethylene fibres. In *Handbook of textile fibre structure*, vol. 1, ed. S. J. Eichhorn, J. W. S. Hearle, M. Jaffe, and T. Kikutani (Oxford; Woodhead, 2009), pp. 352–393.

71. J. L. J. van Dingenen, Gel-spun high-performance polyethylene fibres. In *High-performance fibres*, ed. J. W. Hearle (Cambridge: Woodhead, 2001), pp. 62–92.

72. R. J. Young, Fracture of polymer crystals. In *Developments in polymer fracture*, vol. 1, ed. E. H. Andrews (London: Applied Science, 1979), p. 223.

72. I. M. Ward, *Mechanical properties of solid polymers*, 2nd ed. (London: Wiley-Interscience, 1983).

73. P. J. Lemstra, R. Kirschbaum, T. Ohta, and H. Yasuda, High-strength/high-modulus structures based on flexible macromolecules: gel-spinning and related processes. In *Developments in oriented polymers*, vol. 2, ed. I. M. Ward (London: Elsevier Applied Science, 1987), pp. 39–78.

74. A. E. Zacharides, W. T. Mead, R. S. Porter, Recent developments in ultraorientation of polyethylene by solid-state extrusion. *Chem. Rev.* **80** (1980), 351–360.

75. R. Shishoo, ed. *Plasma technologies for textiles* (Cambridge: Woodhead, 2007).

76. E. Richaud, B. Fayolle, and P. Davies, Tensile properties of polypropylene fibers. In *Handbook of properties of textile and technical fibres*, 2nd ed., ed. A. R. Bunsell, (Duxford: Woodhead-Elsevier, 2018), pp. 515–544.

77. B. Alcock and T. Peijs, Technology and development of self-reinforced polymer composites. In *Polymer composites – polyolefin fractionation – polymeric peptidomimetics – collagens* (Berlin: Springer-Verlag, 2013), pp. 1–76.

78. D. J. Hannant, J. J. Zonsveld, and D. C. Hughes, Polypropylene in cement based materials. *Composites* **9** (2) (1978), 217–226.

79. D. J. Hannant, *Fibre cements and fibre concretes* (Chichester: Wiley-Interscience, 1978), pp 81–98.

80. B. Mobasher, *Mechanics of fiber and textile reinforced cement composites* (Boca Raton, FL: CRC Press, 2011).

81. D. Zhang, Lightweight materials from biofibers and biopolymers. In *Light weight materials,* ed. Y. Yang, H. Xu, and X. Yu (Washington, DC: American Chemical Society, 2014), pp. 1–20.

82. F. K. Ko, Engineering properties of spider silk fibres. In *Natural fibres, plastics and composites*, ed F. T. Wallenberg and N. E. Weston (Boston, MA: Springer, 2004), pp. 27–49.

83. P. Colomban and V. Jauzein, Silk: fibers, films, and composites – types, processing, structure, and mechanics. In *Handbook of properties of textile and technical fibres*, ed. A. R. Bunsell, 2nd ed. (Duxford: Woodhead-Elsevier, 2018), pp. 137–183.

84. F. K. Ko and L. Y. Wan, Engineering properties of spider silk. In *Handbook of properties of textile and technical fibres*, ed A. R. Bunsell, 2nd ed. (Duxford: Woodhead-Elsevier, 2018), pp. 185–220.

85. X. Luo and N. Jin, Fibers made by chemical vapor deposition. In *Handbook of properties of textile and technical fibres*, ed. A. R. Bunsell, 2nd ed. (Duxford: Woodhead-Elsevier, 2018), pp. 929–992.

86. A. R. Bunsell, Small-diameter silicon carbide fibers. In *Handbook of properties of textile and technical fibres*, ed. A. R. Bunsell, 2nd ed. (Duxford: Woodhead-Elsevier, 2018), pp. 873–902.

87. M. Yuan, T. Zhou, J. He, and L. Chen, Formation of boron nitride coatings on silicon carbide fibers using trimethylborate vapour. *Appl. Surf. Sci.* **382** (2016), 27–33.

88. T. F. Cooke, Inorganic fibers: a literature review. *J. Am. Ceram. Soc.* **74** (1991), 2959–2978.

89. D. Wilson, Continuous oxide fibers. In *Handbook of properties of textile and technical fibres*, ed A. R. Bunsell, 2nd ed. (Duxford: Woodhead-Elsevier, 2018), pp. 903–927.

90. J. D. Birchall, J. A. A. Bradbury, and J. Dinwoodie, Alumina fibers. In *Handbook of composites*, eds. W. Watt and B. V. Perov (Amsterdam: North-Holland, 1985), pp. 115–155.

91. T. Yogo and H. Iwahara, Synthesis of α-alumina fibre from modified aluminum alkoxide precursor. *J. Mater. Sci.* **27** (1992), 1499–1504.

92. J. V. Milewski. In *Handbook of reinforcements for plastics*, ed. J. V. Milewski and H. S. Katz (New York: Reinhold van Nostrand, 1987).

93. R. Rothon, Acicular particulate reinforcement. In *Handbook of Polymer–Fibre composites*, ed. F. R. Jones (Harlow: Longman, 1994), pp. 4–7.

94. M. F. L. De Volder, S. H. Tawfick, R. H. Baughman, and A. J. Hart, Carbon nanotubes: present and future commercial application. *Science* **339** (2013), pp. 535–539.

3 Matrices

In this chapter we describe the resins used for the manufacture of composite artefacts. The concept of curing is discussed with respect to the chemistry of typical polymer matrices. The advantages and disadvantages of thermosets and thermoplastics are also discussed.

In the case of thermosets, the importance of thermoplastic and rubber toughening is considered. While we concentrate on polymer matrix materials, ceramic and metal matrices are referred to for completeness.

3.1 Introduction

Most reinforcing fibres have moduli that are at least 10 times the moduli of glassy polymers, so the role of the matrix is to maintain their orientation and transfer loads to the fibres. The rate of transfer of load between the fibres and matrix is a critical aspect, especially for discontinuous fibre systems and at fibre and/or matrix breaks in continuous fibre composites.

3.1.1 Mechanisms of Viscoelasticity in Polymers

The modulus of a polymer is strongly dependent on the mechanism of deformation available [1]. Depending on the molecular structure of the polymer, it can exist at room temperature as either

a. glass
b. retarded high elastic material ('leathery material')
c. rubber
d. viscoelastic fluid.

All polymers can exist in either state, depending on the temperature. Thus, a glassy polymer will go through four stages of viscoelasticity on heating. The glassy and rubbery states can be characterized by unique values of modulus, whereas within the transition regions (b) and (d) the properties are time-dependent. Of these, the glass transition temperature (T_g) is a critical parameter as it defines the service temperature of a rigid structural composite. Figure 3.1 provides a schematic showing the change in relaxation modulus with temperature. The modulus changes over several decades as

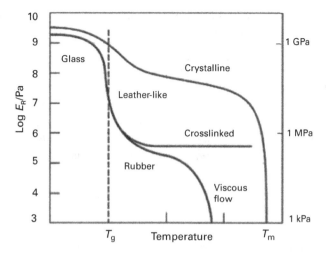

Figure 3.1 The relaxation modulus (E_R) of a polymer as a function of morphology (crystallization and crosslinking) and temperature. An illustration of the mechanical properties at different temperatures relative to the glass transition temperature.

the glassy material becomes rubbery. T_g is defined conventionally as the maximum in tanδ, which occurs at the temperature for maximum damping and is shown in Figure 3.1 at the inflection point of the storage modulus versus temperature curve. In the transition region the material exhibits 'leather-like' properties. The presence of crystallites increases the modulus in an analogous way to the incorporation of particulate fillers. Crosslinking of the polymer chains also enhances the modulus above T_g. Light crosslinking also extends the rubbery plateau (temperature range) and is often referred to as vulcanization. In aerospace engineering the onset of the loss of a glassy modulus is often used to define the value of T_g, which in Figure 3.1 occurs at approximately 5–10 °C lower.

3.1.2 Magnitude of the Modulus (*E*) of a Glassy Polymer

In a polymer glass the only forces resisting deformation are the intermolecular forces between molecular segments. These range from van der Waals through to hydrogen bonding. Crosslinking introduces irreversible covalent bonds in addition to the intermolecular forces. To have a major effect on modulus, crosslinking needs to be extensive. With linear polymers of regular structure, such as polypropylene or polyethylene, crystallization enhances the modulus because of the higher density. In the latter, the modulus can be modelled using composite mechanics. As discussed in Chapter 2, polymer molecules are randomly coiled through rotation about skeletal bonds so that under an applied load the forces of interaction between the molecules carry the load. Therefore, a simple estimate of modulus can be obtained from consideration of a simple frozen hydrocarbon gas, such as methane (CH_4) glass.

The intermolecular energy (ϕ_p) between molecules as a function of distance, R is described by the Lennard–Jones potential:

$$\phi_p = BR^{-n} - AR^{-6}, \tag{3.1}$$

where A and B are constants and $n = 12$ for the simple methane molecule.

As the molecules move together there is an attractive force, which has a maximum value (negative by convention) of ε_o at a fixed distance R_o. When pushed closer a repulsive force develops and the value of ϕ_p becomes progressively more positive. Therefore eq. (3.1) can be normalized to provide the potential energy between two methane molecules:

$$\frac{\phi_p}{\varepsilon_o} = \left(\frac{R}{R_o}\right)^{-12} - 2\left(\frac{R}{R_o}\right)^{-6}. \tag{3.2}$$

For a close-packed structure in which the molecule has 12 neighbours, eq. (3.2) can be simplified in terms of molar volume, V, since $R\alpha\left(\frac{1}{V}\right)^3$:

$$\frac{\phi_p}{\phi_o} = \left(\frac{V_o}{V}\right)^4 - 2\left(\frac{V_o}{V}\right)^2, \tag{3.3}$$

where $\phi_o = 12N_A\varepsilon_o$.

Under compression, the change in molar volume of the glassy methane is given by $dE = PdV$, assuming the change in entropy (dS) is zero. Differentiation gives

$$\frac{dP}{dV} = \frac{d^2E_i}{dV^2},$$

where E_i is the interaction energy. Therefore, the bulk modulus, K, is given by

$$K = -V\frac{dP}{dV} = -V\left(\frac{d^2E_i}{dV^2}\right). \tag{3.4}$$

Thus, eq. (3.3) becomes

$$K = 4\left(\frac{\phi_o}{V_o}\right)\left[5\left(\frac{V_o}{V}\right)^6 - 3\left(\frac{V_o}{V}\right)^4\right], \tag{3.5}$$

where $\phi_o = 12N_A\varepsilon_o$.

Since $V \approx V_o$, eq. (3.5) becomes

$$K = 8\frac{\phi_o}{V_o} = 96\frac{\varepsilon_o N_A}{V_o}. \tag{3.6}$$

For methane, $\varepsilon_o = 0.0127$ eV.

The molar volume, V_o, is given by molecular weight/density. For CH_4, this is:

$$V_o = \frac{16 \times 10^{-3} \text{ kg mol}^{-1}}{547 \quad \text{kg m}^{-3}} = 29 \times 10^{-6} \text{m}^3 \text{ mol}^{-1}.$$

Therefore, $K \approx 4$ GPa.

The bulk modulus, K, is related to Young's modulus, E, according to eq. (3.7):

$$E = 3K(1 - 2\nu), \tag{3.7}$$

where ν is Poisson's ratio. For a polymer, $\nu = 0.3 - 0.4$. Therefore, $E \approx 2.4 - 4.8$ GPa.

The estimate of Young's modulus for glassy methane is ≈ 4 GPa, and the typical maximum value for a glassy polymer will be similar when other deformation mechanisms are absent. Covalent crosslinking between polymer molecules can add resistance to deformation, but the maximum achievable is ≈ 7 GPa. This value has only been approached in highly crosslinked phenolic resins, but these have very low failure strains because ductile mechanisms, which reduce stress concentrations at a crack tip, are not available. Matrix resins such as epoxies have moduli of 2–3.5 GPa and failure strains of 1.5–6% because cooperative skeletal rotation within the bridging network chains provides the mechanism of ductility.

3.1.3   Magnitude of the Modulus of a Rubber

The thermodynamic and statistical analysis of the deformation of a rubber by Treloar [2] provides the magnitude of the modulus. Deformation of a rubber or elastomer involves flow, so a shear modulus, G, is given by:

$$G = \frac{\rho RT}{M_c}, \tag{3.8}$$

where ρ is the density, R is the gas constant, T is the temperature, and M_c is the number-average molecular weight of a network chain.

Unvulcanized or non-crosslinked rubbers deform by a reptation mechanism involving rotation about skeletal bonds leading to non-recoverable flow. To stabilize the flow of rubbers they are vulcanized or crosslinked. Natural and many synthetic rubbers employ sulphur for crosslinking, hence the term *vulcanization*. The degree of crosslinking determines the shear modulus of a rubber, as shown by eq. (3.8). As the degree of crosslinking increases, the length (defined by M_c) of the network chains decreases and G increases.

Real vulcanized rubbers are not perfectly crosslinked, so network defects will exist. These include ineffective network chains such as loops, entanglements, and incomplete network chains or network-ends. Equation (3.8) therefore needs to be corrected:

$$G = \left(\frac{<r_c^2>}{<r_0^2>}\right) \frac{\rho RT}{M_c} \left(1 - 2\frac{M_c}{M}\right), \tag{3.9}$$

where $<r_c^2>$ is the average square end-to-end distance of a network chain and $<r_0^2>$ is the unperturbed value for a free chain. This correction represents the influence of the molecular structure of the network chain. M is the number-average molecular weight of the polymer (before crosslinking) and $(1 - 2\ M_c/M)$ represents the effects of network defects.

A typical value of G for a rubber is ≈ 0.4 MPa, and a low strain elastic modulus (i.e. tensile) would be ≈ 1 MPa. Poly(butadiene) rubber exhibits ideal high elastic behaviour at room temperature, as indicated by zero hysteresis in the load–extension curve. Natural rubber (poly(isoprene)), however, exhibits significant hysteresis. This arises from the energy restriction imposed on the skeletal rotations by the pendant methyl group, but also at high extensions by strain-induced crystallization of the chain. The tensile modulus then has a contribution from loading of the molecular chain bonds.

3.2   Polymeric Matrices for Composites

The fabrication of a composite artefact involves the impregnation of fibres with either a thermosetting resin or a thermoplastic. Both thermoplastics and thermosets are used, but thermosets are preferred because they are liquid and 'monomeric' and have low viscosity so that fibre impregnation is easy and possible at room temperature. They are 'cured' into crosslinked polymers, either thermally or with hardeners and/or catalysts. Thermoplastics are linear polymers that on heating can become viscoelastic fluids for infiltration of continuous fibres or mats, or mixed with discontinuous fibres under pressure. The critical aspect of the latter is the high temperatures required for impregnation, and as a consequence the reproducibility of morphology of the polymer on cooling.

The advantages of a **thermoset** (relative to thermoplastics) are:

- low viscosity for impregnation and wetting of fibres;
- availability of a large number of fabrication processes; and
- the length of the fibres employed in the process is maximized.

The disadvantages of **thermosets** are also worth considering:

- material and artefact are formed simultaneously during curing;
- cast resins have low failure strain (ε_{mu}) and fracture toughness (K_{1c}, G_{1c});
- curing times can be extremely long; and
- the critical length (L_c) of a reinforcing fibre tends to be longer when embedded in thermosets (see Chapter 5).

The advantages of **thermoplastic** matrices (relative to thermosets) are:

- higher ductility;
- no chemistry is involved in the process;
- more rapid composite processing arising from the need for only heating and cooling;
- L_c tends to be lower and short fibre composites can be manufactured by standard thermoplastic moulding techniques such as injection moulding;
- low moisture absorption; and
- good hot/wet props.

These benefits are offset by the following disadvantages:

- high melt viscosity for fibre impregnation;
- high impregnation temperatures;
- crystalline morphology may vary;
- continuous fibre composites present complications to fabrication processes;
- uses mainly short-fibre-reinforced material, which has low mechanical performance arising from the reduced reinforcement efficiency of load transfer to discontinuous and randomly aligned fibres;
- the polymer has a low T_g, which is often offset by the presence of crystallites of high melting point (T_m);
- service temperatures are generally lower for thermoplastics-based structural composites because a small loss of rigidity will still occur at T_g; and
- thermoplastics have a lower modulus (E_m).

3.2.1 Requirements of a Resin Matrix

The following aspects of a resin intended for composite manufacture are desirable:

- low viscosity for impregnation;
- high tensile modulus, E_m (a maximum of 4 GPa for a cured thermoset resin matrix);
- high failure strain, ε_{mu};
- controllable curing for a thermoset; and
- controllable microstructure for a thermoplastic.

Typical **thermosets resins** employed as matrices are phenolics; unsaturated polyester and related resins (vinyl esters, urethane acrylates), which are mainly used in industrial applications; epoxy resins, which are used for structural and high-performance applications; advanced resins, including polyimides, bismaleimides, cyanate esters, and other specialist resins used in high-temperature applications.

Typical **thermoplastics** include most injection mouldable plastics, providing the reinforcements are compatible. The limitation is that the fibres need to be discontinuous in order to flow around the mould (i.e. short fibre composites). More advanced thermoplastics include polyether ether ketone (PEEK), polyether sulphone (PES), and polyetherimide (PEI), which can be used with continuous fibres as well for discontinuous fibre mouldings. The continuous fibres have dominated applications because only high performance can justify their high cost. Polypropene is an interesting matrix polymer; because of its low cost it can be used in mouldings but also as a matrix for continuous glass fibres (continuous and long glass fibres). Its limitations are determined by its compatibility with the reinforcements, but mainly by its low glass transition temperature.

The fabrication of a composite involves the impregnation and wet-out of the fibres, so thermosetting resins are preferred because a large number of manufacturing processes are available for 'long' and continuous fibres.

3.3   Curing of Thermoset Resins

The formation of most polymer composites involves the 'curing' of the resin after the fibres have been impregnated. Curing either occurs by 3D step-growth polymerization; addition copolymerization of unsaturated resin with a reactive diluent or monomer; addition polymerization of cyclic monomers; or addition copolymerization of cyclic monomers.

3.3.1   3D Step-Growth Polymerization

As the name implies, the molecular weight of the matrix builds in a step-wise manner until it becomes infinite as the resin gels into a network. To achieve the rigidity of the matrix for load bearing, further crosslinking needs to occur by diffusion of oligomers to reaction sites, as shown in Box 3.1.

From the above it can be seen that the functionality (the number of reactive groups) of the monomers will determine the extent of reaction of these groups for gelation. With highly functional monomers, gelation occurs at low extents of reaction but the potential degree of crosslinking is higher. To obtain matrices with high moduli and glass transition temperatures, large degrees of crosslinking are required. From the perspective of composite manufacture the above discussion explains the need for post-curing.

The Carothers theory [3] predicts the molecular weight of a polymer obtained by this mechanism as a function of the extent of reaction. For bi-functional monomers, linear polymers are formed and the number-average degree of polymerization of the chain, P_n, is given by:

$$P_n = 1/(1 - \rho), \tag{3.10}$$

where ρ is the extent of reaction of functional groups and $P_n = M_n/M_o$, where M_n is the number-average molecular weight of the polymer from a monomer of molecular weight M_o.

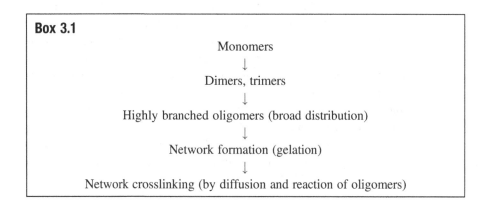

Box 3.1

Monomers
↓
Dimers, trimers
↓
Highly branched oligomers (broad distribution)
↓
Network formation (gelation)
↓
Network crosslinking (by diffusion and reaction of oligomers)

For the purposes of this discussion, we are interested in the gelation point for curing of multifunctional monomers. By applying eq. (3.10) we get

$$P_n = 2/(2 - pf_{av}),$$ (3.11)

where f_{av} is the average functionality of the mixture of monomers. For linear polymerization with bi-functional monomers, $f_{av} = 2$ and eq. (3.11) reduces to eq. (3.10).

At gelation, M_n approaches ∞; thus,

$$p_c = 2/f_{av},$$ (3.12)

where p_c is the critical extent of reaction for gelation or gel point.

From eq. (3.12) we see that the gelation occurs at lower extents of cure for systems with high functionality.

The Carothers theory can be extended to include an imbalance in the concentrations of reactive groups. The Flory statistical theory of gelation considers the probability of reaction of functional groups to predict the gel point. This is beyond the scope of this book, but is of fundamental interest to process engineers modelling matrix curing.

Step-growth polymerization often involves the loss of a small molecule (often water), in which case it is referred to as condensation polymerization. Phenolic resins are an example of condensation or step-growth curing, and demonstrate how curing can be interrupted. It is used here to illustrate the fundamentals or prepreg technology.

3.3.2   Phenolic Resins

Phenolic resins represent the first truly synthetic polymer, which became available at the turn of the twentieth century. Before then the early plastics were modified natural products. The starting reactants are formaldehyde (H_2CO) and phenol (C_6H_6OH). There are two synthetic routes: **resole** and **novolak**.

3.3.3   Resole Resins

When an excess of formaldehyde is used, resoles are formed; the chemistry is illustrated Figure 3.2.

Under alkaline conditions, trihydroxy methyl phenol forms initially, which has a functionality of 3. The phenolic hydroxyl is not involved in the reaction. This product is referred to as **resole** or A-stage resin. Since the resin is monomeric, it has a low viscosity and can be used to readily impregnate the reinforcement. In early applications the reinforcement was paper, which is a discontinuous cellulose fibrous matt. As shown in Figure 3.2, on further heating the trihydroxy methyl phenol monomers react with the loss of water and build up into a crosslinked polymer. The reaction can be interrupted at any stage to provide a resitol or B-stage resin. The resitol properties can be adjusted in this way, but it retains its reactivity and the volatile condensation water is removed in a controlled manner. Thus, a pre-impregnated reinforced layer (or ply)

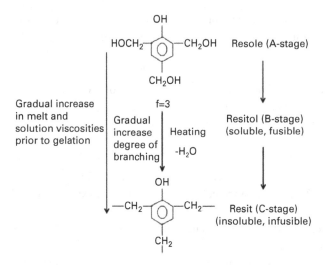

Figure 3.2 Schematic of the chemistry of phenol formaldehyde resole resins (alkali catalysed).

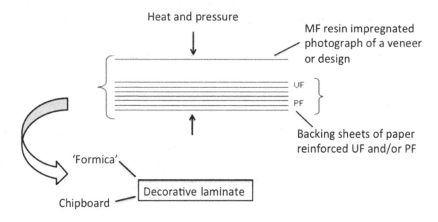

Figure 3.3 Schematic of ply lamination of resole impregnated sheets for decorative laminate manufacture: an early example of 'prepreg' technology.

with residual reactivity can be prepared. A laminate can be prepared by stacking the pre-impregnated plies for consolidation under pressure and heat. Curing occurs by a continuation of the reaction when an insoluble or infusible resit or C-stage resin is formed. This technique was developed to ensure that the volatile condensate could be controlled and voidage limited. Paper laminates were used in the electrical applications and subsequently as a means of manufacturing decorative laminates. Phenolic resins tend to yellow and brown to differing degrees, so water white resins were needed for the latter. These are usually the analogous urea–formaldehyde (UF) or melamine–formaldehyde (MF) resins. Figure 3.3 illustrates the manufacture of a decorative laminate.

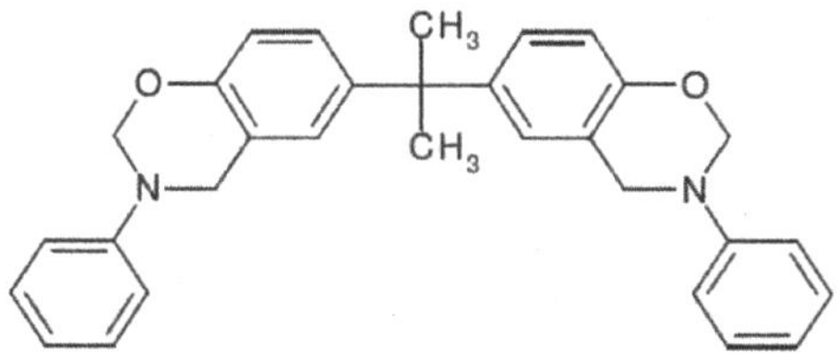

Figure 3.4 Bisphenol-A-based benzoxazine resin (Araldite MT 35600).

Resole resins are used for high-performance laminates from glass and carbon fibres, mainly as textile reinforcements where fire resistance is the main requirement. The main difficulties for structural composites revolves around the need for low voidage, which is not easy to achieve due to the water released during cure. To achieve low voidage, high pressures are employed to prevent bubble formation by keeping the water solubilized. This has the knock-on effect of reducing the matrix T_g by plasticization.

Because of the potential high-temperature and fire resistance of phenolic matrix composites, other related systems have been evaluated. For example bisoxazoline and benzoxazine resins (Figure 3.4) have been developed to meet these disadvantages.

Ishida and Agag [4] have recently reviewed the range of resins available. Figure 3.4 shows the structure of a commercial resin for homo- and copolymerization with epoxy resins (see Section 3.4) into a cured resin with a highly aromatic molecular structure akin to the phenolics.

Resole-type phenolics are also used as a precursor for a carbon or graphite matrix for so-called carbon (fibre)–carbon ceramic composites.

3.3.4 Novolak Resins

The paper laminates described above were the basis of the electrical industry when it developed in the late nineteenth and early twentieth centuries, but the technology was limited to simple shapes. There was a need to develop moulding compounds for moulding into more complex shapes. Baekeland is credited with the development of the technology in 1909 (the first patent was in 1907), which was invented in Belgium and commercialized in the USA. Swinburne, who had developed the resole approach, carried out similar research in the UK, but missed out on the patent by days.

In this approach the phenol is reacted with a deficiency of formaldehyde so that a material with only short linear molecules is formed. Acidic catalysis is employed to control the chemistry. The product is referred to as a **novolak** (Figure 3.5).

The novolaks have 5–7 phenolic groups per chain as a result of the excess phenol used in the reaction. The deficiency in formaldehyde content is addressed by adding hexamethylenetetramine (HMTA), the condensation product of formaldehyde and ammonia. The compound can be thermally shaped in a mould. The degree of flow in the tooling and hence complexity of the shape can be adjusted by pre-reacting the novolak with HMTA in the compounding stage, usually on a hot two-roll mill, to produce a moulding powder. The pressures employed in the moulding process are

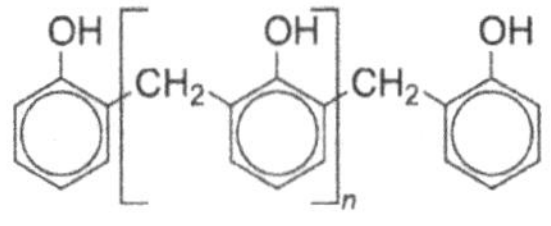

Figure 3.5 Simplified structure of a novolak phenolic resin showing ortho- or 2-substitution. The –CH$_2$– groups can also be attached at para- or 4-position of the phenol ring. n has a value of 2–5.

determined by the extent to which the 'cure' volatiles are lost in the compounding stage. The volatiles released are water and ammonia. Some HMTA residues are also retained in the molecular structure of the resin, which explains the colour that develops during curing. The other critical additive is the reinforcement, which can be either fibrous (e.g. cotton flock, wood flour, or glass fibre) or particulate (e.g. ground minerals or glass beads). The early synthetic plastics utilized composite concepts, as discussed in Chapter 1.

3.3.5   Resins for Composite Manufacture

The phenolic resins have much potential for use for fibre composites, but there are severe disadvantages: (1) curing chemistry is a condensation reaction involving the loss of H$_2$O, which is difficult to control; (2) moulding under high pressure is used to keep the water in solution and prevent bubble formation – therefore, the manufacture of void-free laminates is difficult; and (3) the resin is brittle and not easily toughened. Two resin systems that cure without the loss of water or other volatiles have developed alongside the application of fibre-composite materials. These are the **unsaturated polyester and related resins** and **epoxy resins**. The latter can be cured by a variety of mechanisms, without volatile formation, using step-growth or chain-growth polymerization. The unsaturated polyester resins were responsible for the development of 'glass fibre' or 'fibreglass' reinforced plastics because they could be cold cured using addition polymerization.

3.3.6   Addition Polymerization

The curing of unsaturated polyester, vinyl ester, and urethane acrylate/methacrylate resins, as well as some epoxies employing catalytic systems, occurs by addition polymerization. Addition polymerization is a chain reaction with the following steps:

$$
\begin{aligned}
&\text{Initiation :} && I = 2R^* \\
&\text{Propagation :} && R^* + M = P_1{}^* \\
& && P_1{}^* + M = P_2{}^* \text{ up to } P_n{}^* \\
&\text{Transfer :} && P_n{}^* + SH = P_nH + S^* \\
&\text{Termination :} && P_n{}^* + P_m{}^* = P_{n+m} \text{ or } P_n + P_m,
\end{aligned}
\tag{3.13}
$$

where I is an initiator (often referred to incorrectly as a catalyst); R^* is an active species (free radical (R^*), cation (R^+), anion (R^-), or complex) formed from the dissociation (or reaction) of the initiator; P^* is the propagating active polymer; P_n, P_m, and P_{n+m} are unactive polymers of degree of polymerization n, m, and $n + m$, respectively; SH is a transfer agent, which could be a monomer, polymer, initiator, or additive; S^* is an active species formed from the transfer agent after the transfer of the H atom to the growing active polymer to form an non-active polymer, P_nH. Added transfer agents are used to control the molecular weight of the resulting polymer.

The reason for including this chemistry in the book is to ensure that composite engineers understand how the nature of the curing of resins used as matrices determines the structure and properties. For example, the step-growth systems usually need a post-cure stage to achieve the degree of crosslinking. On the other hand, with addition polymerization each initiated chain grows until it is terminated so that unreacted monomer is always present, which is in contrast to step-growth polymerization. Post-cure is required to ensure that the free monomer concentration is minimized. Furthermore, the differing reactivity of the chain active centres towards co-monomers leads to a range of molecular structures from alternating to homopolymer blocks. The differing fracture properties of unsaturated polyesters and vinyl esters can in part be attributed to this phenomenon.

3.4   Epoxy Resins

Epoxy resins is a generic term for resins that contain the epoxy or epoxide groups that can react with a 'hardener' in a step-growth manner or polymerize catalytically in an addition polymerization.

There are two principal types of epoxy resin, glycidyl ethers or glycidyl amines, represented by the common examples shown in Figure 3.6.

The glycidyl ether DGEBA is available with differing degrees of polymerization and with $n = 0$–34. The mechanical properties of these resins vary significantly across the range because the molecular length between the reactive epoxy groups increases with n. Thus, the crosslink density is reduced, as is the resin modulus, but the failure strain increases. Since DGEBA is difunctional in epoxy groups, the thermomechanical properties T_g, E_m, and G_m are low in magnitude, while the linear thermal expansion coefficient, α_m, will be high because of the low crosslink densities. As a result, these resins are used mainly in industrial composites. The presence of high concentrations of –OH groups in DGEBA resins means they find applications as adhesives.

The cured glycidyl amines such as TGDDM have higher crosslink densities and therefore superior thermomechanical performance. As a result, they find application in aerospace structural composites.

Higher crosslink density can be achieved with multifunctional glycidyl ethers such as the novolak epoxy resin (which is a reaction product of epichlorohydrin and a novolak phenolic resin; epichlorohydrin is the standard reactant for synthesizing epoxy resins). There are also a range of mixed glycidyl ether and glycidyl amines, of which the most common is triglycidyl p-aminophenol (TGAP).

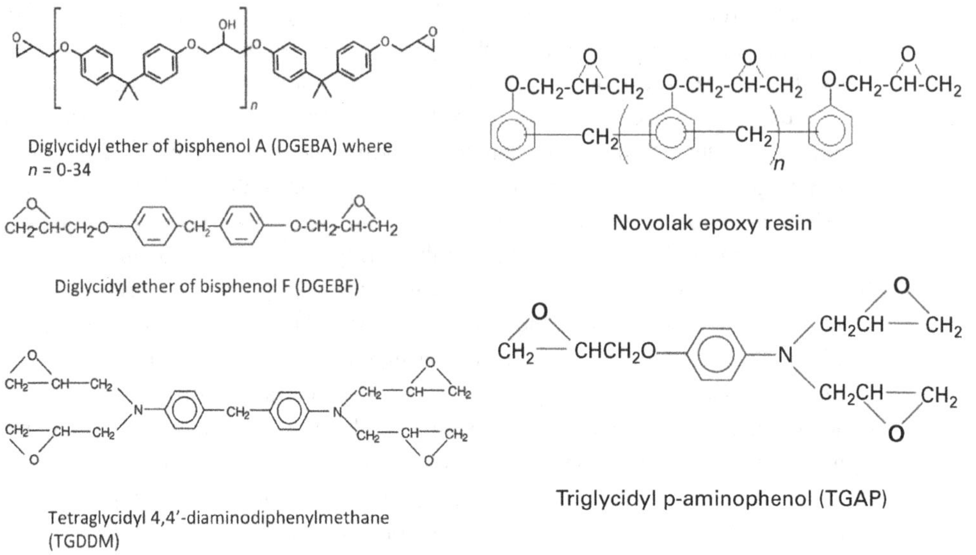

Diglycidyl ether of bisphenol A (DGEBA) where
n = 0-34

Diglycidyl ether of bisphenol F (DGEBF)

Novolak epoxy resin

Tetraglycidyl 4,4'-diaminodiphenylmethane
(TGDDM)

Triglycidyl p-aminophenol (TGAP)

Figure 3.6 The structures of five common epoxy resins: diglycidyl ether of bisphenol A (DGEBA), where $n = 0$–34; diglycidyl ether of bisphenol F (DGEBF); tetraglycidyl 4,4'-diaminodiphenylmethane (TGDDM); novolak epoxy resin; triglycidyl p-aminophenol (TGAP).

3.4.1   Curing of Epoxy Resins

The epoxy group can be polymerized either by a chain polymerization initiated by a *curing agent* or by step-growth copolymerization with a *hardener*.

3.4.1.1   Curing Agents

Often referred to as catalytic curing agents, these molecules open the epoxy group in the initiation step and chain growth occurs by either anionic (negative) or cationic (positive) addition polymerization, as described in Section 3.3.6. The product has a relatively non-polar molecular structure, which is a 'polyether'. A classic curing agent is boron trifluoride, as an etherate or amine adduct, which initiates cationic chain propagation, as shown in Figure 3.7.

Tertiary amines initiate anionic polymerization of the epoxy groups, but this is often employed in combination with a co-curing agent (Figure 3.8). A typical example is the tertiary amine-initiated anhydride systems, such as nadicmethylene tetrahydrophthalic anhydride (NMA). Please note that anhydride curing agents alone require trace concentrations of water in the initiation stage, whereby the anhydride reacts to form the analogous acid which initiates an anionic chain reaction. The cured structure of the network is one of a 'polyester' type. A common tertiary amine used with NMA is benzyl dimethylamine (BDMA).

The cationic homopolymerization of the epoxy groups using a tertiary amine creates a polyether crosslinked structure (Figure 3.9). Addition polymerization tends to provide relatively *low-polar* cured structures so that intermolecular cohesive forces

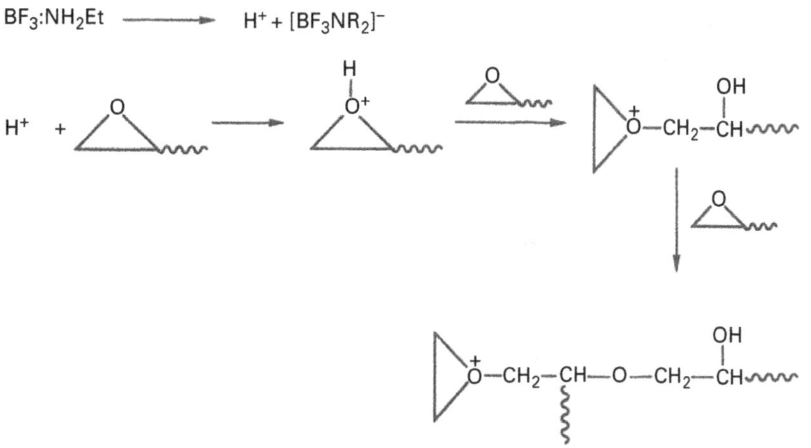

$$BF_3{:}NH_2Et \longrightarrow H^+ + [BF_3NR_2]^-$$

Figure 3.7 The catalytic cure of an epoxy resin using cationic initiator, BF_3 amine adduct, to form a polyether chain crosslink structure.

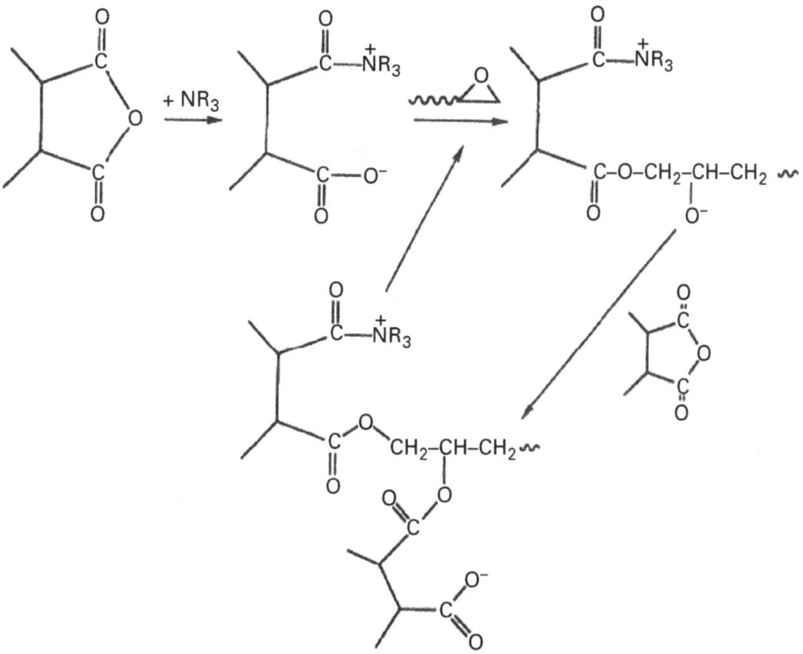

Figure 3.8 Tertiary amine-initiated anionic copolymerization of anhydride ring and epoxy groups during anhydride curing. A polyester chain is formed.

tend to ensure that the thermomechanics of the cured resins have low values unless high degrees of crosslinking are introduced by post-curing at elevated temperatures. We see in the next section that hardeners in general introduce polar groups into the molecular structures of cured epoxy resins.

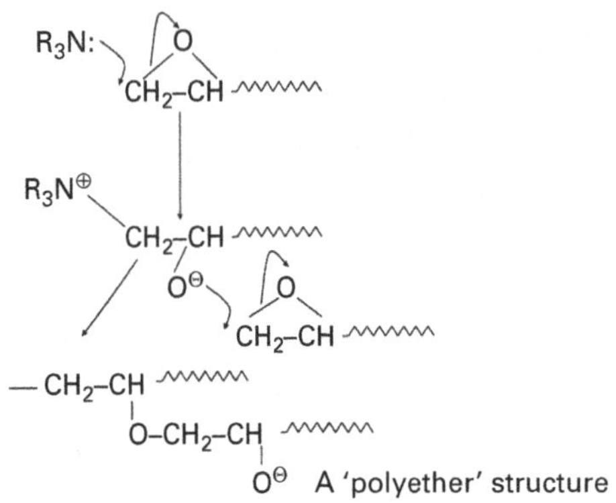

Figure 3.9 Catalytic curing of epoxy resin using tertiary amine-initiated anionic polymerization. A polyether structure is formed.

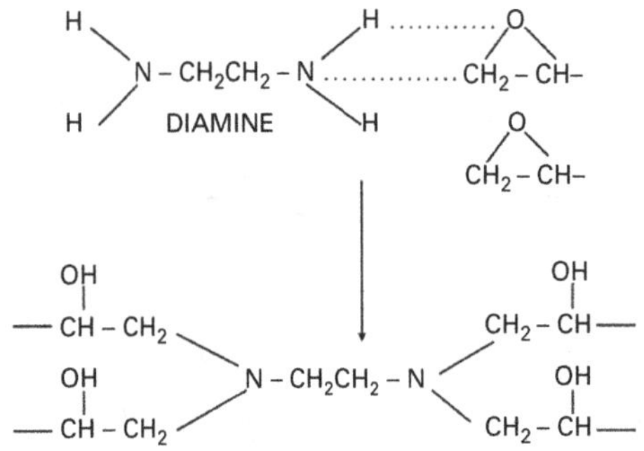

Figure 3.10 Typical cure reaction of epoxy groups with a diamine hardener. The first stage involves a primary amine and subsequently with the remaining secondary amine groups. The latter are less reactive so complete reaction (i.e. full crosslinking) as shown is unlikely.

3.4.1.2   Hardeners

Curing with hardeners occurs by step-growth addition reaction with amines, carboxylic acids, thiols, and cyanate esters. Typical hardeners are either aliphatic multifunctional amines or aromatic diamines. Dicyandiamide (DICY) is a multifunctional hardener that is commonly used as a latent 'curing' agent for epoxy resins and especially for prepreg applications.

A common curing agent for epoxies often used for adhesives is a diamine. Figure 3.10 gives the simplest, ethylene diamine, which is a typical representative of the range of available and common hardeners. A hardener differs from a catalytic curing agent because the mechanism involves step-growth addition reactions, as shown in Figure 3.10.

Cure of epoxy resin using step-growth addition of ethylene diamine (an example of an aliphatic diamine hardener) involves initial reaction with a primary amine, $-NH_2$, followed by a subsequent reaction with the secondary amine, $-NH-CH_2-$,

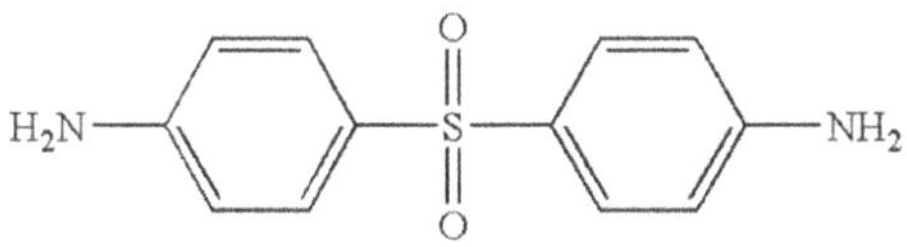

Figure 3.11 1,4 Diamino diphenylsulphone (1,4 DDS): an example of an aromatic diamine hardener.

which is formed. Reaction of the secondary amines with epoxy groups occurs at a lower rate so that post-curing at higher temperatures can increase the degree of crosslinking. This is also accompanied by the formation of β-hydroxyl groups (–CH$_2$CH(OH)–) so that the molecular structure of the crosslinked resin is *polar*, providing hydrogen bonding sites for absorbed water. Therefore, these resins are susceptible to moisture absorption.

Aliphatic diamines such as that shown in Figure 3.10 are highly reactive, enabling curing at room temperature. Extending the aliphatic chain or introducing aromatic groups into the structure can modify the rate of cure. Diamino diphenylsulphone (DDS) is often used as a hardener for high-performance epoxy resins and can be used in prepreg systems for high-temperature autoclave processing. The reactivity of 1,3 DDS and 1,4 DDS (Figure 3.11) also differ, which increases the opportunities for designing epoxy resins with controlled curing.

3.4.1.3 Latent Curing Systems

For prepreg systems, a latent curing system is required so that the pre-impregnated fibre sheet can be stored and the curing of the resin can be thermally initiated. Early systems employed mixed curing agents such as the BF$_3$ adducts in combination with DDS, which becomes active at about 160 °C. The viscosity of the resin and hence the tack of the prepreg could be adjusted by advancing the DDS/ epoxy reaction *in situ*. Dicyandiamide (DICY) is also used to provide latency of cure in a prepreg. Micronized DICY is dispersed into the liquid resin. Since the hardener is not soluble at room temperature, prepreg with improved shelf life can be obtained. The viscosity and tack of the prepreg is adjusted by dissolving a 'thermoplastic' such as polyether sulphone (PES) at temperatures below that of the DICY. In this way the fibres can be impregnated with a low-viscosity resin to maintain their alignment.

The cure temperature is determined by the melting point of DICY, which occurs at 165 °C. High- and low-temperature curing systems are available: 'diuron' or 'monuron' amine accelerators can be added to reduce the curing temperature to 120 °C. Low-temperature systems have clear advantages, such as control of the exotherm during curing. Careful control of the cure schedule is required for the production of good-quality void-free laminates. Thus, a dwell at the temperature of minimum resin viscosity should be included at typically ≈90 °C. A disadvantage of low-temperature cure of DICY-based systems is the increased possibility of residual DICY. There is also a potential for inconsistent resin microstructures, which can be improved by ensuring that the DICY is micronized into a uniform fine particle size. DICY is used at

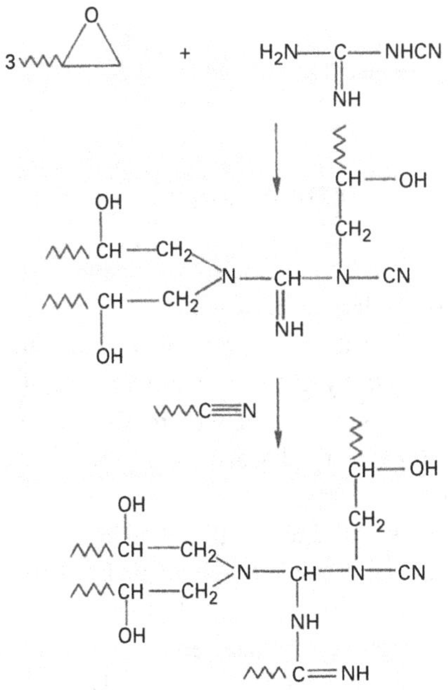

Figure 3.12 The reactions involved in the cure of epoxy resins with dicyandiamide (DICY).

concentrations of 4–6 phr, indicating that the mechanism of cure involves its multi-functionality. Dicyandiamide exists in two molecular tautomers:

$$H_2N—C=N—C\equiv N \rightleftharpoons NH_2—C—NHCN \tag{3.14}$$
$$\quad\quad\;\; | \quad\quad\quad\quad\quad\quad\quad\quad\quad ||$$
$$\quad\quad NH_2 \quad\quad\quad\quad\quad\quad\quad\;\; NH$$

Reaction with an epoxy group is therefore complex (Figure 3.12) and involves the nitrile as well as the primary amine groups. It is also likely to involve other crosslinking mechanisms.

3.4.1.4   Co-curing and Other Curing Agents

There is a large range of curing agents for epoxy resins, many of which are employed for adhesives and/or coatings. It is beyond the scope of this book to provide complete details. Relevant examples used for composite matrices only are included. Full details of epoxy resin systems are given elsewhere [5,6]. Here, we list additional curing agents that are commonly employed for composite manufacture.

Tertiary Amines

For industrial composites, BDMA and tris(dimethylaminomethyl) phenol (Figure 3.13) are often used in combination with other curing agents, such as NMA. BDMA is also used as a co-catalyst with polysulphides and polymercaptans, which tend to have relatively long chain lengths and enable the crosslink density to be controlled.

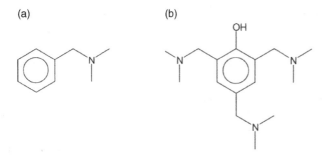

Figure 3.13 Commonly used tertiary amines employed as co-curing agents: (a) BDMA; (b) tris (dimethylaminomethyl) phenol.

Imidazoles

These activators introduce high reactivity and promote curing at lower temperatures. On the other hand, *trisubstituted ureas* or *urons*, which are reaction products of isocyanates with dimethylamine, exhibit outstanding latency of cure and find application in one-pack adhesives.

Carboxylic acids and anhydrides and related carboxylic functional polyester resins are also employed to cure epoxy resins as discussed above.

Sulphur-containing curing agents such as mercaptan-terminated polysulphides can be used as co-curing agents with other long-chain polyamides and cycloaliphatic amines, and introduce high degrees of flexibility because of the low crosslink density. In this way, the extensibility and toughness of the resin can be improved.

Quaternary phosphonium salts such as alkyl and triphenylphosphonium halides are effective latent accelerators for anhydride and phenolic resin-cured epoxy resins. The latter systems are effective at advancing the cure.

Cationic salts such as triarylsulphonium are used as photoinitiators.

3.4.1.5   Reactive Diluents

Diluents are added to epoxy resins as process aids to reduce the viscosity for fibre impregnation. They can be either unreactive or reactive. The former have limited value in composites technology since the properties of the cured resin will be reduced by the presence of a non-bound diluent. Reactive diluents are small molecules that enable the viscosity to be reduced but on curing they are involved in the reaction with the epoxy resin. While the reactive functionality can be other than epoxy, the majority are mono-functional epoxies that can be reacted into the network crosslinks. Examples include glycidyl ethers such as butyl glycidyl ether and phenyl glycidyl ether, or oxiraine compounds such as styrene oxide or octylene oxide.

3.4.1.6   Toughening Techniques

Cured epoxy resins can exhibit high fracture toughness, but this is generally at the expense of the achievable Young's modulus. For composite materials we generally require the matrix to have the highest stiffness, which is obtained from high degrees of crosslinking, but these systems will have relatively low fracture toughness. The fracture toughness of these resins can be improved by adding either mineral fillers, rubber particles, or thermoplastic particles, or introducing a soluble thermoplastic or linear polymer which phase-separates during curing to provide the morphology with the highest fracture toughness.

Mineral fillers are particles that increase the fracture toughness of epoxy resins; the mechanism is considered to arise from (1) crack pinning, (2) bowing of the crack front, and (3) break-away. Thus, K_{1c} increases with particle volume fraction, V_p, reaching a plateau then decreasing; i.e. an optimum V_p can be expected. The effect of particle size is unclear because as the diameter decreases the surface area becomes more significant and influences the structure of the glassy polymer in the interphase region.

For composite materials, recent research has concentrated on increasing the modulus of the matrix resin within a laminate in order to improve the compression strength of the material by preventing fibre buckling. A reduction in the thermal strains arises from the lower thermal expansion coefficient of the 'filled matrix'.

Rubber particles are typically carboxyl-terminated butadiene–acrylonitrile rubber (CTBN), and are employed at <20% by weight. The terminal carboxylic acid groups provide a mechanism of interfacial bonding to the rubber particles. Butadiene rubber has a T_g of −75 °C and therefore exhibits high resilience at room temperature. Butadiene–acrylonitrile rubber, generally referred to as nitrile rubber, has a typical T_g of −38 °C (for a copolymer with 18% acrylonitrile) and is much less resilient at room temperature.

Efficient toughening is achieved with the correct two-phase morphology. Simple dispersion does not provide the required microstructure, so it is necessary to dissolve the rubber in the epoxy for dispersion at the molecular level. Phase separation occurs during curing as the molecular weight of the epoxy increases through crosslinking. The rubber phase is finely dispersed and bonded to the epoxy continuous phase.

In *thermoplastic modification* epoxy resins used as matrices for composites can be toughened using appropriate linear polymers, which phase-separate into a co-continuous morphology. Three differing morphologies are possible, depending on the concentration of the modifier: (1) epoxy continuous phase with dispersed thermoplastic 'particles' (2) co-continuous morphology; and (3) thermoplastic continuous phase with dispersed epoxy 'particles'. To achieve a co-continuous morphology a precise concentration of thermoplastic of defined molecular weight and structure is required. Typically, end-capped linear PES or polyether ether sulphone (PEES) meet these requirements. To understand this we will consider the thermodynamics of polymer solubility.

3.4.2   Compatibility of Polymers

The compatibility of a resin with a linear polymer can be understood using the concept of solubility parameters based on the thermodynamics of mixing described in the Gibbs equation:

$$\Delta G_m = \Delta H_m - T\Delta S_m, \tag{3.15}$$

where ΔG_m, ΔH_m, and ΔS_m are the changes in free energy, enthalpy, and entropy of mixing, respectively.

For complete mixing, ΔG_m has to be negative, while ΔS_m will tend to be positive. For small molecules, ΔS_m will be large and positive so that $-T\Delta S_m$ will readily ensure mixing when there is no interaction between the components and $\Delta H_m = 0$. With polymer–polymer mixing the change in entropy is small and close to zero so that for polymer–polymer compatibility, ΔH_m should be negative. This arises when the two components interact, which can be defined by eq. (3.16):

$$\Delta H_m = v_1 v_2 (\delta_1 - \delta_2)^2 V, \tag{3.16}$$

where δ_1 and δ_2 are the solubility parameters of components 1 and 2, v_1, v_2 are the volume fractions of the two components, V is the total volume of the system, and δ is defined as the square root of cohesive energy density representing the energy of interaction between like molecules. Polar molecules will have high values of δ. For mixing, the interaction between components 1 and 2 must equal that between components (1 and 1) and (2 and 2), otherwise the molecules will prefer themselves. Thus, for mixing $(\delta_1 - \delta_2)$ and hence ΔH_m will be zero. This simply describes the maxim 'like dissolves like'.

The solubility parameter can be readily calculated from its molecular structure according to the simple equation

$$\delta = \rho \sum \frac{G}{M}, \tag{3.17}$$

where ρ is the density, M is the molecular weight, and G is the molar attraction constant for an individual 'molecular' element of the molecule. The sum of the contributions from the structural elements provides the attraction energy density for that molecule.

Equation (3.17) needs a correction factor for polar molecules and a more accurate value needs to include all the molecular interaction components of cohesive energy: dipole, polar, and hydrogen bonding. The total solubility parameter δ_t is given by

$$\delta_1 = (\delta_d^2 + \delta_p^2 + \delta_h^2)^{1/2}, \tag{3.18}$$

where δ_d, δ_p, and δ_h are the dipole, polar, and hydrogen bonding contributions to the solubility parameter, respectively.

Calculation of solubility parameters according to Hoy [7] and van Krevelen and Te Nijenhuis [8] enables the individual contributions to the total solubility parameter to be obtained. The total solubility parameter, δ_t, of DGEBA is in the range 20.6–21.24 MPa$^{1/2}$, depending on the molecular weight. The dipole component, δ_d, decreases from 16.47 to 16.22 MPa$^{1/2}$ with molecular weight, while the polar component increases because of the higher number of polar OH groups. Table 3.1 shows that after curing the value of δ_t tends to be higher, at 22–24 MPa$^{1/2}$. The hydrogen bonding component is fairly constant at 8–10 MPa$^{1/2}$.

Table 3.1. Calculated solubility parameters (δ_t) and the hydrogen bonding components (δ_h) of selected epoxy monomers, hardener (DDS), and typical cured epoxy networks [9] and comparative polymers

Epoxy monomer segment/polymer	Function	δ_t (MPa$^{1/2}$) (calculated)	δ_t (MPa$^{1/2}$) (measured)	δ_h (MPa$^{1/2}$)
DGEBF	Monomer segment	22.36	–	9.75
DGEBA	Monomer segment	21.26	–	8.13
TGAP	Monomer segment	22.92	–	10.34
TGDDM	Monomer segment	21.84	–	8.88
4,4', DDS	Monomer segment	27.18	–	9.18
TGAP/DDS (2:1)	Cured epoxy resin	24.06	–	10.25
TGDDM/DDS (5:1)	Cured epoxy resin	22.30	–	8.95
TGDDM/ TGAP /DDS (19:13:10)	Cured epoxy resin	22.89	–	9.4
DEGBA/NMA (1:2)	Cured epoxy resin	24.0	–	–
PES	Thermoplastic flow control and toughening agent	24.7	23.0	–
PEE)	Thermoplastic flow control and toughening agent	23.5	–	–
Poly bisphenol A sulphone	Comparative thermoplastic	21.7	21.2	–
Polycarbonate	Comparative thermoplastic	20.3	20.4	–
Polybutadiene	Component of nitrile rubber toughening agent (CTBN)	16.8	17.0	–
Polyacrylonitrile	Component of nitrile rubber toughening agent (CTBN)	26.3	26.2	–

The general rule of thumb for identifying compatible pairs is given by:

$$\delta_1 = \delta_2 \pm 2 \text{ MPa}^{1/2}. \tag{3.19}$$

Therefore, for the rubber to dissolve in the epoxy resin it needs a value δ_t in the range 20–26 MPa$^{1/2}$. As shown in Table 3.1, butadiene rubber has a value of δ_t outside of this range. However, we see that polyacrylonitrile has a value of 26.3 MPa$^{1/2}$, so that nitrile rubber, which is an acrylonitrile–butadiene copolymer, will have a closer solubility parameter and compatible with a typical epoxy resin. We see that PES has a similar value to advanced epoxy resins.

Solubility parameters have additive characteristics so that compatibility can be achieved using mixed systems, as indicated by eq. (3.20):

$$\delta_m = x_1\delta_1 + x_2\delta_2, \tag{3.20}$$

where δ_m is the solubility parameter of a mixture of components 1 and 2 at mole fractions of x_1 and x_2.

3.4.2.1 Phase Separation in Polymer Blends

Flory and Huggins' polymer solution theory [3] is a statistical model in which a polymer *segment* is placed, statistically, into a *lattice* of solvent molecules. The assumption is that the segment of a polymer chain has a similar molecular size to the solvent molecule. Equation (3.21) describes the free energy of mixing, ΔG_m:

$$\Delta G_m = RT[n_1 \ln \phi_1 + n_2 \ln \phi_2 + n_1 \phi_2 \chi], \qquad (3.21)$$

where n_1 is the number of solvent molecules, n_2 is the number of segments in the polymer, and ϕ_1 and ϕ_2 are the mole fractions of sites 1 and 2. χ is the Flory interaction parameter, which has both enthalpic (χ_H) and entropic (χ_S) components:

$$\chi = \chi_H + \chi_S. \qquad (3.22)$$

The enthalpic component is related to the solubility parameter according to eq. (3.23):

$$\chi_H = \frac{V_1(\delta_1 - \delta_2)^2}{RT}, \qquad (3.23)$$

where V_1 is the molar volume of the solvent.

The development of eq. (3.21) to understand the criteria for solubility of a polymer is beyond the scope of this book. However, for solubility it is understood that χ has a value of ½ when the entropic and enthalpic components are in balance and the solution can be considered to be a mixture at the molecular level. We also see in eq. (3.23) that χ_H is temperature-dependent. As a result, a polymer–solvent blend will exhibit a temperature above which it has one phase (i.e. a solution), which is referred to as the theta temperature, θ. Below this temperature the polymer will not be soluble. Theta conditions refer to a dilute solution at the theta temperature, T_θ and $\chi = $ ½, and are considered to be *ideal*.

In the context of polymers employed for composite matrices it is relevant to highlight the effect of molecular weight of components on phase separation in polymer blends. The Flory–Huggins theory can be applied to binary polymer–polymer systems, as discussed by Scott [10], who estimated the partial free energies of mixing:

$$\Delta F_{m1} = RT \left[\ln v_1 + \left(1 - \frac{m_1}{m_2} \right) v_2 + m_1 \chi v_2^2 \right], \qquad (3.24)$$

$$\Delta F_{m2} = RT \left[\ln v_2 + \left(1 - \frac{m_2}{m_1} \right) v_2 + m_2 \chi v_1^2 \right], \qquad (3.25)$$

where ΔF_{m1} and ΔF_{m2} are the partial molar free energies of mixing for a polymer-1–polymer-2 binary system. m_1 and m_2 are essentially the degrees of polymerization of the two components at volume fractions of v_1 and v_2.

The critical value of the Flory interaction parameter for solubility/phase separation, χ_c, can be obtained by differentiation and setting $\Delta F_m = 0$:

$$\chi_c = \frac{1}{2} \left[\left(\frac{1}{m_1^{1/2}} \right) + \left(\frac{1}{m_2^{1/2}} \right) \right]^2. \qquad (3.26)$$

For polymer–solvent systems $\chi_c = 0.5$, but for polymer–polymer systems $\chi_c \approx 0.01$. Thus, polymers of infinite molecular weight will tend to be incompatible unless there is a specific interaction (i.e. ΔH_m is negative). In the context of matrices for fibre composites, which are mainly crosslinked network polymers such as epoxy resins, curing causes the molecular weight to increase and eventually become infinite. As a result, any dissolved linear polymer will tend to phase-separate during cure. This is referred to as *reaction-induced phase separation* and is the basis for creating the correct morphology for high fracture toughness when thermoplastics are blended into thermosetting resins. For example, when PES is employed as the flow control and/or toughening agent at a concentration of about 20%, a co-continuous microstructure forms during cure, while below that the PES will phase-separate with a continuous epoxy resin phase. At higher concentrations the epoxy component will be discontinuous. Examination of the thermodynamics above demonstrates that the morphology achieved can be dependent on the cure schedule and the temperatures employed, as well as the molecular weight of the toughening agent.

3.4.3   Cure Property Diagrams

The glass transition temperature, T_g, of a thermosetting resin is not so easy to measure because two major parameters contribute to the formation of a glass: the rates of crosslinking and cooling. As shown in Figure 3.14, a linear polymer will exhibit a linear change in volume with temperature on cooling until the glass transition temperature is reached, when the rate of contraction decreases. On rapid cooling, a glass will form at a higher temperature. A polymeric material consists of randomly coiled molecular chains so that the reorganization of segments in the chain occurs with cooling. To achieve the lowest statistically possible structural arrangement an infinitesimally slow cooling rate is required. Some polymers have a regular structural

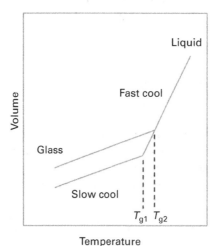

Figure 3.14 Schematic showing the effect of cooling rate on the glass transition temperature, T_g, of a linear polymer.

arrangement at the segmental dimension and will crystallize. Irregularly structured polymers cannot crystallize and form a glass.

Since most polymers do not normally crystallize completely (i.e. 100%), they exhibit melting and glass transition temperatures. Typically, T_g occurs in the range 0.5–0.8 T_m. Thus, a glass with the lowest possible T_g will exhibit the smallest 'unoccupied' volume, which is referred to as *free volume* and defined as the statistical volume for chain–chain crossover, unperturbed by interactions [11]. Therefore, the unoccupied volume in a polymer will consist of free volume and nano-voids [12].

However, a thermosetting polymer such as an epoxy resin differs from a linear polymer in that they are generally non-polymeric, becoming polymeric only during curing. There is an additional shrinkage mechanism, resulting from the crosslinking of the molecules during curing. This results in a different behaviour on cooling, as shown in Figure 3.15. A glass forms at a higher transition temperature, T_{g2}, with a larger unoccupied volume when a resin with a higher extent of crosslinking (degree of cure) is cooled. The cured resin can have a lower density despite a higher glass transition temperature. Thus, the resin in the as-cured state will be in a non-equilibrium state. We see in Chapter 9 that this is responsible for the anomalous moisture absorption of some resin systems, which occurs under thermal cycling.

Gillham and coworkers [13–15] have studied in detail the properties of matrix resins during representative cure schedules using thermal analysis. In particular, torsional braid analysis enables the conversion of a liquid resin into a solid during curing and subsequent cooling. From these studies, time transformation diagrams and cure–temperature property diagrams could be derived.

As discussed in Section 3.3.1, the cure of an' epoxy resin involves the gradual increase in average molecular weight as the relatively low molecular weight

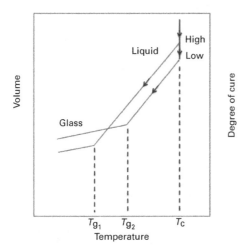

Figure 3.15 Schematic showing the effect of cooling and degree of cure on the glass transition temperature, Tg, of a thermosetting polymer. T_c is the cure temperature.

monomers or oligomers react with the hardeners. The following events occur during a cure schedule with rising temperature:

1. *gelation*
2. *phase separation*
3. *vitrification* (glass formation)
4. *devitrification* (occurs when the temperature exceeds the lowest T_{g0}, the lowest T_g of the system)
5. *degradation* (at high temperatures thermal chain bond scission is possible: resin cure is exothermic and control of cure and hence the actual temperature of the part is essential).

Since cure reactions are exothermic, the temperatures of the part, especially in thick sections and at different positions in the moulding, can vary and be significantly elevated above the cure temperature, T_c. As a result, to avoid excessive rises in temperature from exothermic processes, the initial cure temperature is often kept relatively 'low', then followed by a post-cure at an elevated temperature to achieve full cure. The temperature schedule will depend on the resin system employed.

A time–temperature–transformation (TTT) diagram can be obtained from the experimental determination of the times to gelation, t_{gel}, and vitrification, t_{vit}, under isothermal conditions. Full details are beyond the scope of this text, but a complete description is given elsewhere [5,16–18]. Figure 3.16 gives an example of a TTT diagram.

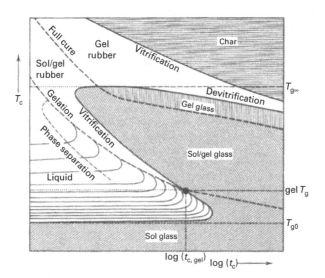

Figure 3.16 Time–temperature–transformation (TTT) diagram. T_c and t_c are the cure temperature and time, respectively. T_g, T_{g0}, $T_{g\infty}$ are the glass transition temperatures at times t_c, t_0 (the resin/curing agent mixture) and the fully cured resin ($t_c = t_\infty$). gel T_g is the T_g at $t_{c,gel} = t_{c,vit}$, the cure times for gelation and vitrification, respectively (the gelation and vitrification lines). *Sol* glass is soluble, whereas *gel* glass is crosslinked and only swells in a solvent. The *liquid* region is represented by iso-viscosity contours (differing by 10) for a homogeneous system. *Char* represents the degradation region [16].

A TTT diagram describes the complexity of the cure of the resin. Initially the cure temperature needs to exceed sol T_g (T_{g0}) for the reactions to proceed. As the complexity of the molecular structure increases with cure, so does the T_g (t_c) (t_c is the cure time at a constant temperature, T_c) until it reaches the cure temperature T_c, when it vitrifies. As T_c approaches T_g (t_c) the molecular mobility decreases and the rate of cure decreases according to a 'second' or higher-order kinetic law, eventually becoming diffusion-controlled. Subsequently, the cure reactions are quenched, despite the presence of reactive functional groups. These occluded reactive sites provide a latent cure mechanism because, as discussed in Section 3.3.3, the cure mechanisms can be reactivated on raising the temperature.

The TTT curve shows the relationship between gelation and glass formation during curing and provides the time and temperature dependence and demonstrates how full cure can be achieved. Thus, it is common practice to employ a post-cure at an elevated temperature. *Phase separation* of a rubber or thermoplastic modifier occurs as the cure progresses, as shown by the increased viscosity identified by a series of isoviscous lines.

Gillham and coworkers [13–15] have discussed the TTT diagrams for epoxy resins. However, Prime [18] has reviewed the application of TTT concepts to the cure of other thermosets, phenolic, amino, allyl, and unsaturated polyester resins.

Wang and Gillham [19] have extended the concept to the cure–temperature property diagrams to represent structural changes (Figure 3.17). In particular, they include β-transitions, which strongly influence the fracture toughness of the cast resin and the elastic modulus. The cure parameter, C, is given by

$$C = \frac{T_g - T_{g0}}{T_{g\infty} - T_{g0}}.$$ (3.27)

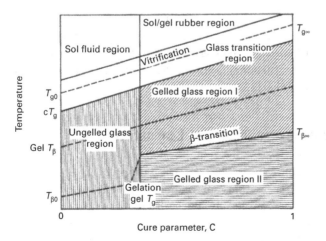

Figure 3.17 Cure–temperature property (CTP) diagram illustrating the influence of the temperature range for the glass–rubber transition and the β transition. cT_g is the onset temperature [16].

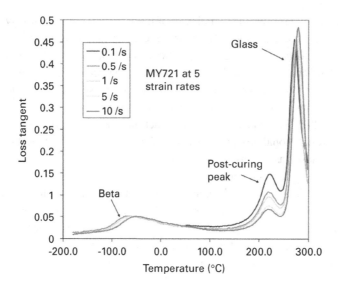

Figure 3.18 Dynamic mechanical thermal analysis (DMTA) spectrum of a DDS-cured TGDDM (MY721). The post-curing peak is present after 180 °C cure but absent after 200 °C cure. The beta peak at T_β is also demonstrated (from J. P. Foreman).

Figure 3.18 shows a typical dynamic mechanical thermal analysis (DMTA) spectrum for a DDS-cured epoxy resin showing the β-transition, which occurs at –50 °C. The T_g appears to have two components after cure at 180 °C. The so-called post-curing peak is not present after cure at 200 °C. The origin of the low-temperature peak is not clear, but could arise from the presence of 'incompletely' reacted components of the MY721 epoxy [20].

The low-temperature beta peak provides a molecular energy absorption mechanism and therefore introduces a degree of fracture toughness to the resin casting. Furthermore, the modulus of the glass is also affected. Group interaction modelling extends eq. (3.1) into a methodology for predicting the thermal and mechanical properties of polymers [21] from the molecular structure. The model has been applied to aerospace epoxy resins [22].

3.5   Unsaturated Polyesters Resins for Lamination

Unsaturated polyester resins (UPE) represent a milestone in the development of the composites industry. These resins are solutions of an unsaturated polyester dissolved in a reactive diluent, which is mainly styrene. The curing of early resins was subject to air inhibition, which limited their introduction. This issue (air inhibition) was solved by ground-breaking work at Scott Bader and Co. Ltd by incorporating paraffin wax in ppm (parts per million) quantities to provide an oxygen barrier at the moulding surface [23]. There are two main classes of resin, casting and laminating, and the latter are most relevant to this discussion.

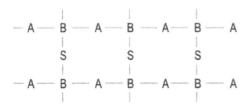

Figure 3.19 Schematic description of a cured unsaturated polyester resin; –A–B– is the polyester, S is the reactive diluent (styrene) built into a copolymer network.

Polyester laminating resins are viscous, pale-yellow liquids. These resins are solutions of unsaturated 'oligomeric' polyester of low degree of polymerization, 8–10 (i.e. MW $\approx$ 2,000) in an unsaturated reactive diluent, which is commonly styrene. These resins cure into a crosslinked network polymer by rapid *addition* copolymerization of the unsaturated groups in the polyester with those of the reactive diluent. To tune the thermomechanical properties of the cured resin, the crosslink density is controlled in two ways:

1. *the spacing of the unsaturated groups* (–C=C–) in the oligomeric polyester by incorporating saturated monomeric units via copolymerization with a saturated acid;
2. *varying the reactivity of the unsaturated groups* to styrene during curing by changing the glycol component in the synthesis, which provides an important means of controlling the rate of cure during composite manufacture [24–27].

The structure of a cured UPE is often represented schematically, as shown in Figure 3.19.

While the cured structure given in Figure 3.19 is commonly used, Figure 3.20 is a better representation because of the average length of the polystyrene bridges ($n \approx$ 2–3). This is a consequence of the kinetics of the copolymerization, which are discussed below.

3.5.1   The Synthesis of Unsaturated Polyester

Figure 3.21 provides the molecular structures of the common monomers employed for the synthesis of polyester.

Maleic anhydride and propylene glycol together with phthalic anhydride are the most common starting materials. The choice of anhydrides, as opposed to diacids, is determined by the more rapid esterification and lower number of moles of H_2O that evolve during the polycondensation, as illustrated in Figure 3.22.

The esterification reaction is a slow process involving the distillation of the water of condensation over 12–20 hours at temperatures up to 200 °C in a nitrogen atmosphere. In batch reactors this is achieved with a nitrogen purge that means the diol often distils out with the condensate. Esterification is acid-catalysed, but sulphuric acid cannot be used because of side reactions that result in coloured products. The other common

(a)

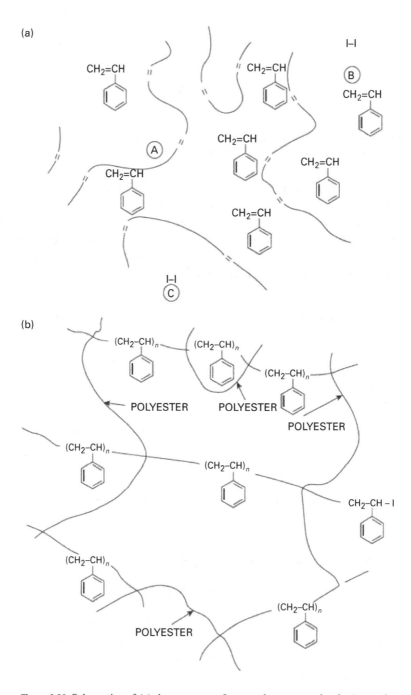

(b)

Figure 3.20 Schematics of (a) the structure of uncured unsaturated polyester resin containing (A) unsaturated PE oligomer, (B) reactive diluent styrene, and (C) 'catalyst' system-initiator fragment 'I'; and (b) the structure of the cured resin illustrating a 'simple' model of the network (n ≈ 2–3) [24,28].

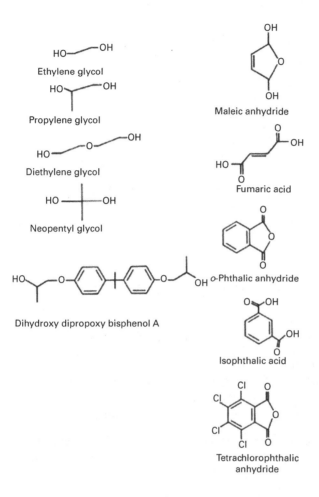

Figure 3.21 Chemical structures of the acids, anhydrides, and glycols used in the synthesis of a UPE resin [26].

esterification catalyst is butyl titanate, but unfortunately it hydrolyses to titania, which results in a white-pigmented resin that is generally unsuitable for the synthesis of laminating resins. Therefore, self-catalysed conditions are employed. The reaction is continued until an acid value of 30–40 mg KOH/g (of resin) is achieved. From 1 mol of diol and 1 mol of diacid this is equivalent to an extent of reaction of $p = 0.95$, which equates to $P_n \approx 20$ (see eq. 3.10), but in reality $P_n \approx 10$ ($M_n = 1{,}400$–$1{,}850$) because a 5–10% excess of diol is used. Therefore, a typical recipe might be:

- 1 mol phthalic anhydride
- 1 mol maleic anhydride
- 2.2 mol propylene glycol (1,2 propane diol).

The chemistry of the synthesis is shown in Figure 3.22.

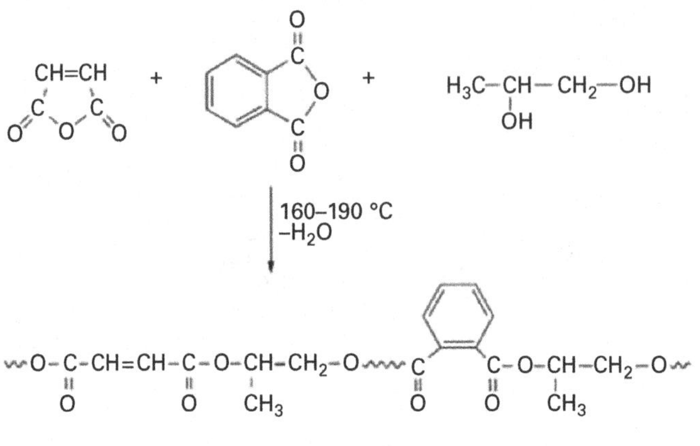

Figure 3.22 Synthesis of unsaturated polyester from phthalic anhydride, maleic anhydride, and propylene glycol.

The excess diol can be lost during synthesis in the following ways:

1. the diol distils as an azeotrope with the water;
2. self-condensation of the diol into volatile cyclic ethers; and
3. self-condensation into higher ether diols (e.g. diethylene glycol from ethylene glycol), which can be incorporated into the resin structure.

The acidity of the mixture and the temperature reached by the reactants determine the structure of the polyester because of the extent of cis–trans isomerization of maleic anhydride.

3.5.1.1   Cis–Trans Isomerization of Maleate to Fumarate During Polycondensation

The most significant aspect of the synthesis that impacts on the reactivity of the resin and the thermomechanical properties of the cured resin is the extent of the cis–trans isomerization of the maleic anhydride into fumaric acid (Figure 3.23), which is incorporated as fumarate unsaturated groups in the polyester chain. The trans –C=C– bonds of the fumarate groups in the UPE chain have a higher reactivity to the styrene diluent in free radical copolymerization. Fumaric acid is also used to synthesize resins with higher reactivity during curing, but the additional costs associated with removing the extra mole of water is not always justified by the benefits. Therefore, controlling the isomerization of maleate to fumarate is an important factor in resin production. Figure 3.24 shows how the degree of isomerization differs with the glycol selected. Propylene glycol is most effective. The cis–trans isomerization is acid-catalysed so the acidity of the component diacid is an important consideration. To promote isomerization in diethylene glycol resins it is essential to include phthalic anhydride in the recipe because ortho-phthalic is the strongest acid (other than maleic).

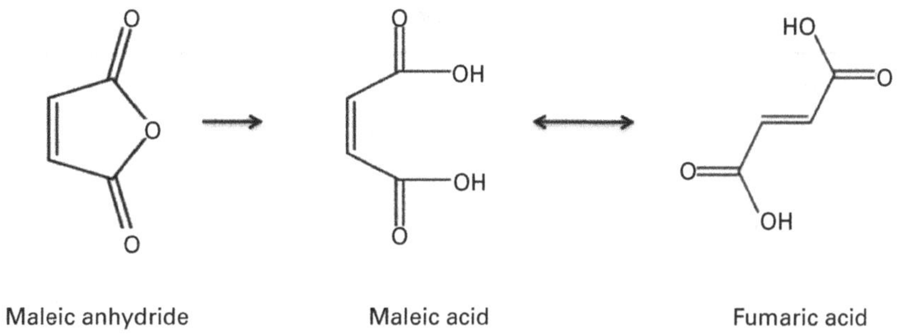

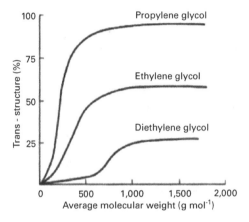

Maleic anhydride   Maleic acid   Fumaric acid

Figure 3.23 Maleic anhydride and subsequent acid-catalysed cis–trans isomerization of maleic acid to fumaric acid.

Figure 3.24 Degree of cis–trans isomerization in glycol maleate polyesters [25,26].

Table 3.2 gives the values of pK_a for the phthalic acids with comparatively strong and weak acids. A strong acid has a low pK_a, so we see that the alternative phthalic acids are less able to promote isomerization.

3.5.2   Formulation of Unsaturated Polyester Ester Resins

The unsaturated polyester with the correct acid value is blended with a reactive diluent, which is mainly styrene (at 30% w/w) in the presence of an inhibitor or free radical scavenger, usually hydroquinone. The latter stabilizes the resin, providing a good shelf life and control over the induction period or 'working time' before gelation, after addition of catalysts and accelerators. In some applications a more complex inhibitor system is employed. Therefore, a typical recipe is given in Table 3.3 [28].

Table 3.2. The acidity of maleic and phthalic acids

Acid	pK$_a$	Strength
HCl	−7	High
Maleic	1.92	
Ortho-phthalic	2.98	
Isophthalic	3.46	
Terephthalic	3.51	
Formic	3.77	
Acetic	4.76	Low

Table 3.3. Typical composition of an unsaturated polyester resin

Component	Function	Polyester structure	Resin composition	Resin composition (parts by weight)
Polyester			2.1 mol	170 parts
Propylene glycol	Diol	1.1 mol		
Maleic anhydride	Unsaturated acid	0.67 mol		
Phthalic anhydride	Saturated acid	0.33 mol		
Styrene	Reactive diluent		0.92 mol	100 parts
Benzyl trimethylammonium chloride	Inhibitor		1.4 mmol	0.26 parts
Hydroquinone	Inhibitor		0.3 mmol	0.034 parts
Quinone	Inhibitor		0.03 mmol	0.0034 parts
Paraffin wax	Oxygen barrier for cure		100 ppm	0.027 parts

The paraffin wax is added as an air inhibitor, which migrates to the free surface to act as an oxygen barrier and prevent reaction with the propagating species involved in the cure of the resin. It prevents the formation of tacky surfaces.

3.5.3   Polyester Formulation

The formulation of the polyester has a number of variants: the nature of the diol or glycol; the choice of unsaturated or saturated acids; and the degree of isomerization of maleate to fumarate esters, which is a function of the choice of glycol (Figure 3.24). Thus, the crosslink density of the cured resin is strongly dependent on the choice of components, synthesis conditions, and the reactivity and concentration of the unsaturated groups.

Maleic anhydride is preferred for general-purpose resins, while fumaric acid-based resins have superior reactivity during cure and hence the achievable thermo-mechanical properties. Resins with high fumarate content achieve high crosslink densities after curing and therefore good temperature and environmental resistance, but tend to be brittle. Fracture toughness and impact strength can be improved by

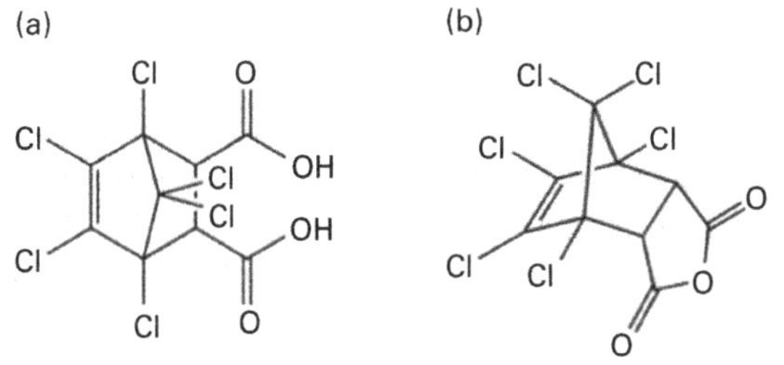

Figure 3.25 (a) Chlorendic or HET acid and (b) chlorendic anhydride used to introduce fire retardancy into unsaturated polyester resins.

optimizing the crosslink density using 'flexible' chain glycols and the ratio of saturated to unsaturated acids in the synthesis. High-performance systems often include polymeric toughening agents. The beauty of some fabrication techniques is that their compatibility can be used to tailor the fracture toughness of the exposed surface in, for example, a pipe.

Isophthalates generally have superior properties to the orthophalates, especially in environments exposed to water, chemicals, and weathering. Residual phthalic anhydride can result from in-service hydrolysis of mono-phthalate chain ends.

Higher temperature resistance, durability, and chemical resistance akin to epoxy resins can be obtained by introducing ethoxylated bisphenol A into the formulation.

Fire retardancy is another important property, which is achieved by incorporating halogenated compounds such as HET acid (chlorendic acid) (Figure 3.25) and/or dibromoneopentyl glycol. HET acid is commonly used to introduce high chlorine content for self-extinguishing properties. However, the thermochemical properties of these resins are inferior to isophthalic resins because the $-C=C-$ is unreactive and fumaric acid is often used to provide the crosslinking mechanism, so careful molecular design is employed for fire-retardant resins.

Fire retardancy can also be achieved or enhanced by adding inorganic fillers such as antimony oxide and/or alumina trihydrate, but at the expense of resin transparency.

3.5.4 Reactive Diluents

Generally, the reactive diluent is styrene, but for translucent sheeting a fraction can be replaced with methyl methacrylate. It cannot be used alone because it does not readily copolymerize with the unsaturated polyester, but provides higher light transmission. Alternative diluents such as vinyl toluene, α-methyl styrene, and diallyl phthalate can be employed in specialist applications. The limitation to the use of styrene is its volatility, and care is needed for safe use. The technological solution is to ensure a low partial pressure of styrene above a moulding during manufacture. This can be achieved in an analogous methodology to air inhibition during curing, whereby ppm quantities of wax or equivalent are added, which migrates to the surface of the moulding.

3.6   Curing of Polyester Resins

These liquid resins can be converted into mechanically stable solid materials without the release of volatiles, and are often referred to as *contact resins* because the composite and the material are formed simultaneously. Curing is achieved by free radical addition copolymerization of the *unsaturated* polyester with the reactive diluent, which is commonly *styrene*. As shown in Figure 3.26, the copolymerization of the maleate groups (and/or fumarate groups) in the polyester with styrene is initiated by a free radical R˙, which is generated either by thermal dissociation of a peroxide 'catalyst' (it is not a true catalyst but an initiator) in hot curing systems, or by reaction with an accelerator in cold curing systems.

Typical peroxide catalysts are shown in Table 3.4. Methyl ethyl ketone peroxide (MEKP) is commonly used for cold curing, while benzoyl peroxide (BP) is employed in hot curing systems. MEKP is available in differing reactivities achieved by altering its concentration in the phthalate ester solution. Hydroperoxide is an impurity arising from the presence of adventitious water, which also affects the reactivity. The Scott Bader handbook [29] differentiates *reactivity*, which controls the gelation and cure times, from *activity*, which is responsible for under-cure of an artefact. Storage of MEKP in dry conditions reduces the loss of its activity. MEKP is commonly used in cold cure formulations with a cobalt accelerator. Free radical formation can also be accelerated for cold curing with dimethyl aniline (DMA). BP is dispersed in a phthalate ester for use as a paste. It is thermally activated above 70 °C. Table 3.4 summarizes the principles for selection of the catalyst.

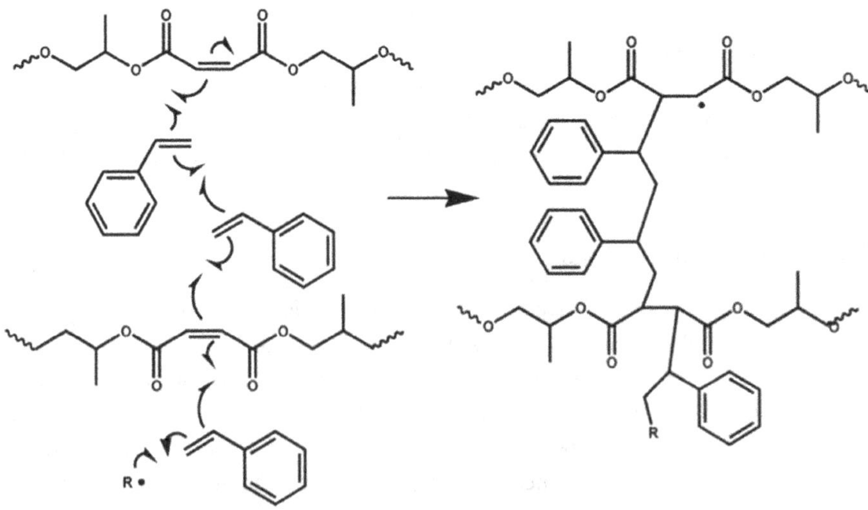

Figure 3.26 Copolymerization of maleate groups in unsaturated polyesters with styrene during curing. The crosslinks involve on average two styrene molecules. R˙ is a free radical generated from the peroxide initiator.

Table 3.4. Peroxides used as 'catalysts' in polyester resin cure

Peroxide 'catalyst'	Form	Reactivity	Critical temperature (°C)	Accelerator	Concentration (%)	Dosing concentration (%)
Benzoyl peroxide (BP)	Paste in a phthalate* plasticizer	Thermal cure	70	Diethyl aniline	50	2–4
Cyclohexanone peroxide (CHP)	Paste in a phthalate** plasticizer	Slow gel + rapid hardening rates	90	Cobalt or vanadium	50	0.5–4
Methyl ethyl ketone peroxide (MEKP) (various reactivity grades may contain hydroperoxide impurity)	Solution in a phthalate plasticizer	Medium + rapid gel + medium hardening rates	80	Cobalt or manganese	60	1–2
Acetyl acetone peroxide (AAP)		Rapid reactivity for RTM and cold press	65		50 (2–4 active O)	1–3
Tertiary butyl peroctoate (TBPO)		Hot-press moulding/pultrusion	70	Cobalt	>50	1–3
Tertiary butyl perbenzoate (TBPB)		Hot-press moulding/pultrusion	90	Cobalt	50	1–3

* Dimethyl phthalate; ** dimethyl phthalate or dibutyl phthalate.

Table 3.5. Accelerators for use in the cure of polyester resins

Accelerator	Solvent	Cobalt concentration (%)	Dosing concentration (%)	Application
Cobalt naphthenate	Styrene	0.5–6.0	0.5–4.0 (100 ppm)	Glass fibre contact moulding
Cobalt octoate	Dimethyl phthalate	0.5–1.0	0.5–2.0	
Vanadium naphthenate/ octoate	Styrene (or naphthenic acid)	2.8–3.2 (by volume)	0.5–2.0	
Dimethyl aniline	DMP/styrene		0.5–2.0	Rapid curing with BP for 'filler' compounds

Table 3.5 gives some typical accelerators. Cobalt naphthenate and dimethyl aniline as solutions in styrene are commonly used. Mixing the catalyst and accelerator can be explosive, so for safety reasons pre-accelerated resins, which have the accelerator added, are available. This is often indicated by the term 'PA'. Gel times should be adjusted by varying the accelerator concentration and not the peroxide catalyst, so it is important to heed manufacturers' recommendations.

3.6.1   Curing Chemistry

It is essential to understand cure mechanisms because the material and artefact form at the same time. Therefore, the manufacturing process can be optimized to achieve consistent properties. Curing involves the initiation of a copolymerization of styrene with the UPE. The rate of reaction is a function of the molecular location of the $-C=C-$, which strongly influences microstructures of cured UPE resin and related vinyl ester resins, and hence their mechanical properties.

Curing involves three stages:

Stage 1 Production of Free Radicals

1. Generation of free radicals from the peroxide (R–O–O–R):
   a. Hot cure:

$$R-O-O-R = R-O^\circ + {}^\circ O- R \text{ or } RO^\circ + {}^\circ OR.$$

   b. Cold cure:

$$R-O-O-R + Co^{2+} = RO^\circ + RO^{\circ -} + Co^{3+}.$$

2. Reaction with inhibitor (hydroquinone, HQ):
   a. HQ controls the shelf life and working time of the catalysed resin:

$$RO^\circ + HQ = ROH + Q^\circ,$$

where Q° is a stable radical unable to initiate curing.

3. Polymer chain growth:
   a. Initiation:

$$RO^o + S = S_1^o,$$

where S is styrene.

   b. Propagation:

$$S_1^o + S = S_2^o,$$
$$S_1^o + P = SP^o \qquad \text{etc.,}$$

where P is polyester.

4. Air inhibition

$$S_{1\to n}^o + O_2 = S{-}O{-}O^o,$$

where $S_{1\to n}^o$ represents a styrene free radical of varying molecular size from degree of polymerization of 1 to n:

$$S{-}O{-}O^o + S = S{-}O{-}O{-}S^o$$
$$S{-}O{-}O{-}S^o + nS + mP = S_nP_mS_n{-}O{-}O{-}S_nP_mS_n{-}O{-}O{-}S_n^o.$$

The rate of reaction of a styrene radical, $S^{\cdot}$, with O_2 is $\sim 10^7$ greater than the reaction with styrene for continued cure. The product is also thermally unstable, releasing free radicals in an uncontrollable mechanism that can initiate further reaction with styrene and O_2.

Stage 2 Copolymerization of UPE Through Reaction of Maleate (M) and Fumarate (F) Groups with Styrene (S)

Figure 3.26 illustrates the copolymerization of the maleate groups with styrene, but in reality we need to consider the relative rates of reaction of the maleate and fumarate groups with styrene, and styrene with styrene [30,31].

The following reactions occur with the given individual rate constants (k):

$$
\begin{array}{ll}
R^o + S = S^o & k_i, \\
S^o + S = S_2^o + nS = S_{n+2} & k_{SS}, \\
S^o + M = M^o & k_{SM}, \\
M^o + M = M_2^o + nM = M_{n+2}^o & k_{MM}, \\
S^o + F = F^o & k_{SF}, \\
S^o + F = F_2^o + nF = F_{n+2}^o & k_{FF}.
\end{array}
$$

The relative rates of styrene-to-styrene radicals, styrene-to-maleate radicals, maleate-to-styrene radicals, and maleate-to-maleate radicals are given by the ratios of rate constants:

$$\frac{k_{SS}}{k_{SM}} = r_{SM} = 6.5,$$

$$\frac{k_{MM}}{k_{MS}} = r_{MS} = 0.005,$$

where r_{SM} and r_{MS} are monomer reactivity ratios for a styrene free radical with maleate groups and a maleate free radical with styrene.

From the values of r_{SM} and r_{MS}, we see that styrene (S) adds to a styrene radical ($S^{\bullet}$) significantly more rapidly than to a maleate ($M^{\bullet}$) radical.

The analogous analysis for fumarate groups is:

$$\frac{k_{SS}}{k_{SF}} = r_{SF} = 0.3,$$

$$\frac{k_{FF}}{k_{FS}} = r_{FS} = 0.07.$$

Thus, fumarate radicals will tend to add styrene, and styrene radicals will tend to add to fumarate unsaturated groups. Thus, an *alternating copolymer* will tend to form.

Methyl methacrylate (m) is an alternative reactive diluent for UPE; the monomer reactivity ratios given below provide understanding [32,33]:

$$\frac{k_{mm}}{k_{mM}} = r_{mM} = 354,$$

$$\frac{k_{MM}}{k_{Mm}} = r_{Mm} = 0.$$

This data shows that methyl methacrylate will tend to *homopolymerize*, with only the occasional maleate links.

Analogously, the reaction of methyl methacrylate with fumarate (F) groups has the following monomer reactivity ratios:

$$\frac{k_{mm}}{k_{mF}} = r_{mF} = 40,$$

$$\frac{k_{FF}}{k_{Fm}} = r_{Fm} = 0.03.$$

Thus, there is still a high probability for methyl methacrylate to *homopolymerize*, but the chain length between fumarate groups will tend to be shorter. Therefore, methyl methacrylate is not an appropriate reactive diluent, but could be included as an additional comonomer.

Stage 3 Diffusion of Monomer

The microstructure of the cured resin is determined in *Stage 2*. Although copolymerization achieves an approximate alternating molecular structure from the two components, the concentration of styrene usually exceeds the molar concentration of maleate and fumarate groups. Therefore, the excess styrene present after initial gelation will diffuse to the free radical sites for further reaction during *Stage 3*. Figure 3.27 shows the consumption of the polyester (P) and styrene (S) components throughout the curing of maleate and fumarate unsaturated polyesters. *Post-curing* is required for full cure. The highest heat distortion temperature (HDT) is achieved with high post-cure temperatures.

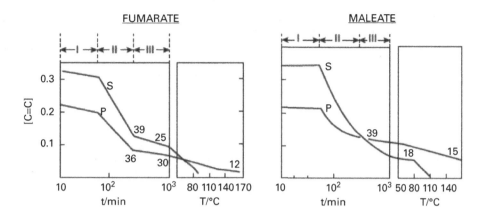

Figure 3.27 The schematic of the decrease in the concentration of unsaturated groups from styrene (S) and maleate and fumarate (P) in the cure of a UPE, illustrating the differing stages: (I) free radical formation; (II) copolymerization; (III) diffusion of monomers during post-cure. The numbers represent the percentage of residual styrene (S) and UPE unsaturated groups [24].

Post-Curing

During chain addition polymerization, polymers form instantaneously so that free monomers are always present unless the reaction is pushed to completion. Therefore, post-curing is usually necessary to achieve the optimum thermomechanical properties. Figure 3.27 shows that, after curing, free styrene is still present in the moulding and that post-curing reduces the free styrene concentration. It is also shown that in the fumarate UPE the residual concentration of unsaturated groups is also lower.

For effective post-curing, the glass transition temperature of the cold-cured resin should be exceeded so that the rate of diffusion of monomers is increased. Post-curing can also involve the decomposition of any residual peroxide into free radicals for further reaction.

Exotherms

The curing of UPE is exothermic, which can be used to manage the degree of post-cure in a 'cold-cured' component. However, uneven exothermic heat dissipation throughout a structure may lead to differential curing and variable thermal residual stresses. Typical exothermic temperatures can reach 200 °C and above, as shown in Table 3.6. The most reactive resins have high fumarate concentrations and exhibit the highest exotherms. This is associated with the presence of the trans –C=C– bonds resulting from the high degree of cis–trans isomerization of the maleate groups in propylene glycol resins (Figure 3.23). Management of the exotherm can be achieved by choosing the correct catalyst/accelerator ratio. A low ratio provides curing without a heat effect. Thus, structures should be designed with heat dissipation in mind so that rapid production is possible.

Table 3.6. The effect of maleate/fumarate isomerization on the gel time and exotherms observed for propylene glycol (PG) and diethylene glycol (DEG) isophthalates (40% styrene) measured using SPI gel test at 82 °C [25]

Acid	Glycol	Gel time (min:s)	Peak exotherm (°C)	Peak exotherm time (min:s)	HDT (°C)	Flexural modulus (GPa)	Flexural strength (MPa)
Fumaric	PG	5:20	210	7:47	100	3.7	126
Maleic	PG	5:20	210	7:44	101	3.7	124
Fumaric	DEG	4:30	202	6:51	52	2.7	106
Maleic	DEG	4:40	183	8:17	43	2.1	84

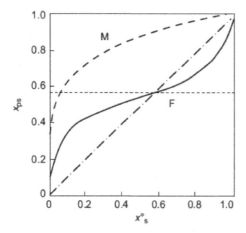

Figure 3.28 Instantaneous copolymer composition of diethyl maleate (M) and diethyl fumarate (F) with styrene (S). x_{ps} is the mole fraction of S in the copolymer. $x^{o}{}_{s}$ is the mole fraction of S in the monomer mixture [30,31,34].

3.7　　Resin Microstructure

The structure of a cured polyester resin is determined by the relative molar concentrations of fumarate groups and the reactive diluent (styrene) and the reactivity ratios (i.e. the rate constants for the copolymerization). The instantaneous copolymer composition can be calculated using the analysis given in Section 3.6.1. Figure 3.28 gives the theoretical instant compositions of diethyl fumarate/styrene and diethyl maleate/styrene copolymers as a function of the composition of the monomer mixture [34].

Assuming that the instantaneous composition of a cured UPE would be similar, it shows that the maleate groups are incorporated less efficiently than the fumarate groups into the molecular structure. Furthermore, a typical UPE employs a mole fraction of styrene of ~0.6 (Table 3.3), which would promote the formation of an alternating copolymer.

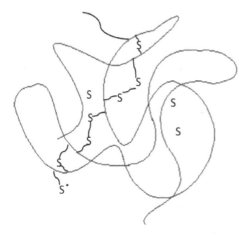

Figure 3.29 A schematic of the conformation of a polyester random coil dissolved in styrene (S) showing the propagation of the polystyryl free radical and the origin of the non-homogeneous microstructure.

The molecular structure of a cured general UPE is considered to have polystyrene crosslinks with an average length of 2–3 monomer units (*n* in Figure 3.20), which reflects the degree of maleate/fumarate isomerization achieved in the polyester synthesis. Further complications arise through the conformation of the polyester molecule in the styrene solution. Styrene has a solubility parameter of 19 MPa$^{1/2}$, whereas a typical polyester has a value of ~21–22 MPa$^{1/2}$, which means that in styrene solution the polyester will exist as a highly conformed coil.

Figure 3.29 shows a schematic of a randomly coiled polyester molecule in styrene solution at the onset of curing. The styryl free radical propagates through adjacent unsaturated groups on the polyester random coil. This results in a 'nodular' microstructure to the cured resin. Funke [35] and Bergman and Demmler [36] have demonstrated by nuclear magnetic resonance (NMR) and infrared (IR) spectroscopy that a cured polyester resin has a defined nodular structure. The development of the microstructure in a UPE is schematically described in Figure 3.30.

3.8   Mechanical Properties of Cast Resins

The mechanical properties of cast UPE resins are a function of the average crosslink density and the 'flexibility' of the molecular chains between the crosslinks. These variables are not fully independent, but provide the options for formulating resins for different applications.

As shown in Table 3.7, increasing the length of the 'diol' chain between the ester groups leads to a decrease in flexural modulus. The rotational flexibility of the 'diol' chain is reduced in the presence of a side group (see EPG vs DEG). Trimethylene glycol (TMG) appears inconsistent with this trend, but the degree of isomerization is

Table 3.7. Flexural properties of cured styrenated UPE with differing diol [25]

Property	PG	DEG	EPG	TMG	EDEG
Chain length (bonds between esters)	3	6	6	4	9
Flexural modulus (GPa)	4.2	3.4	4.1	2.5	2.1
Flexural strength (MPa)	129	119	100	92	83

DEG, diethylene glycol; EDEG, ethylene diethylene glycol; EPG, ethylene propylene glycol; PG, propylene glycol; TMG, trimethylene glycol.

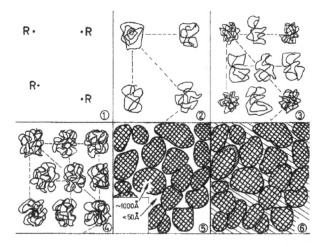

Figure 3.30 Development of the microstructure during the cure of a styrenated PG maleate UPE, established by NMR spectroscopy according to Bergmann and Demmler [36]. 1000 Å = 100 nm = 0.1 μm.

low (and hence so is crosslink density). Furthermore, etherification of TMG can occur during the synthesis so that the length of the 'diol' chain between crosslinks is probably larger.

Aliphatic saturated acids are also used to provide flexibility of the chain between crosslinks. Terephthalic acid can enhance the chain rigidity between crosslinks, but higher synthesis costs means that isophthalic acid is preferred for high-performance resins. Phthalic anhydride is commonly used for general-purpose resins because of lower synthesis costs. Table 3.8 gives the typical properties of cured UPE castings.

The theoretical tensile modulus of an organic glass is calculated to be ≈4 GPa, and most thermosetting resins have similar moduli, depending on the degree of cross-linking and the 'flexibility' of the molecular chain between them. Strength is not a material property because it is dependent on the concentration and dimensions of flaws in the casting. The tensile strength of castings is difficult to measure since flaw-free coupons are difficult to prepare, especially as a result of the microstructure and a high cure shrinkage of ~8% by volume.

Table 3.8. Typical properties of cured styrenated UPE and vinyl ester

Property	UPE	Vinyl ester
Density (kg m^{-3})	1,280	1,120
Hardness (Rockwell M)	110	–
Hardness (Barcol)	40	35
Tensile modulus (GPa)	3.5	3.3
Tensile strength (MPa)	70	70–80
Elongation at break (%)	2.5–4.5	5–6
Fracture toughness, K_{1c} (MN m$^{-3/2}$)	0.49–0.61	0.75
Compressive strength (MPa)	140	–
Refractive index (n_D)	1.57	–
Volume contraction during cure (%)	8	8.3
HDT (°C)	75	100

3.9   Low-Profile Resins

While the addition of reinforcements reduces the volume of resin and hence the curing shrinkage, it is still important to compensate further for the shrinkage to achieve a good surface finish. Low-profile resins contain shrinkage control additives or low-profile additives, which are soluble thermoplastics such as polyvinyl acetate and polystyrene.

3.10   Gel Coat Resins

Gel coat resins are made thixotropic by the addition of 1% by weight of hydrophilic fumed silica. In this way a uniform coating can be applied to the tooling at a thickness of 0.5 mm and gelled prior to the subsequent structural laminating resin and glass fibre. This ensures a good finish to the moulding.

3.11   Related Resins

3.11.1   Vinyl Ester Resins

These resins are designed to achieve the higher performance of epoxy resins while maintaining the flexibility in processing associated with unsaturated polyester resin. A typical vinyl ester is a vinyl-terminated epoxy, which is cured by copolymerization with the styrene diluent using peroxides and accelerators. They are synthesized by reacting an epoxy resin with an unsaturated acid such as methacrylic acid, as shown in Figure 3.31.

Vinyl ester resins have a higher extensibility than standard UPEs, but with a slightly lower tensile modulus. Other vinyl resins based on acrylic acid end-capped

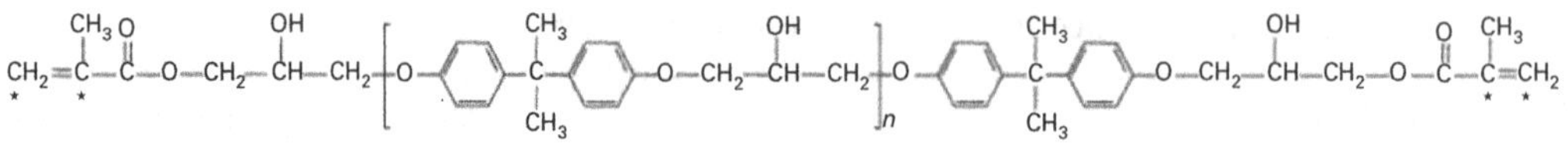

Figure 3.31 Methacrylate end-capped bisphenol A epoxy, an example of a vinyl ester resin. ** indicates the vinyl group.

epoxies are available. Specialist resins based on novolak and tetrabrominated bisphenol A epoxies end-capped with methacrylic acid are also available.

3.11.2   Microstructure of Vinyl Ester Resins

These resins are cured similarly to unsaturated polyester resins; therefore, the analysis given in Section 3.6.1 is relevant to understand the structure of the cast resin [24].

For the methacrylate vinyl ester, the reactivity ratios for the vinyl (V) groups with styrene (S) are:

$$\frac{k_{SS}}{k_{SV}} = r_{SV} = 0.4,$$

$$\frac{k_{VV}}{k_{SS}} = r_{VS} = 0.37.$$

For the acrylic vinyl ester, the reactivity ratios for vinyl (V) groups with styrene (S) are:

$$\frac{k_{SS}}{k_{SV}} = r_{SV} = 0.73,$$

$$\frac{k_{VV}}{k_{SS}} = r_{VS} = 0.72.$$

Since the reactivity ratios for the vinyl group/styrene and styrene/styrene reactions are essentially identical for both systems, the vinyl groups can react either with themselves or with styrene with equal probability. Thus, the chemical structure of the cured resin, which is a copolymer, will be a random arrangement of the monomeric components and a function of the monomer composition. To be random it will consist of blocks of styrene and vinyl ester groups ranging from small ($n \geq 1$) to large. The microstructure should be homogeneous, which is in contrast to the cured polyesters, which have a heterogeneous microstructure, as shown in Figure 3.30. This observation helps explain the higher fracture toughness (K_{1c}) of the vinyl ester resins.

3.11.3   Vinyl Urethanes

Since ester groups can hydrolyse in aqueous environments, especially in the presence of acidic or alkaline conditions, the number of ester groups should be minimized for high durability in aqueous environments. Therefore, hydrolysis of isolated ester groups will not disrupt the network structure significantly. The urethane group is

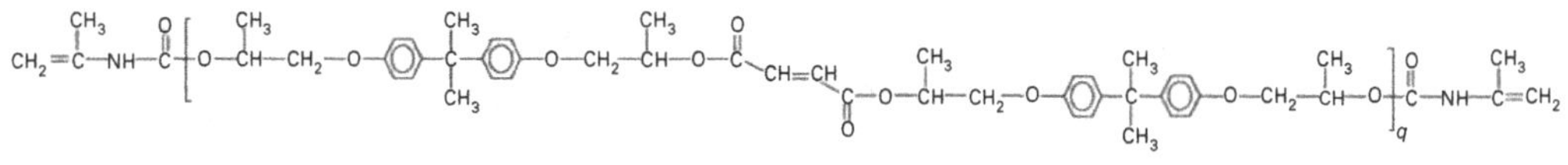

Figure 3.32 Methacrylate end-capped bisphenol A epoxy fumarate, an example of a vinyl urethane [37–39].

more hydrolytically resistant, so vinyl urethanes offer long-term durability. The methacrylate vinyl urethane given in Figure 3.32 is a vinyl end-capped bisphenol epoxy fumarate (originally Atlac 580) that achieves high crosslink density after curing and provides excellent long-term chemical durability in comparison to the isophthalate, bisphenol A polyesters, and vinyl esters. The high crosslink density is reflected in the low fracture toughness ($K_{1c} = 0.45$ MN m$^{-3/2}$). This resin is recommended for long-term durability in *unstressed* applications, but *not* when environmental stress cracking is the most likely failure mechanism, where toughened isophthalate fumarate UPEs are recommended.

3.12   High-Temperature Resin Systems

Many polymeric systems are employed in specialist applications. Here we briefly consider resins for composites used in high-temperature applications.

3.12.1   Thermosetting Polyimides

Polyimides represent the largest class of high-temperature polymers, which are often used for composites. Polyimides are available as thermoplastics or thermosets [40]. We will consider only the main thermosets, PMR and bisimides.

3.12.1.1   PMR Systems

PMR refers to *polymerization of monomeric reactants*, whereby the materials are polymerized *in situ* during the processing of a component. The prepregs are made by impregnating the fibre with a solution of the monomers in methanol. Figure 3.33 illustrates the most common type, PMR-15, which was developed at NASA [41]. The number '15' refers to the molecular weight of the imidized prepolymer, 1,500 g mol^{-1}, which forms from the monomer composition.

The three components shown are:

1. MDA: 4,4′-diamino-diphenylmethane
2. BTDE: benzophenone tetracarboxylic acid dimethyl ester
3. NE: 5-norbornene-2,3-dicarboxylic acid monomethyl ester.

Residual methanol is used to provide the prepreg with tack and drape, and careful cure schedules in an autoclave are employed. The final stage of curing involves

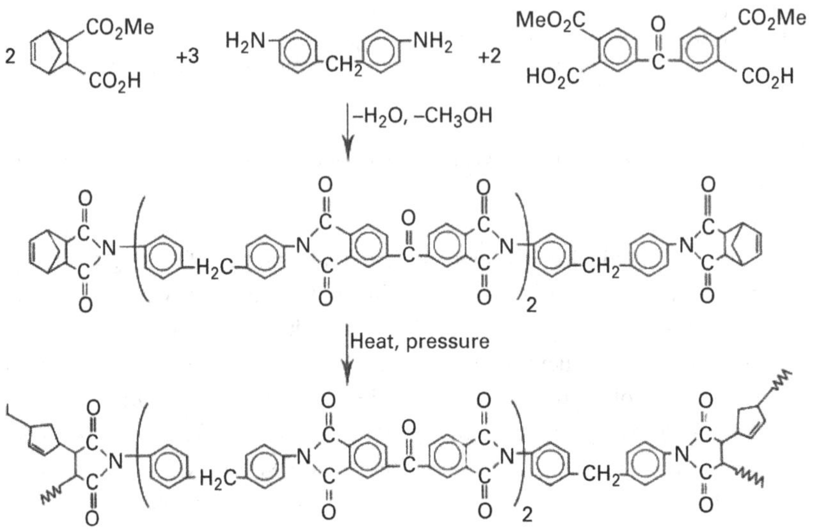

Figure 3.33 The PMR thermosetting polyimide system.

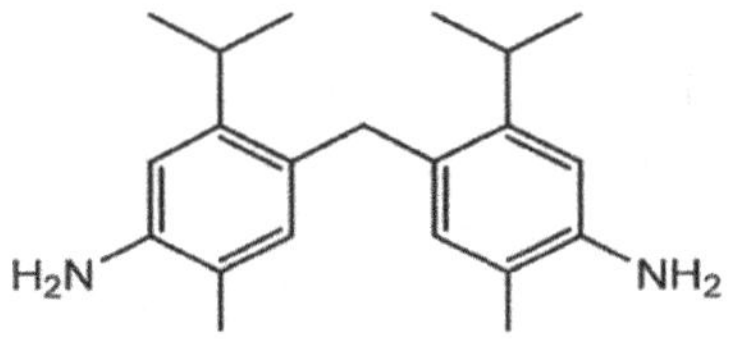

Figure 3.34 CDA: 4,4′-methylenebis-(5-isopropyl-2-methylaniline), a potential replacement for MDA in a low- or non-toxic PMR system.

polymerization through the norbornene end-caps at 316 °C at 200 psi. While this resin is thermally stable and can withstand temperatures up to 350 °C, composites are prone to microcracking resulting from the generation of thermal stresses on cooling. The process appears to be time-dependent because of the loss of volatiles [42].

The major inhibitor to application of PMR resins is the toxicity of the MDA component. Research on replacing MDA is ongoing. For example, Harvey et al. [43] have recently reported the use of a diamine, CDA, with low toxicity for a PMR resin and a T_g approaching that for PMR-15 (Figure 3.34).

3.12.1.2   Bisimides

The most common members of this group are the bismaleimides (BMI). The fundamental structure of a BMI is shown in Figure 3.35, together with a typical example. The cured BMI in Figure 3.35b has a glass transition temperature of 342 °C.

A variety of BMI resins are available, where R in Figure 3.35a can differ by using a range of diamines [43]. Commercial BMIs are available for continuous use at

Table 3.9. Glass transition temperatures of high-temperature matrix polymers (polyimide, cyanate ester, and related resins)

Resin	Type	T_g (°C)	Flexural modulus (GPa)	Flexural strength (MPa)
PMR-15	PMR	340	–	176
Thermid MC-600	BMI	320	4.5	145
Avimid NR 150B2	BMI	340	–	–
Compimide 796	BMI	300	4.6	76
Compimide 796 – toughened	BMI/propenyl copolymer	261	3.7	132
Matrimid 5292	Thermoplastic polyimide	273	4.0	–
PT resin	Phenolic triazine resin	360	4.7	97
AroCy l-10	Cyanate ester	259	2.8	162
AroCy l-10/epoxy blend	Cyanate ester/novolak epoxy	214	–	–
RTX 366	Cyanate ester	175	2.8	121
PBOX	Bisoxazoline-phenolic	>200	–	–
TGDDM-DDS	Epoxy	250	3.6	90

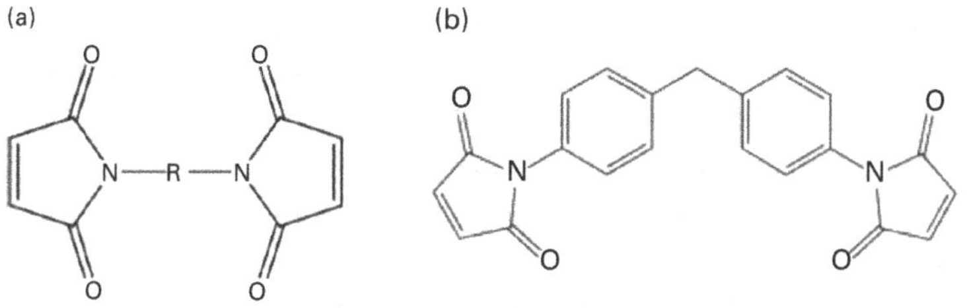

(a)   (b)

Figure 3.35 (a) Structure of a bismaleimide, where R is either an alkyl or aryl chain; (b) a typical component 4,4'- bismaleimidodiphenylmethane [40,44], $T_g = 342$ °C.

temperatures in the range 200–230 °C. To achieve processability for composites a eutectic mixture is often employed. While BMI can be cured by thermal radical polymerization, copolymerization with a range of other monomers provides many formulation options, including toughening with thermoplastic imides. These details are discussed elsewhere [40].

3.12.1.3   Cyanate Resin

Cyanate ester resins have good high-temperature performance, which is only surpassed by that of BMI. Further details can be found elsewhere [45]. The approximate service temperature is related to the T_g of the resin. Table 3.9 provides the T_g of some exemplars of the different resin groups, together with some mechanical properties. The advantages are their rheological properties and the low moisture absorption, but the main disadvantage of cyanate ester resins is the sensitivity of the cyanate groups to hydrolysis. This aspect can be compensated for by copolymerization with epoxy resins [46].

3.12.1.4   Other Potential Matrix Resins

Other high-temperature performance resins such as the bisoaxazoline phenolics, benzocyclobutene (BCB), and phthalonitrile resins are also available. Details are given by Hay [47,48].

3.13   Ceramic Matrices

Ceramics are inorganic materials, and are employed in high-temperature applications – that is, at temperatures $>$1000–2500 °C, which is in excess of that available to metal and polymer matrices. We observed earlier that these materials have moduli greater than typical reinforcements (e.g. alumina at 340 GPa) and exhibit brittle behaviour with failure strains $<$1%. As a result, the role of the reinforcement is to toughen the ceramic. Toughness can be introduced by adding fibrous 'reinforcements', which induce matrix cracking. When the matrix is well bonded to the fibres (i.e. a strong interface) the matrix cracks propagate instantaneously through the fibres without introducing additional energy absorption. However, when the cracks are pinned by the fibres (i.e. with a controlled interface) further cracks form as the load increases. Each time a matrix crack occurs or grows, free surfaces form with energy absorption. Thus, the composite exhibits tough behaviour and the stress–strain curves show an extended displacement as a result of multiple matrix cracking. Figure 3.36 illustrates the extended displacement curve for a fibre composite in comparison to particulate reinforcement [49].

Processing of ceramic matrix composites (CMCs) involves the following stages: (1) impregnating the fibres; (2) consolidation; and (3) formation of the ceramic at high

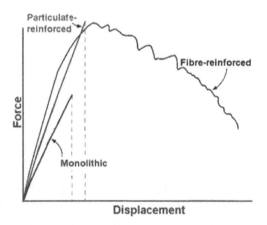

Figure 3.36 Schematic force displacement curves for monolithic ceramic, particulate-reinforced ceramic, and fibre-reinforced ceramic composite. The areas under the curves equate to the energy absorbed through fracture. The fibre composite absorbs energy via multiple matrix cracking [49].

temperature. Generally, melt infusion is impractical because of the high temperatures involved. Therefore, indirect routes are employed in which the fibres are infiltrated with matrix precursors. We will only briefly discuss the available routes.

3.13.1 Slurry Processing

In slurry processing the inorganic powder precursor(s) is dispersed in a carrier liquid, which is often water. For particulate or whisker or short fibre reinforcements a fine dispersion can be prepared ultrasonically then dried. Consolidation is through hot pressing or cold pressing followed sintering. For example, plaster of Paris from $CaSO_4 \cdot 2H_2O$ or ordinary Portland cement involve reaction with the water and subsequent crystal formation for the ceramic matrix to form. A porosity-free matrix is difficult to achieve in this way.

For continuous fibres, the tows are pulled through a slurry of the precursor in an aqueous solution of a polymeric binder to create a 'prepreg', which can be dried and subsequently consolidated by thermal degradation of the binder and hot pressing. Typical matrices formed in this way are glass and glass ceramics. With these matrices, the glass can flow and porosity can be minimized, but that limits the available ceramics.

3.13.2 Liquid Processing

A melt is used to infiltrate the fibres. This could be a polymer precursor, which can be converted into a ceramic during thermal degradation prior to consolidation and firing, such as phenolic resin for carbon, polycarbosilane for silicon carbide, and polysilazane for silicon nitride. An important example is the use of petroleum pitch as a carbon precursor, where the carbon yield can be as high as 80%. In order to achieve high degrees of consolidation and minimum porosity, the pitch normally needs to be converted into a mesophase (or liquid crystal) pitch to maximize the carbon yield during thermal processing. Normally, multiple infiltrations are required to achieve a high-density composite, but the liquid infiltration will not penetrate all the pores left by the loss of volatiles during heat treatment. Full densification is generally not achievable using this approach, and further gaseous infiltration such as chemical vapour infiltration (CVI) or chemical vapour deposition (CVD) can be used to penetrate the remaining pores.

3.13.3 Sol-Gel Processing

In this technique, a monomer reacts with the dispersant, which is mainly water, to provide effectively a solution or a colloid, referred to as a gel, in which the 'particles' are of molecular or nanometre dimensions [50]. As with the slurry technique, the fibres can be infiltrated to prepare a ply, which can be stacked in the manner of prepreg. The advantage is that multiple infiltrations can be used to ensure that the porosity is minimized. The classic example is the use of tetraethoxy

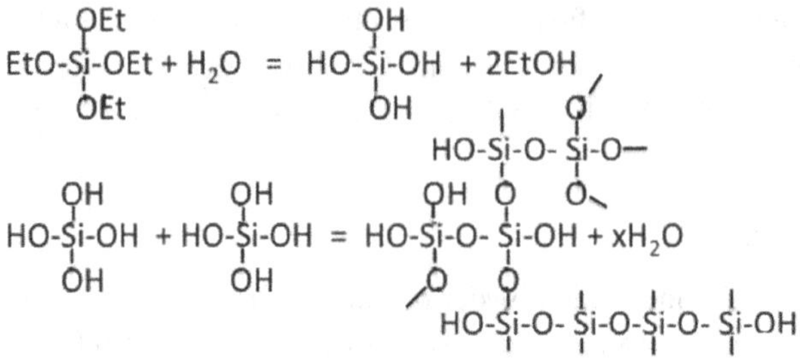

OEt   OH
EtO-Si-OEt + H₂O = HO-Si-OH + 2EtOH
OEt   OH

$$EtO\text{-}Si(OEt)_2\text{-}OEt + H_2O = HO\text{-}Si(OH)_2\text{-}OH + 2EtOH$$

Figure 3.37 Formation of a silica network using sol-gel chemistry from TEOS.

silane (TEOS), which in the presence of an aqueous acid or alkali hydrolyses into a silica sol. The polymerization by condensation of the silanol and ethoxy silane groups causes particles of differing size and structure to form and provides the sol with an increase in viscosity until a gel forms. In this way the infiltrated fibres can have a handleable form for composite manufacture by hot pressing. A number of configurations of molecular structure of the silica are possible, and these can be controlled by TEOS/H_2O ratio as well as the pH. As a result, the silica product will form a matrix that remains either amorphous or crystallizes during hot pressing and firing (Figure 3.37).

A number of oxide-based matrices can be prepared by this method. For example, by mixing silica and alumina sols, mullite can be prepared [51]. Other examples are titanium oxide (titania) or zirconium oxide (zirconia) from titanium or zirconium ethoxides.

3.13.4   Chemical Vapour Infiltration

Chemical vapour deposition is the technique in which chemical reactions in the gaseous state are used to deposit a solid onto a heated surface. Boron and silicon carbide fibres are prepared by deposition onto carrier fibres such as a tungsten wire. For application to composites, the CVD technique is used for CVI of an array of fibres (often in the form of a textile) by gaseous compounds, which are deposited and form the matrix. For example, carbon/carbon fibre composites are prepared in this way from the vapour-phase deposition of carbon using the thermal decomposition of a hydrocarbon. Methane is preferred because it has the highest carbon content for a volatile precursor:

$$CH_{4(g)} \rightarrow C_{(s)} + 2H_{2(g)}.$$

The substrate temperature for CVD and CVI of carbon is generally maintained at 1100 °C, which is the temperature at which the carbon atoms just become mobile. Under isothermal conditions, the infiltration is carried out at low pressures (5–50 torr), whereas with a thermal gradient atmospheric pressures are employed. A pressure gradient technique is used when the porosity deep in the structure needs to be filled.

3.13.5   *In-Situ* Processing

In this method the fibres are infiltrated with a molten metal, which is then reacted with a gas to create the required matrix. In the Lanxide process the fibres are infiltrated by molten aluminium, which reacts with oxygen at the outside surface of the preform to form an alumina–ceramic matrix.

3.14   Metal Matrices

A metallic matrix can be formed using the following techniques.

3.14.1   Solid State

This technique refers to diffusion bonding and powder metallurgy routes. In the former a polymer-bound fibre mat is sandwiched by metal foils to form a stack, which is hot pressed into a consolidated composite. The latter involves mixing powdered metal with short fibres, whiskers, and/or particulates, which are pressed and sintered at elevated temperatures.

3.14.2   Liquid State

Discontinuous fibres or whiskers can be simply mixed into a molten metal for casting into an artefact. Rheocasting employs a more viscous melt, which is stirred at temperatures closer to the melting point so that the reinforcement remains uniformly dispersed.

Continuous fibres can be infiltrated using the technique of *squeeze casting* of the molten metal into a preform. An example is the pressure infiltration of aluminium into random long discontinuous Saffil alumina fibres for locally reinforcing piston crowns for automotive engines.

3.14.3   Deposition

In this technique, spray co-deposition of molten metal and the reinforcement are collected for subsequent processing into an artefact.

3.14.4   *In situ*

In this technique, the reinforcement is a fibre-like crystal introduced into the metal matrix using directional solidification. A movable induction-heating coil is used to directionally crystallize a eutectic alloy to introduce parallel needle crystals, which provide the reinforcement.

3.15   Conclusions

The choice of matrix material is discussed. While the emphasis is on polymeric matrices, we describe metal and ceramic systems for completeness. The fabrication routes described in Chapter 4 demonstrate the range of polymers employed, and we give the reasons for the choice. The metal and ceramic composite matrices cannot be readily divorced from the processing routes, so we use these techniques to explain the choice of material.

3.16   Discussion Points

1. What is the approximate maximum value of tensile modulus of a glassy linear polymer? What molecular force resists deformation?
2. What aspect of the structure of a linear polymer is responsible for a higher modulus than that of a glassy polymer?
3. Thermosetting polymers can have slightly higher modulus than a glassy linear polymer. Why?
4. What are advantages and disadvantages of using a linear thermoplastic matrix or thermosetting matrix?
5. Why have epoxy resins and unsaturated polyester resins commonly been used as matrix resins?
6. What mechanisms are available for curing epoxy and unsaturated polyester resins?
7. Consider the solubility parameter as a method of predicting the compatibility of polymers, and also with diluents.

References

1. J. M. G. Cowie and V. Arrighi Polymers, *Chemistry and physics of modern materials*, 3rd ed. (Bocca Raton, FL: CRC Press, 2007).
2. L. R. G. Treloar, *The physics of rubber elasticity* (London: Oxford University Press, 1958).
3. P. Flory, *Principles of polymer chemistry* (Ithaca, NY: Cornell University Press, 1953).
4. H. Ishida and T. Agag, eds., *Handbook of benzoxazine resins* (New York: Elsevier, 2011).
5. B. Ellis, ed., *The chemistry and technology of epoxy resins* (London: Chapman and Hall, 1993).
6. C. A. May, *Epoxy resin: chemistry and technology*, 2nd ed. (New York: Marcel Dekker, Wiley, 1988).
7. K. Hoy, New values of solubility parameter from vapour pressure data. *J. Paint Technol.* **42** (1970), 76.
8. D. W. van Krevelen and K. Te Nijenhuis, *Properties of polymers*, 4th ed. (Amsterdam: Elsevier, 2009).
9. F. R. Jones, Thermally induced self healing of thermosetting resins and matrices in smart composites. In *Self healing materials: an alternative approach to 20 centuries of materials science*, ed. S. van der Zwaag (Dordrecht: Springer, 2007), pp. 69–93.

10. R. Scott, The thermodynamics of high polymer solutions. V. Phase equilibria in the ternary system: polymer 1—polymer 2—solvent. *J . Chem. Phys.* **17** (1949), 279.

11. R. J. Young and P. A. Lovell, *Introduction to polymers*, 2nd ed. (London: Chapman and Hall, 1991), pp. 241–309.

12. P. J. Flory, *Statistical mechanics of chain molecules* (New York: Interscience, 1989).

13. X. Peng and J. K. Gillham, Time–temperature–transformation (TTT) cure diagrams: Relationship between T_g and the temperature and time of cure for epoxy systems. *J. Appl. Polym. Sci.* **30** (1985), 4685–4696.

14. M. T. Aronhime and J.K. Gillham, Time-temperature-transformation (TTT) cure diagram of thermosetting polymeric systems. *Adv. Polym. Sci.* **78** (1986), 83–113.

15. J. K. Gillham, Formation and properties of thermosetting and high Tg polymeric materials. *Polym. Eng. Sci.* **26** (1986), 1429–1433.

16. B. Ellis, Time-temperature-transformation diagrams for thermosets. In *Handbook of polymer-fibre composites*, ed. F. R. Jones (Harlow: Longman, 1994), pp. 115–121.

17. J-P. Pascault, H. Sautereau, J. Verdu, and R. J. J. Williams, *Thermosetting polymers* (New York: Marcel Dekker, 2002).

18. R. B. Prime, Thermosets. In *Thermal characterisation of polymeric materials*, ed. E. A. Turi (New York: Academic Press, 1981).

19. X. Wang and J. K. Gillham, *J. Coatings Technol.* **64** (1992), 37–45.

20. F. R. Jones, Molecular modelling of composite matrix properties. In *Multi-scale modelling of composite material systems*, ed. C. Soutis and P. W. R. Beaumont (Cambridge: Woodhead, 2005), pp. 1–32.

21. D. Porter, *Group interaction modelling of polymer properties* (New York: Marcel Dekker, 1995).

22. J. P. Foreman, S. Behzadi, S. A. Tsampas, et al. Rate dependent multiscale modelling of fibre reinforced composites. *Plast. Rubber Compos.* **38** (2009), 67–71.

23. B. Parkyn and E. Bader, British Patent 713332 (1954)

24. F. R. Jones, Unsaturated polyester resins, In *Brydson's Plastic Materials*, ed. M. Gilbert, 8th ed. (Oxford: Elsevier, 2017), pp 743–772.

25. H. V. Boenig, *Unsaturated polyesters: structure and properties* (Amsterdam: Elsevier, 1964).

26. A. A. K. Whitehouse, Resin systems. In *Glass reinforced plastics*, ed, B. Parkyn (London: Iliffe-Butterworth, 1970), pp. 109–120.

27. R. A. Panther, Unsaturated polyester resins. In *Handbook of polymer-fibre composites*, ed. F. R. Jones (Harlow: Longman, 1994), pp. 121–127.

28. J. A. Brydson, *Plastics materials*, 5th ed. (London: Butterworth, 1989), pp. 653–696.

29. Scott Bader, *Polyester handbook*. Available at: http://cn.scottbader.com/uploads/files/3381_crystic-handbook-dec-05.pdf.

30. F. W. Billmeyer Jr., *Textbook of polymer science* (New York: Interscience, 1962).

31. F. R. Mayo and C. Walling, Copolymerisation. *Chem. Rev.* **46** (1950), 191–287.

32. W. I. Bengough, D. Goldrich, and R. A. Young, The copolymerisation of methyl methacrylate with diethyl maleate and diethyl fumarate. *Eur. Polym. J.* **3** (1967), 117–120.

33. F. R. Mayo, F. M. Lewis, and C. Walling, Copolymerization. VIII. The relation between structure and reactivity of monomers in copolymerization. *J. Am. Chem. Soc.* **70** (1948), 1529–1133.

34. F. Rodriguez, *Principles of polymer systems* (New York: McGraw-Hill, 1970).

35. W. Funke, Über Vernetzte Makromolekulare Stoffe mit Inhomogener Netzwerkdichte. *J. Polym. Sci. C.* **16** (1967), 1497.

36. K. Bergman and K. Demmler, *Koll. Ztscht. Ztscht. f. Polym.* **252** (1974), 193–206.
37. A. Kandelbauer, G. Tondi, and S. H. Goodman, Unsaturated polyesters and vinyl esters. In *Handbook of thermoset plastics*, ed. H. Dodiuk and S. H. Goodman (San Diego, CA: Elsevier, 2014).
38. R. G. Weatherhead, *FRP technology* (London: Applied Science, 1980).
39. F. R. Jones, Vinyl ester resins. In *Handbook of polymer-fibre composites*, ed. F. R. Jones (Harlow: Longman, 1994), pp. 129–132.
40. D. Wilson, H. D. Stenzenberger, and P. M. Hergenrother, eds., *Polyimides* (Glasgow: Blackie, 1990).
41. T. T. Serafini, P. Delvigs, and G. R. Lightsey, Thermally stable polyimides from solutions of monomeric reactants. *J. Applied Polym. Sci.* **16** (1972), 905–915.
42. M. Simpson, P. M. Jacobs, and F. R. Jones, Generation of thermal strains in carbon fibre reinforced bismaleimide (PMR-15) composites, Pt 3, A simultaneous thermogravimetric mass spectral study of residual volatiles and thermal microcracking. *Composites* **22** (1991), 105.
43. B. G. Harvey, G. R. Yandek, J. T. Lamb, et al., Synthesis and characterization of a high temperature thermosetting polyimide oligomer derived from a non-toxic, sustainable bisaniline. *RSC Adv.* **7** (2017), 23149–23156.
44. R. J. Iredale, C. Ward, and I. Hamerton, Modern advances in bismaleimide resin technology: a 21st century perspective on the chemistry of addition polyimides. *Progr. Polym. Sci.* **69** (2017), 1–21.
45. I. Hamerton, ed., *Chemistry and technology of cyanate ester resins* (Glasgow: Blackie, 1994).
46. S. K. Karad, D. Attwood, and F. R. Jones, Mechanisms of moisture absorption by cyanate ester modified epoxy resin matrices, part III: effect of blend composition. *Composites A* **33** (2002), 1665–1675.
47. J. N. Hay, High temperature resins-thermosetting polyimides. In *Handbook of polymer-fibre composites*, ed. F. R. Jones (Harlow: Longman, 1994), pp. 96–101.
48. J. N. Hay, High temperature resins other thermosets. In *Handbook of polymer-fibre composites*, ed. F. R. Jones (Harlow: Longman, 1994), pp. 101–106.
49. F.L. Matthews and R.D. Rawlings, *Composite materials: engineering and science* (London: Chapman and Hall, 1994).
50. L. L. Hench and J. K. West, The sol-gel process. *Chem. Rev.* **90** (1990), 33–72.
51. J. Wu, M. Chen, F. R. Jones, and P. F. James, Mullite and alumina-silica matrices for composites by modified sol-gel processing. *J. Non Cryst. Solid.* **162** (1993), 197–200.

4 Composites Fabrication

In this chapter the principles of composite manufacture are discussed. The advantages and disadvantages of each method are considered in identifying a process for a particular artefact. Specifically, the need to use sophisticated fibre placement techniques in manufacture is described.

4.1 Introduction

Composite materials are converted into artefacts by combining the reinforcement with a matrix. The matrix, which holds the fibres in the correct orientation, can be ceramic, glass, metal, or polymer. The latter is considered the most important and can be either thermosetting, as in epoxy, phenolic, unsaturated polyester (UPE), and related resins, rubber, or thermoplastic. We will concentrate on the variety of techniques applied to polymer matrix composites (PMCs).

The processing of ceramic matrix composites (CMCs) and metal matrix composites (MMCs) are similar but require high-temperature consolidation. Either high-temperature melt processing, chemical vapour deposition (CVD,) or slurry processing are used to infiltrate the fibres. To limit the concentration of voids, often multiple stages are required, which can employ a mixed set of infiltration steps.

Single-stage impregnation is more easily accomplished for polymers. Many shapes can be manufactured, but for a complex-shaped artefact a moulding compound is employed and, as a consequence, the reinforcing fibres are discontinuous.

The fabrication of fibre composite artefacts is achieved in two principle techniques: *direct impregnation* of the fibres in the form of rovings, textiles, or preforms; or *indirect impregnation* from pre-impregnated moulding materials [1].

4.2 Direct Impregnation

Table 4.1 summarizes the techniques employed for direct impregnation of fibres. In this table the resins identified refer to the differing classes. Thus, UPE is the most commonly used (but also refers to unsaturated polyesters and related resins such as vinyl esters, urethane acrylates, etc.). Epoxy is also a commonly used resin, but for the

Table 4.1. Techniques for direct impregnation of fibres showing typical choices

Process	Resin	Fibres
Hand lay-up	Unsaturated polyester (UPE), vinyl ester (VE), and related resins	Glass fibre
Spray-up	UPE, VE	Glass fibre
Filament winding: • wet • dry winding (of preforms with geodesic shapes) • tape winding of prepreg	UPE, VE, epoxy Epoxy	Glass fibre and sand fillers Carbon, glass, aramid fibre
• Fibre placement • Tow placement • Tape placement	Epoxy and related resins Prepreg (epoxy)	Carbon and glass Carbon, glass, aramid
Pultrusion	UPE, VE, epoxy, phenolic	Carbon, glass, aramid
Centrifugal casting of pipes and vessels	UPE and related resins, hybrid systems	Glass fibre and sand fillers
Resin transfer moulding (RTM) • resin impregnation of a preform • with vacuum assist (VARTM) • Seemann Composites Resin Infusion Moulding Process (SCRIMP)	UPE, VE Epoxy and related resins	Glass fibre Glass, carbon, aramid fibres Non-crimp fabrics (NCFs) Glass fibre Stitched preform with angle configuration
Reinforced reaction injection moulding (RRIM)	Polyurethane	Very short and/or milled glass fibres
Structural reaction injection moulding (SRIM)	Polyurethane	Glass-fibre preform

so-called advanced or high-performance composites a range of bismaleimides and related thermosetting resins are available. Phenolics involve the loss of condensation products during curing, but can be employed for specialist applications. The common denominator is a low viscosity for efficient impregnation of the fibres. It is also essential that the cure kinetics can be controlled to match the chosen process. These methods are summarized in Table 4.1.

To achieve a range of shapes and the required performance, the fibres are in the form of a range of continuous rovings, textiles, discontinuous fibre mats, or preforms. As illustrated in Figure 2.22, there is a huge range of options [1]. Many of these are based on glass fibres, while high-performance fibres such as carbon and aramid are invariably used in continuous form. We should note that the term 'rovings' was originally coined for glass fibres and implies a twist. Glass fibres without a twist are referred to as 'direct rovings'. Carbon and aramid fibres will be used in a continuous form (without a twist) and as textiles. Figure 2.22 provides a summary of the range of fibre forms, but for completeness shows the chopped fibres compounded into moulding materials, which are described in Table 4.2.

Table 4.2. Techniques for indirect moulding of pre-impregnated fibres and compounds with typical resins and fibres

Moulding process	Material	Typical resins	Typical fibres
Autoclave	Prepreg stack	Epoxy Bismaleimide High performance thermoplastic, e.g. polyether ether ketone (PEEK), polyether sulphone (PES)	Carbon, aramid, glass
Vacuum bag	Prepreg stack		Carbon
Inflatable core	Prepreg stack		Carbon
Compression moulding	SMC	UPE	Short glass fibres >20 mm
	DMC	UPE	Short glass fibres 10–20 mm
	Glass mat thermoplastic	Thermoplastic/fibrous paper	Short glass fibres mat
	Moulding powders	Phenolic	1–5 mm glass fibres with glass beads and/or fillers Other fibrous fillers
Injection moulding	Moulding granules	Thermoplastics Thermosets: phenolic and related polymers	1–5 mm glass fibres with fillers Acicular mineral fillers
	Moulding powders	Thermosets mainly Phenolic	1–5 mm glass fibres with glass beads and/or fillers Other fibrous fillers
	DMC	UPE	Glass fibre 10–20 mm

4.2.1   Hand Lay-Up

This approach, in Figure 4.1, mainly uses glass fibres in the form of a chopped strand mat (CSM) or woven rovings, which are placed into a mould on whose surface a gel-coat resin has been applied and gelled. The catalysed liquid resin is impregnated into the fibres by hand using a roller. Further glass fibre and resin can be added to build up to the required thickness. The resin is formulated to cold cure in the specified lay-up time. Consolidation can be facilitated by applying pressure through a polymer film, with or without additional external gas pressure. Hand lay-up is labour-intensive, but is a flexible and economic approach to large mouldings of simple shape using low-cost tooling. As shown in Figure 4.1, this is a common process used for relatively simple shapes in small numbers from UPE or vinyl ester resins.

A gel-coat resin is applied to the tooling at a thickness of 0.5 mm, prior to the structural resin and glass fibre. To ensure a uniform coating the resin is made thixotropic by the addition of ~1% hydrophilic fumed silica. The resin can be

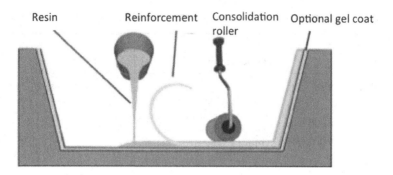

Figure 4.1 Schematic of hand lay-up process for composites [2].

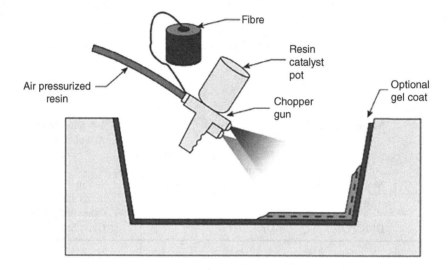

Figure 4.2 Schematic of the spray-up process [3].

formulated for high durability to aqueous environments and may be pre-pigmented for the required finish.

4.2.2 Spray-Up

This is a semi-automated version of hand lay-up in which chopped rovings are sprayed into the mould together with the resin for consolidation, as described above for hand lay-up and shown in Figure 4.2.

4.2.3 Centrifugal Casting

This is similar to spray-up, except that, as shown in Figure 4.3, the fibres and resin are introduced by a lance into a rotating cylindrical mould to produce a pipe. The rate of

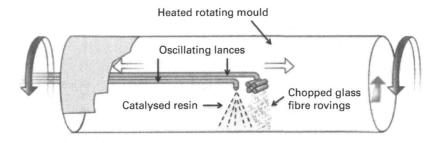

Figure 4.3 Schematic of the centrifugal casting technique, which is a development of spray-up for pipe manufacture, with control over wall construction and optimization of performance.

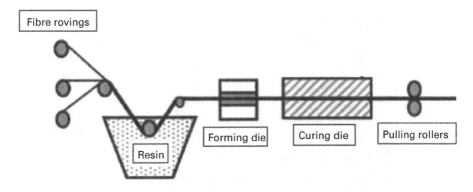

Figure 4.4 Schematic of the pultrusion process for profiles of constant cross-section [4].

rotation provides the centrifugal force for good consolidation. Furthermore, the rate of rotation determines the packing of the fibres so that the fibre volume fraction (V_f) can be adjusted through the wall of the pipe. In this way the rigidity of the pipe can be adjusted. Sand fillers can be added in layers to provide rigidity at the neutral axis of the pipe, where stresses are low under flexure in use, at lower cost. A flexible and/or toughened resin of higher fracture toughness can be applied to the inner surface to provide an integral environmentally resistant liner.

4.2.4 Pultrusion

This technique (Figure 4.4) is used to fabricate continuous lengths of constant cross-section. Pulling the rovings through a resin bath into a heated die, which shapes the section, impregnates the continuous fibres. The resin is hot cured so the skill is to match the rate of cure to the process rate so that the pultrusion has sufficient mechanical strength for release from the tooling on exiting the die. Normally rovings are used that are aligned to the flow direction. This means the product has poor properties in the transverse direction. Incorporating a textile or continuous random

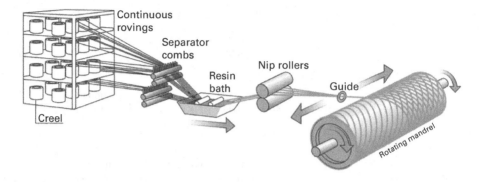

Figure 4.5 Schematic of a wet filament winding process [5].

Figure 4.6 Photograph of the wet winding process for pipes, illustrating consolidation under fibre impregnation.

mat (CRM) outer layer before the die can ameliorate this issue. Alternatively the unidirectional product can be overwound in a subsequent operation.

As shown in Figures 4.5 and 4.6 in the next section, pultrusion and wet filament winding are related in that the fibres are infiltrated with liquid resin and consolidated by pressure applied by the die in the case of pultrusion, and the fibre-tension on contact with the rotating mandrel in the case of filament winding.

4.2.5   Filament Winding

Filament winding can be carried out in two processes:

1. wet winding
2. dry winding.

(a)   (b)

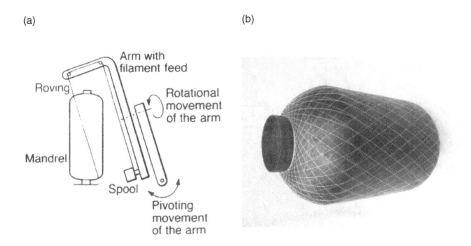

Figure 4.7 Example of planetary filament winding (usually dry): (a) planetary winder; (b) typical fibre wound geodesic shape [6].

In *wet winding*, the rovings are pulled through a resin bath analogously to pultrusion and impregnated on a rotating mandrel, as shown in Figure 4.5. The tension on the fibres aids the consolidation and determines the fibre volume fraction achieved for the walls of a cylinder destined to be a pipe or container (Figure 4.6). Generally, a winding angle of 53° is employed to optimize resistance to internal pressure. Hoop winding achieves an angle of 8°; these pipes are referred to as flexible pipes. To achieve axial stiffness, unidirectional woven rovings are used to provide fibres in the axial direction. These techniques are used to manufacture pipes used for transport of drinking water, sewage, and chemicals. Large-diameter pipes with overwound dished ends are used extensively in the chemical industry.

Dry winding is used to manufacture 3D preforms, which can be impregnated with resin in a separate operation using *resin transfer moulding* (RTM). Often, a polymer foam mandrel is used which is left in place after the resin cures. The strength of this technique lies in the complexity of fibre arrangements, which can be achieved with robotic planetary winders.

An example of a planetary winder is given in Figure 4.7a, while Figure 4.7b illustrates how a geodesic shape can be overwound. It is often necessary to have a 'fixing' technique that prevents the movement of the fibres and ensures that the windings stay in place. An example would be a robot that dispenses an adhesive spot to hold the fibres in position during winding.

4.2.6   Resin Transfer Moulding

In this technique, resin is infused into a fibrous preform enclosed in a closed mould (Figure 4.8). The catalysed resin is pumped into the vented mould where it cold cures before de-moulding. To improve the impregnation and reduce the voids in the

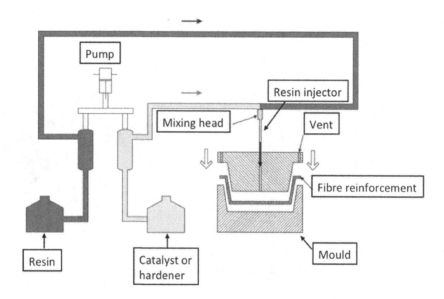

Figure 4.8 Schematic of RTM [7].

composite, vacuum assistance is often used, referred to as vacuum-assisted resin transfer moulding (VARTM). The SCRIMP (Seemann Composites Resin Infusion Moulding Process) process uses a stitched preform with angle configuration. The technique is used for large structural components such as complete car bodies because low clamping pressures mean that tooling costs are low. For advanced composites, non-crimp fabrics (NCFs) are used. In these, layers of fibres at specified angles are stitched so that the fabric configuration is retained during resin infusion and cure. The technique is an attempt to reproduce a laminate structure using a resin infusion approach.

A related technique is reinforced reaction injection moulding (RRIM), in which reactants are rapidly mixed *in situ* at a mixing head for injection into the tooling, which contains the fibrous preform, for fast de-moulding, but is generally not suitable for UPE. This is referred to as structural reinforced reaction injection moulding (SRIM).

4.2.7   Reinforced Reaction Injection Moulding

Reaction injection moulding is a technique that was applied originally to polymerization of ε-caprolactam monomer into a polyamide-6 casting.

The technique described in Figure 4.9 has mainly been applied to polyurethane, which requires the rapid mixing of the two highly reactive monomers, polyol (A) and isocyanate (B), which are transferred immediately into the mould cavity where curing occurs instantaneously. De-moulding is possible after a minimum of 10–20 s. *Reinforced RIM* refers to the equivalent process in which discontinuous fibres are dispersed in one reagent. For polyurethane (PU) the 'milled' glass fibres are dispersed

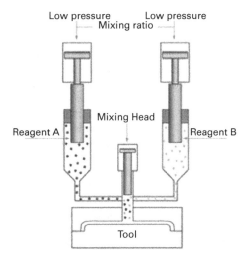

Figure 4.9 Schematic of reaction injection moulding (RIM) [8]. Reinforced RIM has discontinuous fibres or fillers included in one of the reagents, usually the polyol for polyurethanes. Structural RIM has a textile or preform reinforcement included in the tooling [8].

in the polyol. The isocyanate is sensitive to water so inclusion of the reinforcement is not possible.

Mouldings can usually be de-moulded after about 20 s. Extensive research on rapidly curing epoxies and UPE and related resins have attempted to reduce the curing times so the technique could be used more extensively, but this has proved difficult. In any case, fibre volume fractions suitable for high performance cannot be readily achieved. To achieve the mechanical performance associated with composite structures a high V_f of aligned fibres is required. This is achieved in SRIM by including the reinforcement in the form of a preform in the tooling prior to reaction injection.

4.3   Indirect Impregnation

In these fabrication routes the fibres are pre-impregnated with resin or compounded with the resin into a moulding material: prepreg (pre-impregnated fibres sheet); sheet (SMC); dough (DMC); or bulk (BMC); and other moulding compounds. Table 4.2 describes the range of fabrication techniques that use pre-impregnated fibrous materials in the manufacture of composite artefacts.

4.3.1   Moulding Materials

Table 4.3 describes typical moulding materials used for manufacturing composites with a range of properties.

We have identified the requirements of optimum high fibre volume fraction and careful alignment to achieve the mechanical performance, which utilizes the fibre

Table 4.3. Moulding materials with pre-impregnated fibres

Material	Fibre length	Description	Comment
Prepreg	Continuous fibres	Pre-impregnated unidirectional or woven rovings	Not u-polyester Hot press or autoclave (high performance composites)
SMC	Long fibres >30 mm	Equivalent to prepreg for u-polyesters Simple shapes by compression moulding	Achieved using a thickened resin Styrenated resins have a very low viscosity and curing cannot be advanced or easily adjusted
DMC/BMC	Short fibres 10–20 mm	Similar to SMC but very short-fibre dough for moulding complex shapes	Hot press moulding Can be injection moulded
GMT	Short fibres	Glass mat thermoplastic for hot stamping	Short fibrous paper with dispersed powder polymer
SFRTP/SFRTS	Very short fibres <1–5 mm	Short-fibre-reinforced thermoplastics and thermosets Moulding granules and powders	Injection and compression moulding

properties with maximum efficiency. Autoclave and related techniques are used with *prepreg* materials to achieve these aims.

For advanced structural composites the fibres need to be aligned in the principle loading direction. To ensure that the orientation of the fibres is precisely maintained in the artefact, the reinforcements are pre-impregnated with resin prior to moulding. These moulding materials are handleable sheets, referred to as *prepreg* (pre-impregnated fibrous sheet). This is achieved using the low viscosity of the uncured resin for fibre impregnation then increasing the viscosity by advancing the cure to provide a handleable *prepreg* sheet. Not all resin systems are suitable for cure advancement. An alternative process for stabilizing the prepreg involves dispersing a thermoplastic, which dissolves on raising the temperature, increasing the viscosity and providing the required handleability and tack. Prepreg technology is not available with UPE and related resins because of the low resin viscosity and the poor latency of the cure mechanism. The equivalent to prepreg for UPE is a *sheet moulding compound* (SMC). Dough (DMC) or bulk (BMC) moulding compounds are similar to SMC, but employ shorter fibres and fillers in a UPE to create a moulding material in dough form for compression and injection moulding.

4.3.1.1 Prepreg

Prepreg consists of continuous fibres, either unidirectional or in a textile form, which are impregnated with 'molten' monomeric resin (note that early versions used resin solutions, but residual solvent led to defects in the cured material). In order to stabilize fibre orientation, the viscosity of the resin has to be increased by either:

B-staging: the partial reaction of the hardener with the epoxy resin; or
dispersion of a thermoplastic, which dissolves in the resin and increases the viscosity.

These are referred to as flow-controlled prepreg. The quality of a prepreg system is judged by the following characteristics: drape (resulting from the choice of fibre configuration); handleability; tack and degree of 'bleed' (resin excess, as a percentage, over a nominal 40%, i.e. $V_f = 60\%$) in the cured composite. Currently, 'zero' bleed prepreg is common.

A number of variables enable the flow and cure properties of the matrix to be controlled. The curing system needs to have 'latency' so that the prepreg has a long life before the onset of curing. Early systems employed a combination of a catalytic and hardener curing agents to provide B-staging in the prepreg and subsequent curing of the composite. These days a typical epoxy system would employ micronized dicyandiamide (DICY) as the latent curing agent, which becomes active at the melting temperature of 166 °C and could be accelerated by additives such as monuron and diuron. The viscosity control was achieved by a dispersion of a thermoplastic (e.g. polyether sulphone (PES) copolymer), whose molecular weight could be used to control the flow properties of the prepreg. Diamino diphenylsulphone (DDS) also provides sufficient control over cure for use in prepreg systems, but cure temperatures are higher.

The thermoplastic is also chosen to provide the cured resin with a microstructure that optimizes its fracture toughness. Other resin systems are also used to manufacture prepreg, as considered in Chapter 3. The manufacture of prepreg is illustrated in Figure 4.10. There are a number of variants, including solution casting of the resin film, but efficient removal of the solvent proved troublesome. In some early prepreg systems residual solvent was responsible for voidage in cured composites.

A latent curing system is employed so that the prepreg can be stored prior to conversion into an artefact. Most systems involve epoxy resins, although more

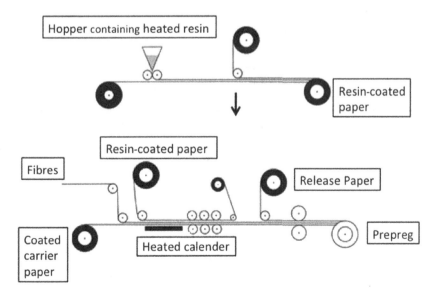

Figure 4.10 Schematic of the manufacture of prepreg [9].

advanced resin systems are available for specialized applications. Thermoplastics such as PEEK and PES (known as advanced polymer composites, or APC) can also be used to create 'prepreg', which has the advantage of thermoplastic fusion for creating laminates.

4.3.1.2 Sheet Moulding Compound

Ionomer formation with the unsaturated polyester carboxyl end-groups is used to increase the molecular weight of the UPE and hence the viscosity of the styrenated resin. This is achieved by the addition of 1% MgO, as shown in Figure 4.11. The catalysed UPE is compounded with a filler, usually stearate-coated calcium carbonate, a 'low-profile' or shrinkage additive and release agent. A typical formulation is given in Table 4.4. The resin slurry is deposited onto a polyolefin release and backing film. As shown in Figure 4.12, glass fibres are deposited onto one of the films of resin slurry and sandwiched with the second layer and consolidated between a series of impregnating rollers. The rolled up sheet is then 'aged' for several days until the resin has thickened by ionomer formation. Ionomers self-assemble into higher molecular weight polymers, which increase the viscosity of the styrene solution. Usually,

Table 4.4. Typical formulations of SMC, DMC, and BMC [10–12]

Component	Nature	Concentration (parts by weight)	
		SMC	DMC/BMC
Resin	Unsaturated PE	66.6	66.6
Low-profile additive	Shrinkage control additive	33.3	33.3
Cure system: catalyst	Tert-butyl perbenzoate (TBPB)	1.25	1.5
Cure system: inhibitor	Hydroquinone	0.01	–
Thickener	Magnesium oxide (MgO)	1.00	–
Release agent	Zinc stearate ($Zn(O_2CC_{17}H_{35})_2$)	2.00	3.5
Filler	Calcium carbonate ($CaCO_3$)-stearate coated	180	160
Glass-fibre reinforcement	E-glass chopped strand (hard-sized) (25–50 mm)	93	–
	E-glass chopped strand (hard-sized) (6 mm)	–	46

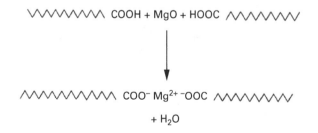

Figure 4.11 Schematic of the formation of an ionomer between added magnesium oxide and the carboxyl end-groups of a UPE [10].

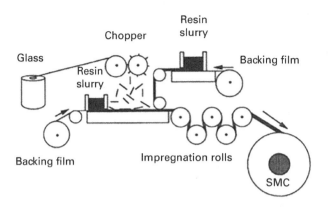

Figure 4.12 Schematic diagram of the manufacture of SMC [10].

chopped strands of 25 or 50 mm length are used, as described in Chapter 2. Continuous fibres in a random configuration can also be used to achieve either highly curved mouldings or better mechanical performance. It is also usual to use hard-sized chopped strands so that the reinforcing element is a bundle. This provides the composite with higher impact strength because a large surface area is created when these elements fracture by defibrillation, absorbing energy. Sheet moulding compound materials are processed into artefacts by hot press or compression moulding at 140–150 °C under pressures of 0.5–5.0 MPa. De-moulding is possible after 1–3 min. To achieve a good surface finish, in-mould coating is employed. Typically, a tensile modulus of 9–12 GPa and tensile strength of 65–80 MPa can be achieved. A notched Izod impact strength of $1-1.2$ kJ m^{-2} can be anticipated.

4.3.1.3   Dough or Bulk Moulding Compound

DMC and BMC refer to dough or bulk moulding compounds, which are dough-like thermosets, which can be hot press, compression, or injection moulded. The terms are interchangeable, with DMC often used in the UK and BMC commonly used in the USA. Table 4.4 compares the differing recipes for SMC and DMC. In contrast to SMC, DMC/BMC have a lower volume fraction of random chopped strands of length of 6–12 mm. They have a greater degree of flow than SMC and can be used for more complex-shaped parts. Both hot press moulding and injection moulding can be used. The latter has the advantage of cycle times of 20–50 s. Injection moulding machines need to be equipped with a hydraulic-assisted feed. Plug flow occurs with injection moulding of these thermosets. The orientation of the fibres leads to an anisotropy in mechanical properties. As with thermoplastics, tooling design is critical to optimum performance.

4.4   Autoclave Moulding

In order to ensure the maintenance of fibre angle and alignment in structural components, laminates are prepared from stacks of prepreg. Consolidation requires control

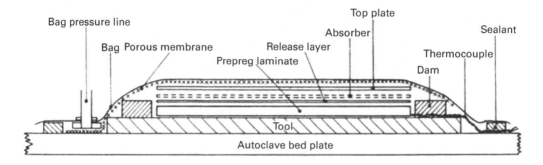

Figure 4.13 Schematic of a bagged-up prepreg stack or laminate with a description of the components employed [13].

Figure 4.14 A 'bagged-up' artefact and autoclave pressure vessel.

over the flow of the resin without disturbing the fibres and introducing voids. Autoclave moulding at high pressure is used to achieve these aims. Figure 4.13 illustrates the arrangement of the prepreg stack within the bagging arrangement.

For structural composites, the prepreg is laid up into a stack of plies of the designed arrangements of fibre angle. A typical lay-up would have the following configuration: $0°_2/\pm45°/90°_2/\pm45°_2/0°_2$. The prepreg stack is placed between release cloths and an absorber layer, or a 'bleed' pack to absorb excess resin, which is bagged up as shown in Figure 4.14. The bag pressure line is used to evacuate the bag, analogously to

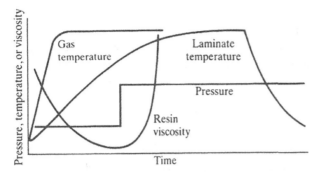

Figure 4.15 Typical cure cycle for an autoclave cure of laminate showing the effect of temperature and pressure on the viscosity of the resin [13].

'vacuum packed food'. A pressure of 1 atm can be achieved in this way, and higher pressures through the autoclave. Early prepreg used ≈5% excess resin, but modern systems employ zero bleed, which ensures that fibre orientation is maintained more precisely. However, some bleeding of the resin helps the removal of air during curing and consolidation of the prepreg into a void-free laminate. The other variables to aid consolidation are the cure cycle and temperature and pressure profiles.

Figure 4.15 shows a typical cure cycle, illustrating changes in temperature and pressure. The resin viscosity is critical since a 'low' viscosity is required for removal of entrapped air before the pressure is applied for consolidation. The bagged-up stack as shown in Figure 4.14 is evacuated to aid removal of air and any volatiles, which also applies a pressure of 1 atm. Some prepreg systems, such as the BMIs and phenolics, use a much higher autoclave pressure to ensure that any entrapped volatiles remain soluble in the polymer and do not lead to voidage in the composite. Thus, control over the cure schedule and pressure and temperatures profiles is essential to the manufacture of laminates with minimum defects and voids. The advantages of this technique are the quality of the structure and maintenance of fibre alignment, but we see that the overall manufacturing cycle is time-consuming, especially the time in the autoclave, which can be as long as 24 h. A typical autoclave and bagged-up stack is shown in Figure 4.14, which illustrates the scale of the process.

The prepreg stack is placed on the tool surface and covered by a porous release layer. The absorber takes up excess resin, which exudes from the stack. Many systems have a 5% bleed designed into the prepreg, but modern systems with zero bleed are available. A close-fitting dam surrounds the prepreg. The top or caul plate is optional, but provides a more even finish. To avoid contamination, the cutting and stacking of the prepreg is done in clean-room conditions. A bagging or membrane film is used to surround the 'stack' and is sealed with a peripheral sealant. A porous membrane is used to ensure an even pressure when the bag is evacuated. The vacuum is applied, analogously to vacuum packing via the 'bag pressure line' port and valve. The bagged stack is placed in a pressure vessel (autoclave), as shown in Figure 4.14. A pressure of 1 atm is achieved through evacuation of the bag and additional pressure is supplied through the gas in the autoclave. Typically, the pressures cover the range 100–1,000 kPa. The pressure used in the autoclave is a function of the temperature and the temperature-dependence of the resin viscosity, so that each prepreg system is provided with recommended cure conditions. A typical cure cycle is given in Figure 4.15, demonstrating the accuracy required to ensure good quality and performance. A free-standing post-cure can also be employed to shorten the process.

The preparation of a stack of prepreg requires control over the fibre angle in the individual plies, which can only be achieved by accurate cutting and laying of prepreg. The other issue is the laying of the prepreg onto curved surfaces, which can be done using tailoring techniques. These are time-consuming hand lay-up methods, so automatic placement techniques are necessary. There are many approaches to combining placement techniques with an automatic lay-up process. Figure 4.16 provides an example in the form of a 'tape placement' head. More recently, analogous 'tow and fibre' placement techniques have been developed (Figure 4.16b). Usually, the tow and/or filament is coated with a compatible size or matrix resin.

With well-designed prepreg systems, the use of an autoclave can be avoided; this technique is referred to as 'vacuum bagging', and can be seen in Figure 4.14.

Resin systems have been developed that have sufficient flow at 1 atm (101.325 kPa or 0.1 GPa) pressure so that the evacuated 'bagged-up' prepreg stack can be consolidated in an oven without the need for the additional pressure normally achieved in the autoclave.

Complexity and the time-consuming nature of autoclave moulding has led to extensive studies on 'out-of-autoclave' processing to speed up the process while maintaining the accuracy of fibre placement. For example, NCFs have through-thickness stitching to stabilize the fibre arrangement for RTM. Out-of-autoclave processing is a subject of continuing research in manufacturing.

Alternative approaches involve placement of *cut prepreg or tape* onto a tooling surface using a robot. The control has developed into a finer-scale process known as

(a) (b)

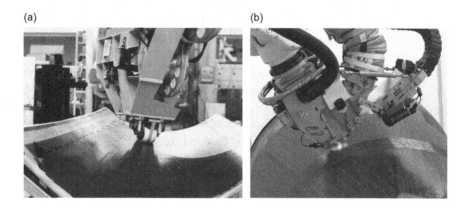

Figure 4.16 Techniques for placing prepreg and fibres onto a curved surface: (a) tape laying of prepreg; (b) automatic dry fibre placement (courtesy of Coriolis Composites®).

tow placement. Robotic placement of tows and impregnated tows of fibres can be used to provide accuracy to the fibre arrangement in the composite structure. Current developments have moved to an even finer scale of fibre placement. Figure 4.16 compares early tape placement with modern dry fibre placement.

4.5   Inflatable Core Moulding

This technique is essentially an inverse vacuum bagging technique in which a hollow core is inflated under pressure against the prepreg stack held inside the tooling. The artefacts prepared in this way are structures that are subjected to bending loads in use. The hollow core provides rigidity in flexure to the structure. The most common application is in sports racquets, such as for tennis and badminton. Bicycle frame parts can also be made this way. Figure 4.17 shows how this technique is employed.

4.6   Hot Compaction

This is a technique akin to hot pressing, and is used for sintering MMC and CMC materials. It is also applied to thermoplastics to manufacture self-reinforced

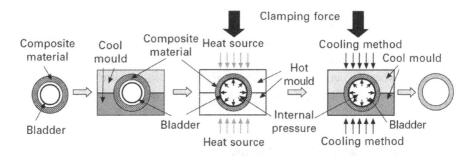

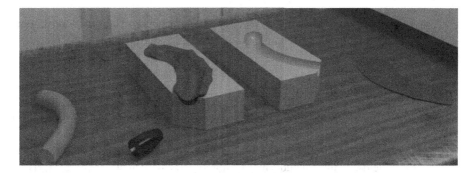

Figure 4.17 Inflation core moulding. (a) Schematic of the inflatable core moulding process [14]. (b) Typical components of bladder moulding, with the prepreg placed in the bottom tool followed by a silicone elastomer bladder and inflation valve (courtesy of Fenner Precision UK).

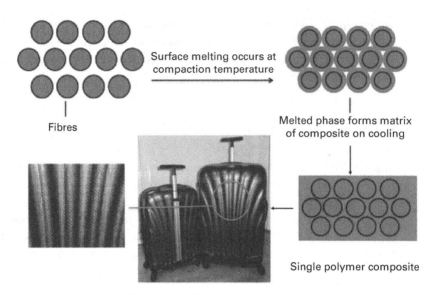

Figure 4.18 Schematic of hot compaction, in which the molten surface of polypropylene fibres forms the matrix. The example product from hot compaction of Curv® textile polypropylene, a lightweight suitcase (Cosmolite). The inset shows the fibre textile pattern (pers. comm., P. Hine).

composites in which the matrix and reinforcing fibres are made from the same polymer, either polyethylene or polypropylene. During the process the surfaces of highly oriented fibres are selectively melted to produce the matrix phase. There are two approaches: one employs heated platens which rapidly melt the fibre surfaces when the compaction pressure is applied; the second uses a fibre with a co-extruded outer surface polymer of lower melting point (Figure 4.18).

4.7   Other Moulding Techniques

Composite artefacts can also be manufactured using the conventional plastics moulding techniques compression moulding and injection moulding.

4.7.1   Compression Moulding

Compression moulding involves placing a charge into the cavity of the mould and applying pressure, as shown in Figure 4.19. Typical materials are the phenolic moulding powders DMC and SMC. The former employ compounded novolak systems, which include hybrid fibrous and particulate reinforcements, and require clamp breathing (a brief opening of the tooling) to allow curing volatiles to exit. Hot pressing can also be used to form laminates without full bagging and autoclave, in analogy to lamination of Formica kitchen laminates.

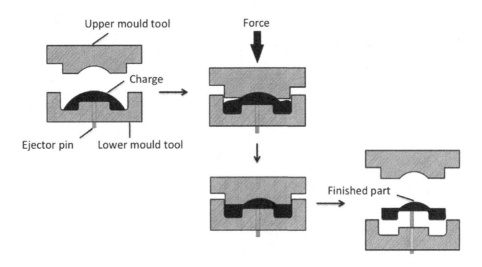

Figure 4.19 Schematic of compression moulding [15].

4.7.2   Injection Moulding

Injection moulding is a standard plastics moulding process whereby the polymer granules are plasticized and homogenized in the barrel as the 'melt' moves along the screw (Figure 4.20). Once the 'melt' is homogenized, the screw moves forwards and injects the 'melt' into the mould, under pressure. The size of the machine is determined by the clamping pressure required to keep the mould closed during injection. For thermosets it is often necessary to use clamp breathing (a brief opening of the mould) and/or a vented barrel to release any volatiles that are formed during the process [16–18].

The main advantage of injection moulding is that it is a rapid manufacturing process; however, for composites we can only employ short fibres because the 'melt' has to flow within a 3D mould cavity and through a narrow gate. Table 4.3 shows that thermoplastics (SFRTP) and thermosets (SFRTS) have fibres with length >1–5 mm, but during plasticization the longer fibres are broken down further. To maximize fibre length in the moulding, the fibres are usually in bundle form with an average length greater than that required for the moulding. The flow of the 'melt' induces a fibre orientation and careful tooling design is required to avoid anisotropy in the moulding. Figure 4.21 shows how the discontinuous fibre adapts to the flow and leads to anisotropy in a moulding, as shown in Figure 4.22.

The moulding in Figure 4.22 will have analogous micromechanics to a 0°/90°/0° continuous fibre composite, with brittle failure in the central material and ductile failure in the 'surface' material.

Figure 4.23 illustrates the flow of the 'melt' around holes and inserts in the tooling and the formation 'weld' lines, which can lead to premature fracture of the part. This also illustrates how mould design can direct the flow to reduce such defects. Software

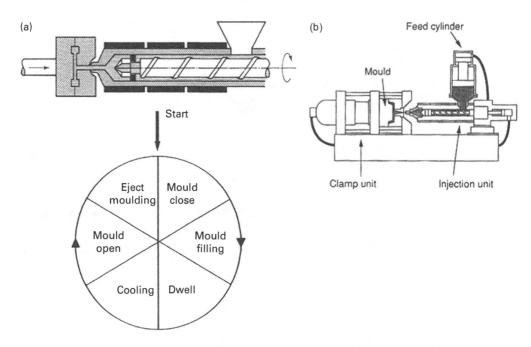

Figure 4.20 Schematic of injection moulding equipment. (a) Typical screw injection moulding machine together with an illustration of the moulding sequence; (b) the modified hopper feed employed for DMC or BMC [11,16].

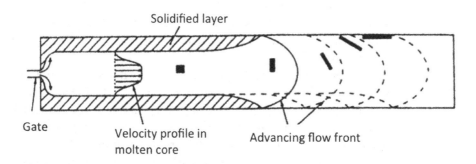

Figure 4.21 Schematic of the flow of a short-fibre/polymer melt in a simple gated mould, illustrating the change in fibre orientation [17,18].

is available to model flow in the presence of multiple gates. With composites, it is possible for the fibres to be circumferentially aligned around a hole to provide higher resistance to torque. This can best be achieved with the correct melt viscosity and aided by the inclusion of particulate reinforcements, which aid flow of the fibres.

Dough moulding compounds are thermosetting UPE materials with longer fibres than fibre reinforced thermoset (FRTS), providing higher performance. The material

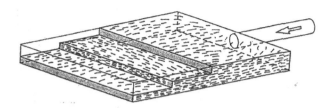

Figure 4.22 Illustration of anisotropy; 'sandwich' fibre orientations in an injection moulding. The fibres are aligned parallel to the mould fill direction at the tooling surfaces, but normal to the flow direction in the centre section [19].

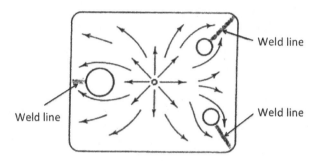

Figure 4.23 Diagram illustrating the flow direction of a melt around a centre-gated plate mould with three circular holes showing the 'knit' or 'weld' lines that form where the melt fronts join [19].

has a dough-like consistency and therefore requires a modified hopper with pressurized flow into the screw, as shown in Figure 4.20b. The nature of the dough is that plug flow occurs in the mould, and anisotropic mouldings are more probable than with conventional thermoplastics.

As shown in Figure 4.23 melt flows around inserts but the melt fronts do not mix, but align as flow continues. Thus, the fibres will be parallel, leading to a weak region in the moulding. There are a number of techniques to address the anisotropy of fibre distribution within a moulding:

1. *Multi-gated tooling* for injection-moulded artefacts to avoid weld lines. The gating of the tooling should be designed by flow simulation software to minimize the opportunity for weld line formation.
2. *Include particulate fillers* to modify the flow of the melt within the mould cavity.
3. *Fibre management* techniques. Control of the fibre orientation can be achieved using a *live-feed* device to apply a shear stress to the melt.

Multi-live-feed was developed by Bevis and Allan [20,21] to provide shear to the melt in the tooling, as shown in Figure 4.24. In this way, the fibre orientation can be optimized within the component. The optimum strength of components can be achieved with this technology.

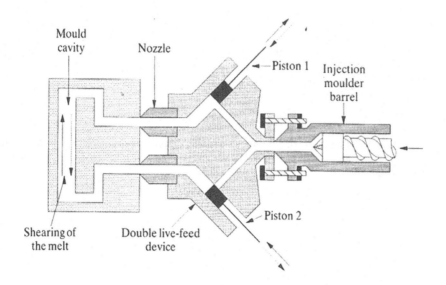

Figure 4.24 Schematic of live-feed device used to selectively provide optimum fibre orientation in the tooling by applying either shear or compression to the melt. With pistons 1 and 2 out of phase, the fibres become oriented to the flow direction. Pistons 1 and 2 in phase or applying compression to the melt will lead to fibre orientation at 90° [20,21].

Figure 4.24 shows a drawing of a two-live-feed device attached to an injection moulding machine between the mould cavity and the barrel. The technique is referred to as *Scorim*. A similar *extrusion* technique has been referred as *Scorex*.

Shear can be applied to the melt in the mould using pistons 1 and 2.

Three modes are possible:

1. Pistons 1 and 2 are pumped backwards and forwards with a phase difference of 180° to align fibres through the thickness.
2. Pistons 1 and 2 are pumped backwards and forwards at the same frequency and in phase to align the fibres along the length.
3. Pistons 1 and 2 are held to apply a static pressure and compact the melt.

Four live-feed devices can be incorporated between two injection screws in order to provide the option of aligning the fibres in a more complex quasi-isotropic arrangement.

4.8   Selection of Manufacturing Method

The range of potential manufacturing processes is reviewed in Table 4.5. Choice is determined by the requirements of the moulding:

1. performance: high or low;
2. shape: simple or complex;

Table 4.5. Summary of composite Manufacturing Routes with choice of resin and typical applications

Process	Reinforcement	Vol (%)	Method	Common resin	Typical application
1. Hand lay-up	CSM or woven mat (G)	15–30	Resin impregnation of dry fibre with brush and roller	UPE Vinyl ester	GRP boat hull
2. Spray-up	Chopped rovings (G)	15–30	Direct application of chopped rovings and resin	UPE	As above
3. Autoclave	Continuous fibres (G, C, A)	60	Forming of stacked prepreg in a vacuum bag in a pressure vessel	Epoxy BMI PMR PEEK	Aircraft wing or rotor blade
4. Filament winding	Continuous rovings, occasionally particulate fillers (e.g. sand) (G, C)	50	Wet resin carried by fibres onto a mandrel at predetermined angles. Variants: prepreg tape, woven fibre tape Dry winding of preform	UPE Vinyl ester Epoxy Phenolic	Pipes and containers
5. Pultrusion	Continuous, occasional CSM and fillers (G, C)	60–80	Wet resin impregnated roving or cloth formed with a hot die	UPE Vinyl ester Epoxy	I-beam or rods
6. Centrifugal casting	Chopped rovings (G) and particulate fillers (e.g. sand)	40–60	Chopped fibres (G) and resin sprayed into rotating mould	UPE and related resins	Pipes
7. Compression Moulding	Chopped or continuous fibres and fillers (G, C, A)	10–60	SMC, DMC wet resin plus CSM preform, prepreg, phenolic moulding powder Pressed to shape in a closed mould	Polyester Epoxy Phenolic	Medium- sized complex mouldings (e.g. lorry body panels)

Table 4.5. (cont.)

Process	Reinforcement	Vol (%)	Method	Common resin	Typical application
8. Hot stamping of thermoplastics	Continuous and discontinuous (G)	15–50	Forming of FRTP sheeting with fully impregnated fibres	Polypropylene Nylon Advanced thermoplastics; PEEK, PES	Medium-size thermoplastic mouldings (e.g. car fascias)
9. Resin injection or transfer moulding (RTM)	Continuous or discontinuous fibres. Polymer foam inserts (G, C, A)	25–30	Wet resin injected into woven or CSM preform, within a closed mould	UPE Vinyl ester	GRP body panels
10. Reinforced reaction injection moulding (RRIM)	Short-milled glass fibres or fibre preform (G)	5–20	Fibre–polyol dispersion and isocyanate mixed at rapid mixing head	Polyurethane (nylon 6)	Large complex parts with high impact resistance (e.g. vertical automobile panels)
11. Injection moulding	Short fibres (G)	5–20	Conventional injection Moulding of pre-compounded granules DMC, BMC, FRTS, FRTP	UPE Thermoplastics Phenolics moulding powders	Small to medium domestic and automotive parts

A, aramid fibres; C, carbon fibres; G, glass fibres.

3. production rate: high or low;
4. anisotropic properties: acceptable or not;
5. cost;
6. choice of fibre configuration:
   a. continuous simple shapes;
   b. woven continuous fibres;
   c. fibre mats, simple shapes but more complex curvature;
   d. dispersed short fibres, complex shapes with high production rates possible.
7. compromise: speed, complexity, performance.

How do we choose from the available methods given above and in Tables 4.1 and 4.2? Examination of Figure 4.25 shows the range of mechanical performances that can be achieved with glass fibres as a function of the form of the reinforcement.

Table 4.5 summarizes the manufacturing routes for composites, with suggestions of suitable materials and applications. Of particular relevance are the typical fibre volume fractions that can be achieved. Since V_f mainly determines the rigidity, the selected technique will be determined by the required performance of the moulding. There is a compromise between the mechanical properties, the complexity of shape, and the rate of production. This aspect is best illustrated by reference to the range of glass-fibre composites manufactured from available reinforcements (Figure 4.25). We see that the maximum achievable fibre volume fraction (V_f) is a function of the nature of the reinforcement and the moulding process. Thus, with short or discontinuous fibres the maximum V_f is $\approx 20\%$, which determines the mechanical properties that can be achieved. The performance is also affected by the fact that in a 3D moulding the orientation is random and the fibres have a distribution of length.

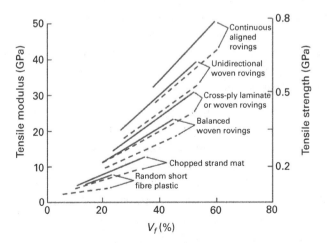

Figure 4.25 Map of the mechanical properties of E-glass-fibre composites from differing types of reinforcement, illustrating the relationship between fibre orientation and volume fraction and their effect on performance (adapted from [22]).

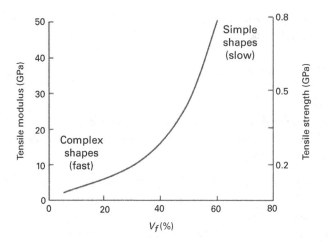

Figure 4.26 Mechanical properties of glass-fibre composites as a function of reinforcement type, illustrating the compromise between performance, complexity of shape, and rate of production (see Figure 4.25 for details).

Low-volume-fraction materials are prepared from techniques such as injection or compression moulding, which can produce complex-shaped artefacts relatively rapidly, but at a cost to the performance. For high performance, control over fibre geometry is essential and these techniques are relatively slow and only for 'simple' shapes. Autoclave moulding can take 24 h. This explains why carbon fibre is not used to the same degree in discontinuous fibre composites. The reduced performance of discontinuous carbon fibre reinforcement may not be justified by the increased cost of the moulding.

Figure 4.26 is a schematic of the spread of mechanical performance of glass-fibre composites derived from Figure 4.25, illustrating the compromise between *performance*, *complexity of shape*, and *rate of production*.

Thus, a complex-shaped artefact manufactured by a moulding technique will probably have a fibre volume fraction of $\leq 20\%$, and we can summarize the need to consider performance, shape, and production rate:

1. injection mouldings (3D); V_f up to $\approx 20\%$;
2. hand lay-up moulding with CSM (2D); V_f approx. 15–30% random in-plane;
3. hand lay-up or wound moulding from balanced woven roving; V_f approx. 20–45%;
4. hot pressed cross-ply laminates; V_f approx. 20–50%;
5. pultrusion from unidirectional rovings; V_f approx. 30–55%;
6. hot pressed sheet using unidirectional aligned continuous fibres; V_f approx. 40–60%;

With higher-performance fibres, such as carbon fibre, most fabrication routes will expect to achieve V_f of 60%. Thus, the curve in Figure 4.26 would be extended to achieve a modulus of ≈ 140 GPa and strength of ≈ 2 GPa (high-strength carbon fibre). In reality a range of curves for each fibre could be constructed.

Aligned continuous fibres provide the highest performance and maximum practical V_f of 0.6. Thus, from the simple law of mixtures analysis, the highest modulus of a glass-fibre composite is $0.6 \times E_f \approx 46$ GPa; a carbon fibre (PAN-based) composite ≈ 135–210 GPa. Examination of Figure 4.25 shows that employing random discontinuous fibres would reduce the modulus significantly. Given the higher cost of carbon fibre, it becomes clear that there is only limited benefit from using randomly oriented discontinuous carbon fibre. As a result, most carbon fibre systems employ continuous fibres either as 'rovings' or as textiles. Glass-fibre composites dominate the discontinuous fibre systems.

A number of textile arrangements are available, as illustrated in Figure 2.22. These provide control over the fibre orientation in the manufactured part. The choice is determined by the drape required to manufacture a curved component. In shapes requiring a large displacement of the reinforcement, CRM can be employed. Sheet moulding compounds have discontinuous random or continuous random fibres in a thickened resin compound, which can be compression moulded into relatively complex curved components such as lorry cabs. For more complex 3D shapes, discontinuous fibres are dispersed into the resin to make a DMC or BMC. Alternatively, thermoplastic systems with dispersed short fibres can be more rapidly injection moulded into a complex 3D shape. As shown in Figure 4.26, there is a compromise between the achievable mechanical properties of the composite and complexity of the shape of the artefact and the rate of production. For high performance, carbon fibre prepreg can be moulded into simple shapes using relatively slow processing, such as an autoclave. On the other hand, a 3D moulding would normally require discontinuous reinforcements so that injection moulding can be employed but mechanically the performance will be lower. Process costs for a high-performance structure are determined by labour requirements for lay-up, as well as the running costs for an autoclave in addition to the cost of materials. As a result, out-of-autoclave processing using stitched unidirectional fibre mats for RTM impregnation is being developed. The main issue is to ensure that the fibre orientation can be precisely retained in the moulding.

Selection of the process for manufacturing composites requires an understanding of this concept. Clearly the choice of matrix polymer dictates performance, but also which method is applicable. As mentioned above, out-of-autoclave moulding has the potential to increase the rate of production of structural parts. One solution is the use of continuous fibre thermoplastic prepreg in which tape laying is combined with thermal welding of the matrix.

Table 4.5 illustrates the range of processing options, but the beauty of composite materials is the flexibility of individual techniques for specific shapes. For example, curved pultrusions have been formed by placing the uncured pultrudate onto a large-diameter wheel. This technique was used by Goldsworthy [23,24] to manufacture simple constant cross-section automobile springs. Graduated leaf-spring equivalents required a prepreg technique. Also, combining pultrusion with subsequent introduction of DMC or SMC enables non-constant cross-section artefacts to be manufactured.

4.9   Conclusions

In this chapter we have described the principal methods for the manufacture of PMCs. The advantages and disadvantages of individual techniques are discussed and the selection criteria provided with respect to the performance, shape, and production rate of the components. The manufacturing options for a composite artefact are discussed in detail. The different moulding techniques and materials are described to provide an overview of the choices needed. There is a compromise between performance, shape, rate of process, and costs, which needs to inform the design process.

Non-polymeric matrices such as ceramics and metals can be processed in analogous techniques to those described here. The major factors to be considered are the higher temperatures involved and the protection of the fibres from reaction with the matrix at these temperatures. We have indicated some of these aspects in Chapter 3.

4.10   Discussion Points

1. What is the advantage of employing discontinuous fibres in a composite?
2. Which fibres are preferred for complex 3D mouldings and why?
3. Why is it unlikely that carbon fibres are employed in these mouldings?
4. What is the main reason for using autoclave moulding with advanced fibres such as carbon fibre?
5. Why is out-of-autoclave processing becoming more probable? What technique is used with reinforcement and what is the reason? Which impregnation methods are best used?
6. Discuss the selection of a process for (1) a high-performance artefact or (2) a rapidly manufactured artefact with a 3D moulding shape.

References

1. F. R. Jones, Glass fibres. In *High performance fibres*, ed. J. W. S. Hearle (Cambridge: Woodhead, 2001), pp. 191–238.
2. M. Raji, H. Abdellaoui, H. Essabir, et al. Prediction of the cyclic durability of woven-hybrid composites. In *Durability and life prediction in biocomposites, fibre-reinforced composites and hybrid composites,* ed. Moh. Jawaid, Moh. Thariq, N. Saba (Oxford: Woodhead, Elsevier, 2019) pp. 27–62.
3. K. Balasubramanian, M. T. H. Sultan, and N. Rajeswari, Manufacturing techniques of composites for aerospace applications. In *Sustainable composites for aerospace applications*, ed. Moh. Jawaid and Moh. Thariq (Oxford: Woodhead, 2018) pp. 55–67.
4. C. Acquaha, I. Datskova, A. Mawardib, et al., Optimization under uncertainty of a composite fabrication process using a deterministic one-stage approach. *Comput. Chem. Eng.* **30** (2006), 947–960.

5. J. P. Nunes and J.F. Silva, Sandwiched composites in aerospace engineering. In *Advanced composite materials for aerospace engineering,* ed. S. Rana and R. Fangueiro (Oxford: Woodhead, Elsevier 2016) pp 129–174.

6. V. Middleton, Filament winding. In *Handbook of polymer-fibre composites*, ed. F. R. Jones (Harlow: Longman, 1994), pp. 154–160.

7. D. Dai and M. Fan, Wood fibres as reinforcements in natural fibre composites: structure, properties, processing and applications. In *Materials, Processes and Applications*, ed. A. Hodzic and R. Shanks (Oxford: Woodhead, 2014), pp. 3–65.

8. X. Hu, E. M. Wouterson, and M. Liu, Polymer foam technology. In *Handbook of manufacturing engineering and technology*, ed. A. Nee (London: Springer, 2015) pp. 125–168.

9. O. Diestel and J. Hausding, Pre-impregnated textile semi-finished products (prepregs). In *Textile materials for lightweight constructions*, ed. C. Cherif (Berlin: Springer, 2016) pp. 361–379.

10. A. G. Gibson, Sheet moulding compounds. In *Handbook of polymer-fibre composites*, ed. F. R. Jones (Harlow: Longman, 1994), pp. 200–205.

11. A. G. Gibson, Bulk moulding compounds and dough moulding compounds. In *Handbook of Polymer-Fibre Composites*, ed. F. R. Jones (Harlow: Longman, 1994), pp. 150–153.

12. F. R. Jones, Unsaturated polyester resins. In *Brydson's plastics materials*, ed. M. Gilbert, 8th ed. (Oxford: Elsevier, 2017), pp. 743–772.

13. F. R. Jones, Autoclave moulding. In. *Handbook of polymer-fibre composites*, ed. F. R. Jones (Harlow: Longman, 1994), pp. 139–144.

14. M. D. Wakeman and J-A. E. Månson, Composites manufacturing: thermoplastics. In *Design and manufacture of textile composites*, ed. A. C. Long (London: Elsevier, 2005), pp. 197–241.

15. R. A. Tatara, Compression moulding. In *Applied plastics engineering handbook: processing, materials, and applications* ed. M. Kutz (Oxford: Elsevier, 2017) pp. 291–320.

16. M. J. Folkes, Injection moulding: thermoplastics. In *Handbook of polymer-fibre composites*, ed. F. R. Jones (Harlow: Longman, 1994), pp. 165–168.

17. M. J. Folkes and D. A. M. Russell, Orientation effects during the flow of short-fibre reinforced thermoplastics. *Polymer* **21** (1980), 1252–1258.

18. M. J. Folkes, *Short fibre reinforced thermoplastics* (New York: Research Studies Press, 1982).

19. M. G. Bader, Short fibre reinforced thermoplastics. In *Handbook of polymer-fibre composites*, ed. F. R. Jones (Harlow: Longman, 1994), pp. 274–278.

20. P. S. Allan and M. J. Bevis, Plastics. *Rubber Proc. Appl.* **7** (1987), 3.

21. P. S. Allan, M. J. Bevis, Injection moulding: fibre management by shear controlled orientation. In *Handbook of polymer-fibre composites,* ed. F. R. Jones (Harlow: Longman, 1994), pp. 171–176.

22. F. R. Jones, Fibre-reinforced plastic composites. In *Aluminium alloys: contemporary research and applications*, ed. A. K. Vasudevan and R. D. Doherty (San Diego, CA: Academic Press, 1989) pp. 605–647.

23. W. B. Goldsworthy and D. Dawson, Composites, fabrication. In *Encyclopedia of polymer science and technology*, 4th ed. https://onlinelibrary.wiley.com/doi/book/10.1002/0471440264.

24. W. B. Goldsworthy, Continuous manufacturing processes. In *Handbook of composites*, ed. G. Lubin (New York: SPE, van Nostrand, 1982), pp. 479–490.

5 Mechanical Properties of Composite Materials

This chapter describes the mechanical performance of a fibre composite. A number of variables that control deformation and fracture are discussed: continuous or discontinuous fibres; fibre angle; fibre length; the transfer of stress between matrix and fibre at a short fibre and/or a fibre-break; and the role of the matrix. Individual components can fracture independently and control the micromechanics; the redistribution of stress after these events is discussed.

5.1 Continuous Fibre Reinforcement

Figure 5.1 shows a unidirectional continuous fibre composite with parallel continuous fibres illustrating that the elastic modulus (or Young's modulus) differs with the direction of the applied stress. Therefore, four principle moduli are defined:

$$E_1(or\ E_l), E_2(or\ E_t), E_3(=\ E_t), E_\theta,$$

where E is the elastic modulus under tension and G is shear modulus. σ, ε, τ, and γ are the stress, strain, shear stress, and shear strain, respectively. The subscripts 1 (l), 2 (t), and 3 (t) represent the loading direction with respect to the fibres (Figure 5.1). Within this chapter, f, m, and c refer to the properties of the fibre, matrix (resin or other continuous phase), or composite. u refers to the ultimate property, so that strength is defined as σ_{cu} and failure strain as ε_{cu}.

5.1.1 Longitudinal Modulus

With a tensile load parallel to the fibres and assuming a perfect interfacial bond, the strain in the composite (ε_c) equals the strain in the matrix (ε_m) which equals the strain in the fibres (ε_f). Therefore:

$$\varepsilon_c = \varepsilon_m = \varepsilon_f. \tag{5.1}$$

Since

$$\sigma_f = E_f \varepsilon_1 \ \text{and} \ \sigma_m = E_m \varepsilon_1 \ \text{and} \ E_f > E_m, \tag{5.2}$$

$$\text{then} \quad \sigma_f > \sigma_m. \tag{5.3}$$

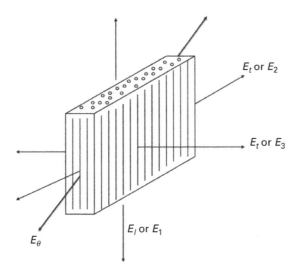

Figure 5.1 A schematic of a unidirectional continuous fibre composite with the loading angles shown. E_1 or E_l is the elastic (or Young's) modulus in the longitudinal or 1-direction. E_t or E_2 is the elastic modulus in the transverse or 2-direction. E_t or E_3 is the elastic modulus through-thickness, which is equivalent to E_2. E_θ is the elastic modulus at a θ angle direction to the fibres.

Consider a load, P_c, across the cross-sectional area (CSA), A_c, of the composite:

$$P_c = \sigma_1 A_c,$$
$$P_c = P_f + P_m,$$
$$P_f = \sigma_f A_f, P_m = \sigma_m A_m,$$
$$P_c = \sigma_f A_f + \sigma_m A_m,$$
$$\sigma_1 = E_1 \varepsilon_1.$$

Therefore, eq. (5.3) becomes

$$E_1 = \frac{E_f A_f}{A_c} + \frac{E_m A_m}{A_c}, \tag{5.4}$$

where (A_f/A_c) defines the volume fraction of the fibres (V_f) and (A_m/A_c) is the volume fraction of the matrix (V_m):

$$V_f = A_f/A_c, V_m = A_m/A_c.$$

Therefore, eq. (5.4) becomes

$$E_1 = E_f V_f + E_m V_m. \tag{5.5}$$

Since $V_m = 1 - V_f$, eq. (5.5) can be written as

$$E_1 = E_f + E_m(1 - V_f). \tag{5.6}$$

Equation 5.5 (or 5.6) is known as the law of mixtures and assumes Poisson's ratio of the matrix (v_m) equals Poisson's ratio of the fibre (v_f). This is strictly incorrect but introduces an error of $<1\%$, which is within the experimental error in the measurement.

The law of mixtures can also be used to estimate Poisson's ratio of the composite:

$$v_{lt} = v_{12} = v_f V_f + v_m(1 - V_f), \tag{5.7}$$

where v_f, v_m, v_{lt}, or v_{12} are Poisson's ratios of the fibre, matrix, or the composite, which is loaded in the longitudinal (l) or 1-direction and records the contraction in the transverse (t) or 2-direction.

5.1.2 Transverse Modulus

With a tensile load at 90° to fibres and assuming a perfect interfacial bond, the stress in the composite (σ_c) equals the stress in the matrix (σ_m) which equals the stress in the fibres (σ_f):

$$\sigma_c = \sigma_m = \sigma_f \tag{5.8}$$

Corresponding strains are:

$$\varepsilon_f = \frac{\sigma_2}{E_f},$$

$$\varepsilon_m = \frac{\sigma_2}{E_m}.$$

Thus,

$$\varepsilon_2 = V_f \varepsilon_f + V_m \varepsilon_m, \tag{5.9}$$

$$\varepsilon_2 = \frac{V_f \sigma_2}{E_f} + \frac{V_m \sigma_2}{E_m},$$

$$\sigma_2 = E_t \varepsilon_2.$$

Therefore,

$$\frac{1}{E_t} = \frac{V_f}{E_f} + \frac{V_m}{E_m}. \tag{5.10}$$

Rearrangement and substitution for $V_m = (1 - V_f)$ gives

$$E_t = \frac{E_f E_m}{E_f(1 - V_f)} + E_m V_f. \tag{5.11}$$

Equation (5.10) can be improved by taking into account Poisson's ratio of the matrix (v_m):

$$E_t = \frac{E_f E'_m}{E_f(1 - V_f)} + E'_m V_f, \tag{5.12}$$

$$E'_m = \frac{E_m}{(1 - v_m^2)}. \tag{5.13}$$

Unfortunately, the assumption that $\sigma_c = \sigma_m = \sigma_f$ is invalid because the fibres are not normally packed in a uniform manner. The empirical Halpin–Tsai equation (eq. 5.14) generally provides a more accurate estimate of E_t:

$$E_t = \frac{(1 + \zeta\eta V_f)}{(1 - \eta V_f)} E_m, \tag{5.14}$$

where

$$\eta = \frac{\left(\dfrac{E_f}{E_m} - 1\right)}{\left(\dfrac{E_f}{E_m} + \zeta\right)}$$

and ζ is the packing geometry factor, which is generally determined experimentally. It takes a value of ~0.2 for fibrous reinforcements. Equation (5.14) is closely related to the Nielsen equation developed for particulate-filled materials.

5.1.3   Halpin–Tsai Equations [1]

The Halpin–Tsai equations are usually presented in a generic form:

$$M = \frac{(1 + \zeta\eta V_f)}{(1 - \eta V_f)} M_m, \tag{5.15}$$

where M is the composite modulus, E_t (or E_2), G_{12}, or v_{23}; M_f is the corresponding modulus or Poisson's ratio of the fibre, E_f, G_f, v_f, and M_m are the corresponding modulus or Poisson's ratios of the matrix, E_m, G_m, v_m:

$$\eta = \frac{\left(\dfrac{M_f}{M_m} - 1\right)}{\left(\dfrac{M_f}{M_m} + \zeta\right)}. \tag{5.16}$$

Table 5.1 shows the extremes of properties that are achievable in unidirectional carbon, glass, and aramid fibre composites. This illustrates the phenomenon of anisotropy in mechanical properties, which has to be dealt with in the design of durable structures using these materials. It also shows that in the fibre direction the modulus and strength are relatively high and reflect the choice of fibres. σ_{cu} and ε_{cu} are the failure stress (i.e. strength) and strain of the composite.

5.1.3.1   Angular Modulus (E_θ)

As shown in Table 5.1, the in-plane tensile modulus has two extreme values as defined in Figure 5.2. These values can be estimated according to eqs (5.5) and (5.14). At angles other than $0°$ and $90°$, equations can be obtained from a consideration of the resolved stresses acting in different directions (Figure 5.3). E_x and E_y are the moduli in the x- and y-directions, respectively, of laminae that act at $90°$ to each other and are

Table 5.1. Typical properties of unidirectional composites in the 1- (0°) and 2- (90°) loading angles

Laminate	0° CFRP	90° CFRP	0° GFRP	90° GFRP	0° AFRP	90° AFRP
E (GPa)	127 ± 3	8.3 ± 0.3	42 ± 1	14 ± 0.5	83	5.6
σ_{cu} (GPa)	1.7 ± 0.1	0.039 ± 0.003	0.92	0.056	1.31	0.039
ε_{cu} (%)	1.16 ± 0.04	0.48 ± 0.05	2.2	0.5	–	–
v	0.29 ± 0.004	0.019 ± 0.002	0.27	0.09	0.34	–
V_f	0.62 ± 0.02	0.62 ± 0.02	0.55 ± 0.05	0.55 ± 0.05	0.62 ± 0.03	0.62 ± 0.03

AFRP, Kevlar 49; CFRP, high-strength carbon fibre; GRP, E-glass fibre.

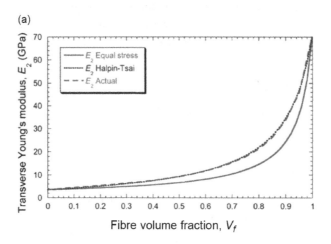

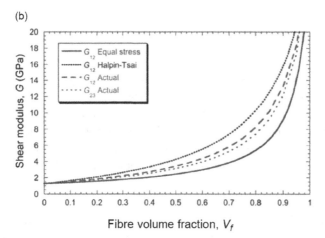

Figure 5.2 Comparison of the predicted (a) transverse and (b) shear moduli of a glass-fibre composite from the constant or equal stress model and Halpin–Tsai equations. E_2 is predicted well by the Halpin–Tsai equation, whereas G_{12} overestimates the actual value [2].

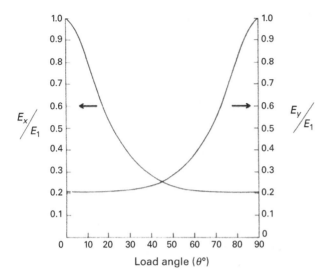

Figure 5.3 Predicted angular moduli of a unidirectional composite illustrating the relationship between E_x and E_y [3].

rotated in-plane from the longitudinal (or 1-) and transverse (or 2-) axes. They therefore define E_θ relative to the fibre direction. Equation 5.17 is obtained from laminate analysis as described elsewhere:

$$\frac{1}{E_x} = \frac{1}{E_1}c^4 + \left(\frac{1}{G_{12}} - \frac{2v_{12}}{E_1}\right)s^2c^2 + \frac{1}{E_2}s^4$$

$$\frac{1}{E_y} = \frac{1}{E_1}s^4 + \left(\frac{1}{G_{12}} - \frac{2v_{12}}{E_1}\right)s^2c^2 + \frac{1}{E_2}c^4$$

$$\frac{1}{G_{xy}} = 2\left(\frac{2}{E_1} + \frac{2}{E_2} + \frac{4v_{12}}{E_1} - \frac{1}{G_{12}}\right)s^2c^2 + \frac{1}{G_{12}}(s^4 + c^4)$$

$$v_{xy} = E_x\left[\frac{v_{12}}{E_1}(s^4 + c^4) - \left(\frac{1}{E_1} + \frac{1}{E_2} - \frac{1}{G_{12}}\right)s^2c^2\right],$$

(5.17)

where $c = \cos\theta$ and $s = \sin\theta$. G_{12} and v_{12} are the in-plane shear modulus and Poisson's ratio on the 1- and 2-coordinates. G_{xy} and v_{xy} are the analogous properties on the xy coordinates.

5.1.3.2   Poisson's Ratios

Figure 5.4 shows schematically the changes in dimensions of an aligned unidirectional composite and the definitions of the three constants.

Figure 5.5 shows how the predicted values of the three Poisson's ratios are influenced by the volume fractions of fibres. The non-linear behaviour of v_{21}, v_{31} is attributed to the constraint of longitudinal shrinkage by the higher-modulus fibres, while for v_{32} and v_{23} the strains involved in the loading of resins are higher.

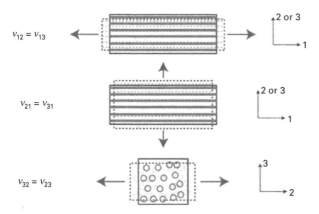

Figure 5.4 The definitions of Poisson's ratios that operate in an aligned unidirectional composite under different loading directions. Loading is in the (a) 1-direction; (b) 2-direction; and (c) 3-direction [2].

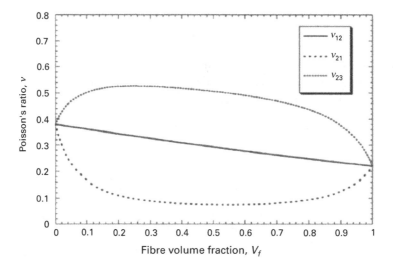

Figure 5.5 Predicted values of Poisson's ratios of a glass-fibre–epoxy composite as a function of fibre volume fraction [2].

5.2   Transverse Strength

The transverse or 90° strength of a unidirectional fibre composite is difficult to predict because it is a function of:

a. interface strength;
b. matrix strength;
c. transverse strength of the fibre; and
d. stress concentration factors.

Clearly one of (a)–(c) will dominate, depending on whichever is weakest. The locations of stress concentrations will also have a major impact on failure micromechanics. Therefore, we need to consider the extremes of interfacial fibre–matrix bonding.

5.2.1   Poorly Bonded Fibres

Consider a square array of poorly bonded fibres, which will behave as holes (Figure 5.6). The volume fraction of fibres, V_f, in a square array is given by

$$V_f = \frac{\pi}{4} \left(\frac{r}{R}\right)^2,$$ (5.18)

When the fibres touch, $R = r$; therefore,

$$V_{f(\mathrm{max})} = 0.785.$$

The spacing between fibres is given by

$$S = 2\left[\left(\frac{\pi}{4V_f}\right)^{1/2} - 1\right] r.$$

Therefore, the introduction of fibres reduces the CSA of the matrix by:

$$\frac{S}{2r} = \left[1 - 2\left(\frac{V_f}{\pi}\right)^{1/2}\right].$$

Therefore, the strength of the composite is given by:

$$\sigma_{2u} = \sigma_{mu}\left[1 - 2\left(\frac{V_f}{\pi}\right)^{1/2}\right].$$ (5.19)

From this analysis, the strength of the composite is given by eq. (5.19), which is plotted in Figure 5.7. A composite of typical V_f of, say, 0.5 will have a transverse strength (σ_{2u} or σ_{tu}) that is $\approx 20\%$ of the matrix strength (σ_{mu}).

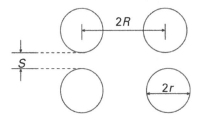

Figure 5.6 Square array of fibres of diameter, $2r$ separated by the distance between the centre of the fibres, $2R$. S is the interfibre spacing.

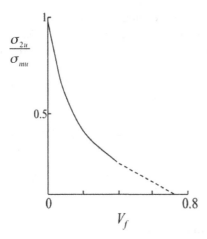

Figure 5.7 The predicted effect of the volume fraction of fibres on the transverse strength for a square array of poorly bonded fibres. $V_{f\,max}$ is 0.785 when the fibres touch and strength falls to zero.

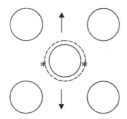

Figure 5.8 Model array of poorly bonded fibres showing the regions of stress concentration. The stress concentration is defined by use of the factor K' (see text); * marks the position of maximum tensile stress.

This model is unrealistic on two accounts: (1) the unrepresentative square array of fibres and (2) exclusion of stress concentrations associated with the poorly bonded fibres. Bailey and Parvizi addressed these deficiencies in a simplistic model [4], which is described in Figure 5.8.

In this model (1) V_f is better represented by $V_f \approx \frac{\pi r^2}{(S+2r)^2}$ and (2) the load is carried by the matrix and enhanced by a stress concentration factor, K':

$$\sigma_2 = V_m \sigma_m$$
$$\sigma_{m(\text{max})} = K' \sigma_m,$$

where K' is the stress concentration factor.

Failure occurs when

$$\sigma_m^{max} = \sigma_{mu}$$

$$\therefore \qquad \sigma_{2u} = \frac{\sigma_{mu} V_m}{K'} \qquad\qquad (5.20)$$

$$\text{or} \quad \varepsilon_{2u} = \frac{\varepsilon_{mu} E_m V_m}{K' E_t}. \tag{5.21}$$

K' has a value of 2.0 according to Bailey and Parvizi [4], who used the estimate of Chamis [5]. Equation (5.21) gives a value of $\varepsilon_{2u} = 0.6\%$ for a glass-fibre–epoxy composite with $V_f = 0.5$.

5.2.2 Well-Bonded Fibres

Under load at 90° to the fibres there is a *strain magnification* in the matrix as a result of the difference in the moduli. The Kies [6] model uses a strain magnification factor (SMF) to define the strain acting on the resin:

$$\text{SMF} = \frac{\text{Strain in resin between fibres}}{\text{Strain in tranverse ply}},$$

$$\text{SMF} = \frac{(2r + S)}{2r E_m E_f + S}. \tag{5.22}$$

For a square array of fibres,

$$S = r\left[\left(\frac{\pi}{V_f}\right)^{1/2} - 2\right].$$

Since

$$\text{SMF} = \frac{\varepsilon_{mu}}{\varepsilon_{2u}},$$

$$\varepsilon_{2u} = \frac{\varepsilon_{mu}}{1 + 2\left[\left(\frac{\pi}{V_f}\right)^{1/2} - 2\right]^{-1}},$$

$$\sigma_{2u} = \frac{\sigma_{mu} E_t}{E_m\left[1 + 2\left[\left(\frac{\pi}{V_f}\right)^{1/2} - 2\right]^{-1}\right]}.$$

Therefore,

$$\sigma_{2u} \approx 0.5\sigma_{mu} \tag{5.23}$$

Thus, the transverse strength of the composite (σ_{2u} or σ_{tu}) is approximately half the matrix strength.

If debonding precedes failure, then a stress concentration will occur at position X in Figure 5.9 and should be included in the model.

For a transverse ply at a stress of $\sigma_t = \sigma_2$, the maximum local stress in the matrix at X is $\sigma_m^{max} = K\sigma_t = K\sigma_2$, where K is the stress concentration factor.

Composite failure will occur when σ_m exceeds σ_{mu}; therefore, we can define transverse strength by σ_m^{max}. Therefore, $\sigma_m^{max} = \sigma_{mu}$, but $\sigma_{2u} = \frac{\sigma_{mu}}{K}$ and $\sigma_{mu} = \varepsilon_{mu} E_m$. Therefore,

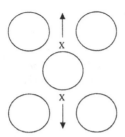

Figure 5.9 A more representative fibre array used in the prediction of transverse strength of composite containing well-bonded fibres. The stress concentrations are located at position X.

$$\varepsilon_{2u} = \varepsilon_{tu} = \frac{\varepsilon_{mu} E_m}{K E_t}. \tag{5.24}$$

The experiments of Bailey and Parvizi [4] show that Kies theory is an overestimate of $\sigma_{2u} (or \ \sigma_{tu})$ and the incorporation of the stress concentration factor brings the predictions closer to the measured values. The Kies model for a glass-fibre–epoxy composite provides $\varepsilon_{2u} \approx 0.65\%$ for $V_f = 0.5$, while eq. (5.24) gives $\varepsilon_{2u} \approx 0.5\%$ for $V_f = 0.5$. The measured value is $\varepsilon_{2u} \approx 0.55\%$ for a composite with $V_f = 0.5$.

5.3   Longitudinal Strength

The strength of a $0°$ (unidirectional aligned) fibre composite depends on the individual failure strains of the two components. Therefore, we need to consider the cases for brittle and extensible matrices. For polymer matrix composites (PMCs), the fibre failure strain (ε_{fu}) < matrix failure strain (ε_{mu}), whereas for brittle matrix composites such as ceramic matrices (and some brittle polymers) the fibre failure strain (ε_{fu}) > matrix failure strain (ε_{mu}).

With a perfect interfacial bond in a composite under an axial load the matrix and fibre will deform simultaneously so that a constant strain operates:

$$\varepsilon_c = \varepsilon_m = \varepsilon_f,$$
$$E_1 = E_f V_f + E_m V_m. \tag{5.25}$$

In terms of stress:

$$\sigma_1 = \varepsilon_1 E_f V_f + \varepsilon_1 E_m (1 - V_f),$$
$$\sigma_1 = \sigma_f V_f + \sigma_m (1 - V_f).$$

The strength depends on which element fails first. We assume that the failure strains of the fibres and matrix have fixed values. – this is unrealistic and its consequences are discussed elsewhere, but it is a necessary assumption for the development of simple models.

Thus, there are two cases: (1) $\varepsilon_{fu} < \varepsilon_{mu}$ (ductile matrix composites) and (2) $\varepsilon_{fu} > \varepsilon_{mu}$ (brittle matrix composites).

5.3.1   Ductile Matrix Composites

In this case the *fibres fracture first* because the matrix has a higher failure strain and the load carried by the fibres is transferred to the matrix. At low V_f the extra load on the matrix does not lead to failure. The strength of a uniaxial composite in the 0° direction is given by

$$\sigma_{1u} = \sigma_{mu} V_m = \sigma_{mu}(1 - V_f). \tag{5.26}$$

At high V_f the matrix is unable to carry the extra load so the composite fails. The strength is given by:

$$\sigma_{1u} = \sigma_{fu} V_f + \sigma'_m(1 - V_f), \tag{5.27}$$

where σ'_m is the stress on the matrix when the fibres fail:

$$\sigma'_m = \varepsilon_{fu} E_m.$$

Thus, at high V_f simultaneous fracture of the matrix and fibres occurs, whereas at low V_f multiple fracture of the fibres takes place. The latter is often referred to as fibre fragmentation. Fragmentation ceases when the load transferred from the fibres to the matrix causes the interface to fail in shear. This phenomenon is used to assess the strength of the fibre–matrix interface and will be discussed elsewhere. Figure 5.10 provides these mechanisms schematically.

V'_f is the minimum volume fraction of fibres above which the micromechanisms of fracture change. Thus, below V'_f multiple fibre fracture occurs (fragmentation), whereas above V'_f the fibres and matrix fracture simultaneously (i.e. a single failure). V'_f is obtained using eqs (5.26) and (5.27):

$$\sigma_{mu}(1 - V_f) = \sigma_{fu} V_f + \sigma'_m(1 - V_f),$$

and thus

$$V'_f = \frac{(\sigma_{mu} - \sigma_m)}{(\sigma_{fu} + \sigma_{mu} - \sigma'_m)}. \tag{5.28}$$

Typical values of V'_f are 0.006 for glass-fibre composites and 0.03 for carbon-fibre composites, so this analysis describes the basis of the behaviour of CFRP and GRP. How these micromechanics operate in real materials is discussed later.

As shown in Figure 5.10, a critical volume fraction of fibres (V_{crit}) is required for reinforcement, which is given by using eq. (5.27) for the matrix strength:

$$\sigma_{mu} = \sigma_{1u} = \sigma_{fu} V_f + \sigma'_m(1 - V_f), \tag{5.29}$$

$$V_{crit} = \frac{(\sigma_{mu} - \sigma'_m)}{(\sigma_{fu} + \sigma'_m)}, \tag{5.30}$$

where $\sigma'_m = \varepsilon_{fu} E_m$.

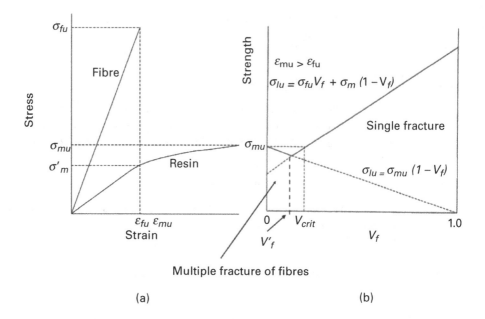

Figure 5.10 (a) Schematic stress–strain curves of the fibres and matrices for $\varepsilon_{fu} < \varepsilon_{mu}$ and $E_f > E_m$. (b) Schematic of variation of fracture strength of a $0°$ unidirectional lamina with fibre volume fraction V_f.

Since V_{crit} is well below the actual volume fraction employed in structural composites, failure will involve fibre fracture. In reality, the strengths of the individual filaments differ so that composite fracture involves the accumulation of fibre-breaks. It is important to understand how the fibre-breaks interact. Figure 5.11a illustrates random fracture of filaments without interaction, whereas Figure 5.11b shows how fibre-breaks can interact and lead to failure.

5.3.1.1   Variable Strength of Fibres

Weibull statistics [7] are commonly used to describe the distribution of individual filament strengths. Figure 5.12 illustrates how the distribution can be ranked.

A cumulative probability distribution function, $P(\sigma)$, can be obtained from the data:

$$P(\sigma) = 1 - \exp\left[-L(\sigma_{fu}/\sigma_o)^m\right], \tag{5.31}$$

where σ_{fu} is the strength of an individual filament of length L, m is the Weibull modulus, and σ_o is the characteristic strength of the population. σ_o is also referred to as a shape parameter. It should not be confused with the average strength of the population ($\bar{\sigma}_{fu}$). Figure 5.13 provides an example of the cumulative probability distribution function given in eq. (5.31).

(a)

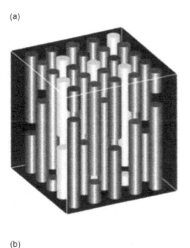

(b)

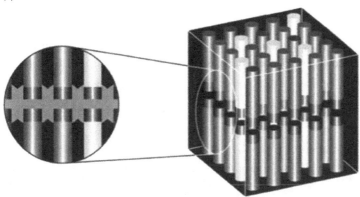

Figure 5.11 Schematic showing (a) random filament fracture within a $0°$ composite; (b) failure crack propagation resulting from the interaction of filament fractures. The strength of the interfacial bond between the fibre and matrix determines whether stress transfer occurs in the plane of fibre fracture.

Taking logarithms of eq. (5.31) enables values of m and σ_o, which describe the distribution function, to be estimated:

$$\ln\left\{1/L \ln\left[1/(1-P(\sigma))\right]\right\} = m \ln \sigma_{fu} - m \ln \sigma_o. \qquad (5.32)$$

A plot of eq. (5.32) is linear, with a slope of m, and $\ln \sigma_o$ can be obtained when $\ln\left\{1/L \ln\left[1/(1-P(\sigma))\right]\right\} = 0$. Figure 5.14 gives example plots for two different strength distributions. It is clear from this figure that when $m_1 > m_2$ there is a narrower distribution in fibre strength.

5.3.1.2 Strength of an Unimpregnated Fibre Bundle

A simple model for a unidirectional composite is an unimpregnated fibre bundle. Coleman [9] has reported eq. 5.33:

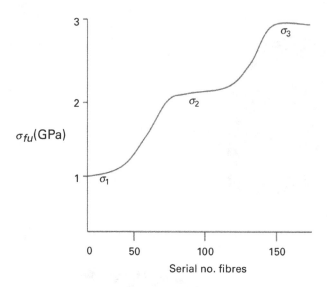

Figure 5.12 Schematic of the ranking of the strength of individual glass filaments within a bundle showing the presence of three strength populations [8].

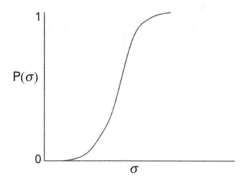

Figure 5.13 Example of a cumulative probability distribution function.

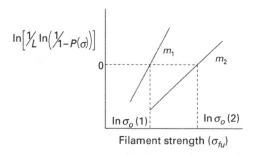

Figure 5.14 Schematic Weibull plots for two populations of fibres (1, 2). m_1, m_2 and $\sigma_o(1)$, $\sigma_o(2)$ are the Weibull and characteristic strengths of the two populations.

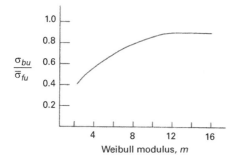

Figure 5.15 The strength of a bundle of unimpregnated fibres (σ_{bu}) as a function of the Weibull modulus for the strength distribution.

$$\frac{\sigma_{bu}}{\bar{\sigma}_{fu}} = \left(\frac{1}{m\varepsilon_1}\right)^{1/m} \left(\frac{1}{\Gamma\left(1+\frac{1}{m}\right)}\right),$$

(5.33)

where $\bar{\sigma}_{fu}$ is the average strength of the filaments in the bundle and Γ is the gamma function.

For typical reinforcing fibres where $m = 2$–5 the bundle strength is given by eq. (5.36):

$$\sigma_{bu} \approx (0.5 - 0.6)\bar{\sigma}_{fu}.$$

(5.34)

Figure 5.15 shows how bundle strength depends on the Weibull modulus, m. The strength of an unimpregnated bundle is about half the mean strength of the fibres. Therefore, the matrix has an important role to play in the performance of a so-called reinforced material.

5.3.1.3   Strength of an Impregnated Bundle

A $0°$ composite can be modelled as an impregnated bundle. As implied by Figure 5.11, there are two potential models of failure, which describe the fracture process:

1. cumulative weakening (resulting from an increase in filament breaks); and
2. fibre-break propagation.

Cumulative Weakening Model

In this model the load carried by an individual fibre is transferred to the matrix through shear when it breaks, so that the stress around the ends of the broken filaments is modified. Therefore, at a fibre-break there is an ineffective length, L_c, which does not carry a load. This is illustrated in Figure 5.16. Progressive fracture of the filaments leads to progressive weakening of the material. Rosen [10] provides a prediction of composite strength.

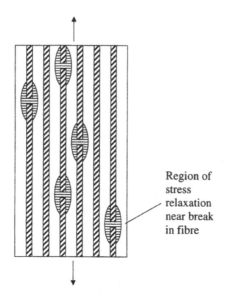

Figure 5.16 Schematic illustrating the cumulative weakening model for 0° strength (from Rosen [10,11]).

The strength of the composite can be taken as that of the bundle, provided that fibre-breaks do not interact, over the ineffective length L_c:

$$\sigma_1 = \sigma_f V_f + \sigma_m (1 - V_f).$$

Thus, eq. (5.35) can be derived:

$$\sigma_{1u} = \sigma_{lu} = \underbrace{\left(\frac{L}{L_c m \varepsilon_c}\right)^{1/m} \frac{1}{\Gamma\left(1 + {}^{1}/_{m}\right)} \bar{\sigma}_{fu} V_f}_{\text{Fibre- weakening}} + \underbrace{(1 - V_f)\sigma_{mu}}_{\text{Matrix-contribution}} , \qquad (5.35)$$

where L is the length of the specimen.

For typical values of the ineffective length, when $m = 5$ the composite strength is $\sigma_{1u} \approx 4\bar{\sigma}_{fu}$ and when $m = 10$, the composite strength is $\sigma_{1u} \approx 2\bar{\sigma}_{fu}$.

This shows how the matrix contributes to the fracture behaviour. In contrast to 'dry' bundles, impregnated bundles and composites have higher strengths, which is a function of the distribution of strengths in the fibre population, as well as their interaction with the matrix. We will see later how the stress transfer between fibre and matrix involves the quality of the interface or interphase.

Fibre-Break Propagation Model

This is the case for a strongly bonded fibre–matrix interface so that a fibre-break propagates through the matrix and adjacent fibres, as shown schematically in

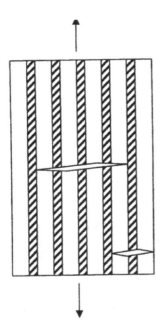

Figure 5.17 Schematic of the fibre-break propagation model for 0° strength.

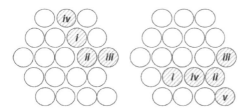

Figure 5.18 The sequence of fibre failure (numbered) in two examples of a fibre composite layer with different random fibre failure distributions. White and shaded fibres represent intact and broken fibres, respectively [12].

Figure 5.17. This is a result of the stress concentration on the adjacent matrix and fibres in the plane of the first fibre-break.

5.3.1.4   0° Strength of Ductile Matrix Composites: A Review

In reality, the cumulative weakening still occurs through fibre fracture. At each fibre-break the stress on the fibre is redistributed onto the matrix and neighbouring fibres, increasing the probability of fracture. Figure 5.18 shows two examples of the fibre failure sequence in typical composite layers. Initially, stable groups of broken fibres form in the layer that withstand the overload imposed by the broken fibres, which then become unstable at higher applied strains.

Shear yield in the matrix or interphase or debonding of the fibre–matrix interface dissipates the additional stress across a larger volume of material and reduces the

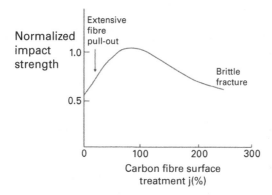

Figure 5.19 Effect of surface treatment on the interfacial bond strength of high-strength carbon fibres with epoxy resins on the impact strength of composites [13]. The optimum surface treatment is defined as 100%. Thus, overtreatment leads to a strong interface and brittle behaviour.

stress concentration on the adjacent matrix and fibres. Fracture surfaces will exhibit varying degrees of fibre pull-out. The composite strength is a complex function of fibre, interface, and matrix properties.

The consequences of these phenomena are shown in Figure 5.19, which describes the development of oxidative surface treatment of carbon fibres to optimize composite impact strength.

5.3.2   Brittle Matrix Composites

5.3.2.1   Case 2a ($E_f > E_m$)

This is the case for PMCs. As above, we assume that the failure strains of the fibres and matrix have fixed values, which is a necessary assumption for the development of simple models. As noted above, this is unrealistic. Since $\varepsilon_{fu} > \varepsilon_{mu}$ (or $\varepsilon_{mu} < \varepsilon_{fu}$), the matrix fractures first and the load carried by the matrix is transferred to the fibres.

At low V_f the fibres will not be able to support the additional load and the composite will fracture simultaneously. Thus, the strength of the composite is given by an analogous equation to eq. (5.27):

$$\sigma_{1u} = \sigma'_f V_f + \sigma_{mu}(1 - V_f),$$ (5.36)

where σ'_f is the stress on the fibres when the matrix fractures:

$$\sigma'_f = \varepsilon_{mu} E_f.$$

With $E_f < E_m$, the matrix carries a small proportion of the load so that at high V_f the additional load on the fibres is insufficient to cause fibre fracture, so the strength is given by an analogous equation to eq. (5.26):

$$\sigma_{1u} = \sigma_{fu} V_f.$$ (5.37)

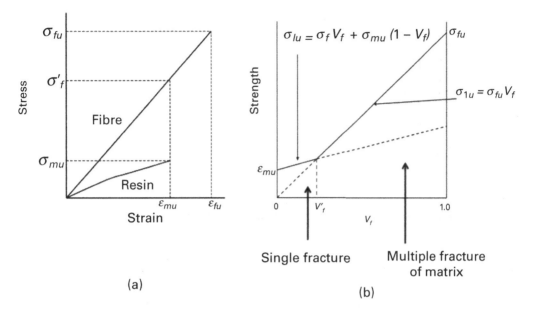

Figure 5.20 (a) Schematic stress–strain curves of the fibres and matrices when $\varepsilon_{fu} > \varepsilon_{mu}$ and $E_f > E_m$. (b) Schematic of variation of fracture strength of a 0° unidirectional lamina with fibre volume fraction, V_f.

By equating eqs (5.36) and (5.37), we can estimate the value of V'_f above, for which there is an enhanced reinforcing effect:

$$\sigma'_f V_f + \sigma_{mu}(1 - V_f) = \sigma_{fu} V_f$$

$$V'_f = \frac{\sigma_{mu}}{(\sigma_{fu} - \sigma'_f + \sigma_{mu})} a. \tag{5.38}$$

V'_f is the minimum fibre volume fraction for enhanced reinforcement and also represents the point at which the micromechanics change.

The response of a unidirectional brittle matrix composite under an axial load is illustrated in Figure 5.20.

Below V'_f, fracture involves a single failure, whereas above V'_f multiple matrix cracking occurs (providing the fibre–matrix interface remains intact). It represents a change in micromechanics from *single failure to multiple matrix cracking*. A typical value of V'_f can be calculated from eq. (5.38). For glass-fibre composites ($E_f = 70$ GPa), $V'_f \approx 0.11$.

Since typical commercial composites will have $V_f \approx 0.4 - 0.7$, which is above V'_f, we can expect glass-fibre composites under load to exhibit multiple matrix cracking. Multiple matrix cracking is illustrated in Figure 5.21. Under an applied stress, the deformation of the composite will exceed the matrix failure strain so that the material will behave in a 'tough manner'.

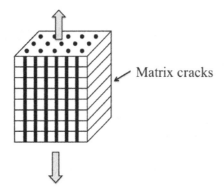

Figure 5.21 Schematic illustrating the phenomenon of multiple matrix cracking in a 0°
composite under a uniaxial load.

5.3.2.2   Case 2b ($E_f < E_m$)

This a specific case for ceramic materials reinforced with fibres of lower modulus but
high failure strain. An example is the use of polymeric fibres for reinforcing cement [14].

In this case, the load carried by the fibres is much less because $E_f < E_m$ and
$\sigma_f(= \varepsilon_f E_f)$ is lower. On matrix fracture, $\sigma'_f(= \varepsilon_{mu} E_f)$ is also lower. Since

$$\sigma_{1u} = \sigma'_f V_f + \sigma_{mu}(1 - V_f),$$

σ_{1u} will decrease with V_f. Thus, V'_f represents the minimum volume fraction of fibres
at which the micromechanics changes. Thus, a critical fibre volume fraction for
reinforcement, V_{crit}, is given by using eq. (5.37) for matrix strength:

$$\sigma_{mn} = \sigma_{1u} = \sigma_{fu} V_f, \tag{5.39}$$

$$V_{crit} = \sigma_{mu}/\sigma_{fu}. \tag{5.40}$$

This is illustrated schematically in Figure 5.22b.

Below V'_f, fracture involves a single failure; above V'_f, multiple matrix cracking
occurs. This behaviour provides brittle matrices such as cement or ceramics with
toughness as a result of the surface area generated during multiple matrix cracking.

5.4   Stress Transfer at Interfaces

To understand further the nature of multiple fibre (fragmentation) and multiple matrix
fracture in composite materials we must understand the role of interfaces. In the case
of fibre fragmentation and multiple matrix cracking, it is the fibre–matrix interface that
is important, whereas for laminate multiple transverse cracking (often referred to as
matrix cracking) it is the ply interfaces that are critical. In all of these we need to
consider the stress transfer mechanics. There are two extreme situations: (1) weak
adhesion (i.e. poor interface) and (2) strong adhesion (i.e. good interface). The
constant shear model (Kelly–Tyson) is used when a weakly bonded interface exists,

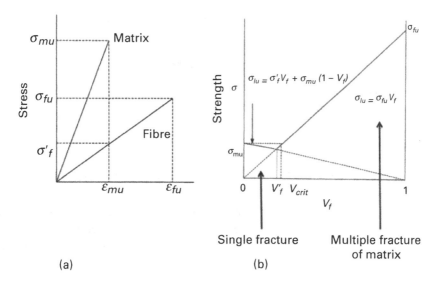

Figure 5.22 (a) Schematic stress–strain curves for the fibres and matrices where $\varepsilon_{fu} > \varepsilon_{mu}$ and $E_f < E_m$. (b) Schematic of variation of fracture strength of a 0° unidirectional lamina with fibre volume fraction V_f.

and the shear lag (or Cox) model for a perfect bond. Modelling of intermediate bonded interfaces is not so straightforward. For a full description the reader is referred to the text *Strong Solids* [15].

5.4.1 Constant Shear Stress Model

This is often referred to as the **Kelly–Tyson model** and was developed to describe stress transfer in an elastic fibre–ductile metal matrix composite with the reloading of tungsten wires [16]. It has been adopted for PMCs in which the interface is poorly bonded so that a frictional interface exhibits constant shear stress. Figure 5.23 illustrates the development of stress in a fibre under a constant interfacial shear stress.

The following assumptions are implicit in the constant shear model:

1. poor fibre–matrix interface (or fibre elastic–matrix plastic);
2. the interfacial shear stress, τ_i, is constant (for MMC, τ_y = constant); the subscript i refers to interface and y the yield strength of the matrix;
3. interface fails in shear;
4. absence of end-adhesion; and
5. the stress builds up linearly to a plateau in the fibre.

The stress on the fibre, σ_f, is given by the force on the fibre, P_f, acting on the CSA of the fibre of radius r_f:

$$\sigma_f = \frac{P_f}{\pi r_f^2}. \tag{5.41}$$

The shear stress τ is given by the interfacial area (i.e. the cylinder):

$$\tau = \frac{P_t}{2\pi r_f L_t},$$ (5.42)

where L_t is the transfer length.

From the force balance, σ_f is given by

$$\sigma_f = \frac{2\tau L_t}{r_f}.$$ (5.43)

Thus, the stress on the fibre increases linearly over the transfer length because the shear stress is constant.

The maximum stress σ^{max} that the fibre can resist is its strength, σ_{fu} when the fibre breaks. Thus, when $\sigma_f = \sigma_{fu}$ the fibre breaks; examination of Figure 5.23 shows that a continuous fibre would fragment until it reached its critical length, L_c, which is given by:

$$L_c = 2L_t.$$ (5.44)

On rearranging eq. (5.43), L_c is given by

$$L_c = \frac{\sigma_{fu} r_j}{\tau}.$$ (5.45)

Equation (5.45) is also represented by a critical aspect ratio, (L_c/d_f):

$$\frac{L_c}{d_f} = \frac{\sigma_{fu}}{2\tau_i}.$$ (5.46)

Supercritical fibres can continue to fragment until they are too short to further break. Subcritical fibres cannot be fractured further. This is the basis of the

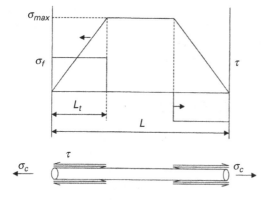

Figure 5.23 The constant shear stress, or Kelly–Tyson, model for describing stress transfer between matrix and fibre. σ_f is the tensile stress in the fibre and σ_{max} the maximum achievable. L_t is the transfer length over which the fibre is reloaded to σ_{max} – i.e. the stress carried by a continuous fibre or by the fibre before it fractured. σ_c is the applied stress on the composite. τ is the shear stress operating at the interface.

fragmentation test, which is discussed in Section 5.9.1 for the estimation of the interfacial bond strength. When τ_i exceeds τ_{iu}, the interface will fail and the fragmentation process ceases so that a value of τ_{iu} can be calculated from the average fragment length.

5.4.2   Cox Theory

This model was developed by Cox [17] to describe the behaviour of a cellulose fibre in paper. It is assumed that:

1. the fibre is elastic;
2. the fibre is perfectly bonded to an elastic matrix; and
3. the modulus of the fibre is much larger than the modulus of the matrix.

Under an axial load and since $E_f \gg E_m$, the fibre and matrix will experience locally different axial displacements so that a series of shear strains will be produced in the matrix in planes parallel to the fibre axis. This is illustrated in Figure 5.24.

Figure 5.25 illustrates the assumptions behind this model. The discontinuous filament is embedded in a cylinder of matrix and loaded axially. Under a composite strain, ε, the load on the fibre at a distance x from the fibre-end is P_f. Let the displacement of the distance x from the fibre-end be u. At the same point, in the absence of fibre, the displacement would be v.

Thus, the rate of load transfer from matrix to fibre is given by

$$\frac{dP_f}{dx} = H(u - v), \tag{5.47}$$

where H is a constant dependent on the geometric arrangement of the fibres and the moduli of the fibres and the matrix.

Differentiation of eq. (5.47) gives:

$$\frac{d^2 P_f}{dx^2} = H\left(\frac{du}{dx} - \frac{dv}{dx}\right), \tag{5.48}$$

$$\text{but } \frac{du}{dx} = \text{strain in the fibre} = \frac{P_f}{E_f A_f} = \varepsilon_f$$

$$\text{and } \frac{dv}{dx} = \text{strain in the matrix} = \varepsilon_m.$$

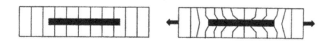

Figure 5.24 Schematic of the stress transfer between the matrix and a perfectly bonded fibre under an axial load.

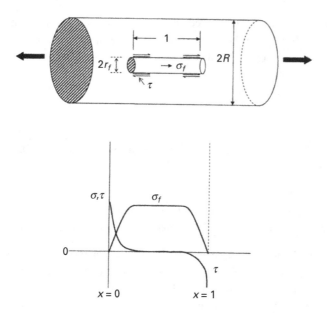

Figure 5.25 Schematic of the stress transfer between a perfectly bonded elastic fibre in an elastic matrix according to the Cox theory.

Therefore, eq. (5.48) becomes

$$\frac{d^2 P_f}{dx^2} = H\left(\frac{P_f}{A_f E_f} - \varepsilon_m\right),$$ (5.49)

where A_f is the CSA of the fibre.

The solution to this differential equation is:

$$P_f = E_f A_f \varepsilon_m + S \sinh \beta x + T \cosh \beta x,$$

where $\beta = \left(\frac{H}{A_f E_f}\right)^{1/2}$ and S and T are constants.

To evaluate S and T we set $P_f = 0$ at $x = 0$ and $x = 1$:

$$\therefore \quad \text{for} \quad 0 < x < l/2,$$

$$P_f = E_f A_f \varepsilon \left[1 - \frac{\cosh\beta\left(\frac{1}{2} - x\right)}{\cosh\beta\left(\frac{1}{2}\right)}\right]$$ (5.50)

or

$$\sigma_f = E_f \varepsilon \left[1 - \frac{\cosh\beta\left(\frac{1}{2} - x\right)}{\cosh\beta\left(\frac{1}{2}\right)}\right],$$ (5.51)

where ε is the maximum possible value of strain and εE_f is the maximum stress. Therefore, the average stress on the fibre, $\bar{\sigma}_f$, is given by integration of eq. (5.51):

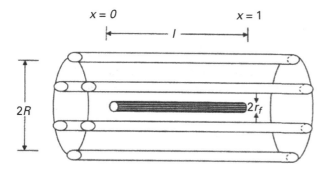

Figure 5.26 Arrangement of a discontinuous fibre within a uniaxial composite to define the cylinder of matrix enclosed by a series of parallel fibres so that the value of β can be determined.

$$\bar{\sigma}_f = \frac{E_f \varepsilon}{l} \int_0^1 \left[1 - \frac{\cosh\beta\left(\frac{1}{2} - x\right)}{\cosh\beta\left(\frac{1}{2}\right)} \right] dx$$

$$\bar{\sigma}_f = E_f \varepsilon \left[1 - \frac{\tanh\left(\beta\frac{l}{2}\right)}{\left(\beta\frac{l}{2}\right)} \right]. \tag{5.52}$$

This assumes that no load is transferred across the end faces of the fibre.

5.4.2.1 Estimation of β

β can be estimated by considering a composite consisting of a set of parallel fibres of constant length and circular cross-section of radius r_f and a centre-to-centre separation of $2R$.

We consider that shear strains will exist on all planes parallel to the fibres under an axial load. This results from the higher displacements in the lower-modulus polymeric matrix. At the fibre surface (i.e. the interface), we can write:

$$\frac{dP_f}{dx} = -2\pi r_f \tau_i, \tag{5.53}$$

where τ_i is the interfacial shear stress.

From eq. 5.49,

$$\frac{dP_f}{dx} = H(u - v)$$

$$H = \frac{2\pi r_f \tau_i}{(u - v)}. \tag{5.54}$$

We consider equilibrium forces

$$2\pi r_f \tau_i = 2\pi R \tau_r, \tag{5.55}$$

where τ_r is the shear stress acting on planes parallel to the fibres. Since at

$$R = 0 \quad \tau_i = \tau_r,$$

eq. (5.55) becomes

$$\tau_r = \frac{r_f}{R}\tau_i.$$ (5.56)

Since $\gamma = \frac{\tau_r}{G_m}$ and $\gamma = \frac{dw}{dR}$, where G_m is the shear modulus of the matrix and γ is the strain in the matrix, w is the actual displacement in the matrix close to the fibres.

At the fibre–matrix interface, assuming no slippage, then $w = u$:

$$\frac{dw}{dR} = \frac{\tau_i r_f}{RG_m},$$ (5.57)

$$\Delta w = \int_{r_f}^{R} dw = \frac{\tau_i}{G_m}r_f \ln\frac{R}{r_f}.$$ (5.58)

The model assumes a perfect interfacial bond so that there is no slippage at the interface; therefore, the actual displacement close to the fibre, w will equal u. Thus:

$$\mathbf{u} = \mathbf{w}.$$

However, at a distance R from the central fibre axis, matrix displacement is unaffected by the presence of a fibre so that $w = v$:

$$\Delta w = v - u = -(u - w).$$

Combining eq. (5.58) with (5.54), we get eq. (5.59):

$$H = \frac{2\pi G_m}{\ln\dfrac{R}{r_f}}.$$ (5.59)

From eqs (5.50) and (5.54) we obtain an expression for β:

$$\beta = \left[\frac{2\pi G_m}{E_f A_f \ln\left(\dfrac{R}{r_f}\right)}\right].$$ (5.60)

For a hexagonal packing arrangement of the fibres (a close approximation to that achieved in practice) we can estimate a value for R/r_f.

$$\ln\left(\frac{R}{r_f}\right) = \frac{1}{2}\ln\left[\frac{2\pi}{3^{1/2}V_f}\right]$$

$$= 0.64\ln V_f.$$

From inspection of eq. (5.60) we see that β is a function of the shear modulus of the matrix and tensile modulus of the fibre:

$$\beta = f\left(\frac{G_m}{E_f}\right)^{1/2}.$$ (5.61)

This shows that stress transfer is more rapid in high-modulus matrices.

Table 5.2. The estimates of the ratio of maximum shear stress to maximum fibre tensile stress $(\tau_{max}/\sigma_{f\,max})$ for glass- and carbon-fibre composites. A square array has been used to calculate R/r_f.

Fibre	G_m/E_f	V_f	$\tau_{max}/\sigma_{f\,max}$
Glass	0.017	0.6	0.25
Carbon	0.005	0,6	0.13

5.4.2.2 The Variation of Shear Stress along a Fibre

Equations (5.47) and (5.51) can be combined to obtain a relationship that describes the shear stress in the matrix at the interface, τ_i. Since $P_f = \pi r_f^2 \sigma_f$,

$$\tau = E_f \varepsilon \left\{ \left[\frac{G_m}{E_f 2\ln\left(R/r_f\right)} \right]^{1/2} \left[\frac{\sinh \beta \left(1/2 - x\right)}{\cosh \beta \left(1/2\right)} \right] \right\}. \tag{5.62}$$

τ is the shear stress in the matrix at the interface and can be considered to be τ_i; it is at a maximum at the fibre ends. Thus, τ_{max} occurs at $x = 0$ and at $x = 1$.

Equation (5.62) can be rewritten in the form:

$$\frac{\tau_{max}}{\sigma_{f\,max}} = \left[\frac{G_m}{2E_f \ln\left(R/r_f\right)} \right]^{1/2} \coth \beta \left(1/4\right), \tag{5.63}$$

where $\sigma_{f\,max} = E_f \varepsilon$.

Table 5.2 shows that in order to utilize the maximum stress the fibres can resist (σ_{fu}), the matrix needs to withstand high shear stresses, so polymeric matrices will tend to fail in shear because they have a relatively low shear strength.

5.4.3 Conclusions

The above analysis shows:

1. The regions at the end of a discontinuous fibre do not carry a full load. Therefore, the average stress on the fibre is less than that on a continuous fibre under an external load.
2. Reinforcing efficiency decreases as the length decreases.
3. Reinforcing efficiency depends on the quality of the interface – that is, the interfacial bond strength. With a poor interface the shear stress, which develops at the interface under an axial load, will easily exceed the interfacial shear strength and the fibre will debond. Stress transfer to the fibre will effectively be lost.
4. Therefore, the shear stress which is generated at the fibre ends can initiate the following phenomena:
   a. debonding of the interface $(\tau > \tau_{iu})$;
   b. cohesive failure of the matrix $(\tau_{iu} > \tau)$;
   c. cohesive failure of the fibre $(\tau_{fu} > \tau)$;
   d. shear yielding of the matrix $(\tau > \tau_y)$.

5. The analysis is not exact and ignores end-adhesion, so may underestimate shear stress concentrations at the end of the fibre.

5.5   Multiple Matrix Cracking

We have already discussed the criteria for multiple matrix cracking in unidirectional brittle matrix composites. In this analysis we identified a minimum fibre volume fraction, V_f', above which multiple cracking of the matrix occurs. After the first matrix crack forms the additional stress on the fibre is transferred back into the matrix over a transfer length that is dependent on the quality of the interface. There are *two* cases that determine the extent of matrix cracking and hence the crack spacing: (1) unbonded fibres and (2) bonded fibres.

5.5.1   Unbonded Fibres

In the case of unbonded fibres (Figure 5.27), reloading of the matrix at a matrix crack occurs via a constant shear stress arising from friction at the fibre interface. Further loading causes the matrix to fracture into a set of parallel cracks spaced x' and $2x'$ apart. The transfer of load from fibre to matrix can be modelled by a simple force balance described by Kelly–Tyson [15,16] (Section 5.4.1). In this case,

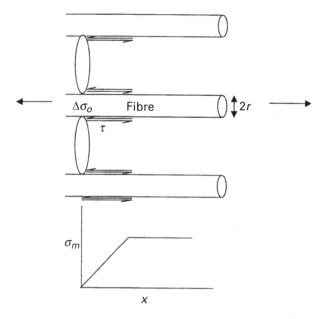

Figure 5.27 The description of the reloading of the matrix through stress transfer at the fibre interface after the first matrix crack. Case 1: poor interfacial bonding and friction so that the shear stress (τ) is constant along the transfer length.

$$\frac{d\sigma_m}{dx} \propto \tau.$$

The area of interface of a fibre is $2\pi r dx$; the area of interface per unit length of a composite of fibre volume fraction V_f is given by $(2V_f/r)dx$. Therefore,

$$\frac{d\sigma_m}{ds} = 2\frac{V_f}{V_m}\frac{\tau}{r}, \tag{5.64}$$

where V_m is the matrix volume fraction.

For a second (and third) crack to form the matrix stress needs to reach the maximum value (i.e. strength $= \sigma_{mu}$). The distance x' over which sufficient stress σ_{mu} is transferred for the matrix to fracture is given by integrating eq. (5.64):

$$x' = \frac{V_m}{V_f}\frac{\sigma_{mu}}{2\tau}r. \tag{5.65}$$

Provided fibre deformation is elastic then the additional stress upon the matrix will lead to a series of parallel cracks spaced between x' and $2x'$ apart, but the average crack spacing will be closer to x'.

5.5.2 Bonded Fibres

In the case of bonded fibres the matrix cracks are pinned by the fibres, as shown in Figure 5.28. The additional load on the fibres does not lead to debonding of the interface, but is transferred back to the matrix via shear in the interfacial matrix. The shear stress is not constant but decreases with distance from the matrix crack. The Cox or shear lag model can be used to describe the rate of stress transfer. While we can use eq. (5.62), a modified shear lag model was used by Bailey et al. [18] to predict the crack spacing that would be observed. They used an exponential decay of the additional stress placed on the fibres, $\Delta\sigma_o$, in the plane of the first matrix crack over distance y:

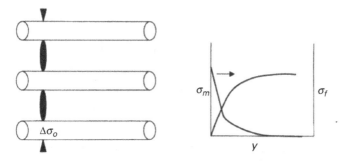

Figure 5.28 The description of the reloading of the matrix through stress transfer at the fibre interface after the first matrix crack. Case 2: good interfacial bonding so that the shear stress (τ) decays exponentially along the transfer length.

$$\Delta\sigma = \Delta\sigma_o \exp\left(-\Phi^{1/2}y\right),\tag{5.66}$$

where

$$\Phi^{1/2} = \left(\frac{2G_mE_c}{E_fE_mV_m}\right)^{1/2}\frac{1}{r\left[\ln\left(\frac{R}{r}\right)\right]^{1/2}}$$

and E_c is the modulus of the composite in the longitudinal direction.

After the first matrix crack, an additional stress $\Delta\sigma_o$ is thrown onto the fibres in the plane of the crack. It is equal to the stress on the fibres at the failure strain of the matrix:

$$\Delta\sigma_o = \frac{\sigma_c}{V_f} - \varepsilon_{mu}E_f,\tag{5.67}$$

where σ_c is the applied stress on the composite. Therefore, the shear stress at the interface, τ, is given by

$$\tau = \frac{r}{2}\Delta\sigma_o\Phi^{1/2}\exp\left(-\Phi^{1/2}y\right).\tag{5.68}$$

Rewriting eq. (5.64) in terms of load gives:

$$\frac{dF}{dy} = \frac{2V_f\tau}{r}.\tag{5.69}$$

Therefore, the load, F, transferred to the matrix at a distance l from the crack is given by

$$F = V_f\Delta\sigma_o\left[1 - \exp\left(-\Phi^{1/2}l\right)\right].\tag{5.70}$$

When $\Delta\sigma_o$ arises from a matrix crack, one crack forms. An additional load on the composite will cause further matrix cracking into blocks between l and $2l$. The crack spacing can be estimated from eq. (5.70):

$$F = \sigma_{mu}V_m,$$

$$\Delta\sigma_o \gg \frac{\sigma_{mu}V_m}{V_f},$$

$$l = \frac{1}{\Phi^{1/2}}\frac{\sigma_{mu}V_m}{\Delta\sigma_oV_f}.\tag{5.71}$$

Comparison of eqs (5.71) and (5.65) from the *constant shear model* shows that the crack density for the bonded case is higher than that for the non-bonded case:

$$x' = \frac{V_m\sigma_{mu}}{V_f}\frac{\sigma_{mu}}{2\tau}r.\tag{5.72}$$

Multiple matrix cracking will occur until the crack spacing is such that the reloading of the matrix blocks between the cracks to σ_{mu} is not possible. As a result, crack spacing for the bonded case is lower than for the non-bonded case.

Equation (5.73) also provides a value for the maximum interfacial shear stress, τ_{max}:

$$\tau_{max} = \frac{\Delta\sigma_o}{2}\Phi^{1/2}r. \tag{5.73}$$

5.5.3   Multiple Matrix Cracking in Cross-Ply Laminates

The reader is referred to Chapter 6, where the progression of damage by multiple transverse cracking of the 90° plies is discussed. The analysis in Section 5.5.2 provides the fundamentals of this phenomenon. In this case, the 90° plies are analogous to a brittle matrix and stress transfer occurs at the ply–ply interface, and similarly early transverse cracks are pinned while at higher loads the shear stresses generated lead to delamination.

5.6   Discontinuous Fibre Reinforcement

Continuous fibres need to be accurately placed to utilize their strength efficiently. We saw in Section 5.1.3.1 that the tensile modulus decreases rapidly with loading angle while the shear modulus increases. As a consequence, the failure mechanism of a unidirectional fibre composite changes with loading angle from fibre fracture at 0° to shear failure at 5–20°. A full analysis is given in Chapter 6. Table 5.3 summarizes the outcomes of this study. The most crucial point is that at an angle >3–5° shear failure dominates. Discontinuous or short-fibre composites are manufactured in moulding processes in which either the fibres flow together with the matrix polymer into a mould or the resin flows into a fibrous preform. In these materials the fibres are often randomly distributed in the matrix.

We need to consider the effect of loading angle and length on predicted performance. As a result, we need to include length and angular efficiency factors. We can use the law of mixtures prediction for modulus (eq. 5.5), modified to take into account the length and the angle of discontinuous fibres in the material:

Table 5.3. Summary of failure mode of a unidirectional fibre composite as a function of loading angle (θ)

Loading angle (θ)(°)	Failure mechanism
0	Longitudinal tensile fracture through fibre-breaks
5–20	Shear failure
20–45	Mixed mode of shear and tensile
45–90	Transverse tensile failure

$$E_c = \eta_l \eta_\theta V_f E_f + V_m E_m, \tag{5.74}$$

where η_l is the length efficiency factor and η_θ is the angle efficiency factor.

The value of η_θ is difficult to estimate because measurement of an average of θ of an array of fibres at a myriad of angles has practical limitations. Thus θ_{ave} (and η_θ) is often inferred from fitting the experimental data to an expanded version of eq. (5.74). We concentrate on estimating η_l.

5.6.1   Elasto-plastic Matrices

5.6.1.1   Length Efficiency Factor (η_l)

To understand how the length of a fibre influences its reinforcing efficiency, we should understand the concept of critical fibre length. Most discontinuous or short-fibre composites are referred to as reinforced plastics, based on thermoplastics where both matrix yielding and a limited interfacial bond operate. Therefore, the constant shear model is more appropriate than the shear lag model. This model enables a critical fibre length, L_c, to be defined.

5.6.1.2   Critical Fibre Length

The maximum stress σ_{max} that the fibre can resist is its strength, σ_{fu}, when the fibre breaks. Thus, when $\sigma_f = \sigma_{fu}$ the fibre breaks; examination of Figure 5.29 shows that a continuous fibre would fragment until it reached its critical length, $L < L_c$, which is given by:

$$L_c = 2L_t. \tag{5.75}$$

Rearranging, L_c is given by

$$L_c = \frac{\sigma_{fu} r_f}{\tau}. \tag{5.76}$$

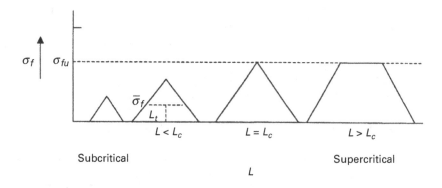

Figure 5.29 Schematic illustrating the fracture of an embedded filament under a constant interfacial shear stress. A supercritical fibre can fracture because it can be loaded to σ_{fu}. The critical fibre length ($L = L_c$) represents the minimum that can be loaded to a maximum stress. Subcritical fibres are of insufficient length to be loaded to fracture.

Equation (5.76) is also represented by a critical aspect ratio, (L_c/d_f):

$$\frac{L_c}{d_f} = \frac{\sigma_{fu}}{2\tau_i}. \tag{5.77}$$

Supercritical fibres can continue to fragment until they are too short to further break. Subcritical fibres cannot be fractured further.

We can see from Figure 5.29 how supercritical fibres $(L > L_c)$ can break into smaller fibres, which may be subcritical $(L > L_c)$ or supercritical, in which case we need to estimate the average stress on the fibres. From the geometrical conditions provided in Figure 5.29, the average stress carried by fibres, σ_{fave}, of differing length can be estimated. For $L < L_c$,

$$\sigma_{f(ave)} = \frac{\tau L}{2r_f}, \tag{5.78}$$

where $\sigma_{f(ave)}$ is the average stress in the fibre scaled from the force balanced with $L = 2L_t$.

For $L > L_c$,

$$\sigma_{f(ave)} = E_f \varepsilon_e \left(1 - \frac{L_c}{2L}\right), \tag{5.79}$$

where $E_f \varepsilon_c$ is the peak stress in the fibre.

5.6.1.3 Modulus of Aligned Discontinuous Fibre Composite

For an aligned discontinuous composite, $\eta_\theta = 1$. Therefore, in eq. (5.74) we define η_l as the fraction of average stress on the discontinuous fibre to that of a continuous fibre:

$$\eta_i = \frac{\sigma_{f(ave)}}{\sigma_{f(max)}}. \tag{5.80}$$

Therefore, for $L > L_c$, eq. (5.80) becomes

$$\eta_l = E_f \varepsilon_c \frac{\left(1 - {L_c}/{2L}\right)}{E_f \varepsilon_c}.$$

Therefore,

$$\eta_l = 1 - \frac{L_c}{2L}. \tag{5.81}$$

With $\eta_l = 1$, eq. (5.74) becomes

$$E_c = \left(1 - \frac{L_c}{2L}\right) V_f E_f + V_m E_m. \tag{5.82}$$

For $L < L_c$, since

$$\sigma_{f(ave)} = \frac{\tau L}{2 r_f}, \tag{5.83}$$

Table 5.4. Reinforcing efficiency factors of short fibres as a function of length L, according to eq. (5.81)

Fibre length (L)	Length efficiency factor (η_l)
$100\,L_c$	0.99
$10\,L_c$	0.95
L_c	0.5
$<L_c$	<0.5

Table 5.5. Typical values of critical fibre length (L_c) in different polymeric matrices

Fibre	Polymer	L_c (mm)
Glass	Epoxy	0.5–0.8
Glass	Vinyl ester	1–1.5
Glass	Polyamide	≈ 0.1
Carbon	Epoxy	≈ 0.5

then eq. 5.80 becomes

$$\eta_1 = \frac{\tau L}{2r_f \varepsilon_c E}. \tag{5.84}$$

Therefore,

$$E_c = \frac{\tau L}{2r_f \varepsilon_c} E_f V_f + E_m V_m. \tag{5.85}$$

Figure 5.29 shows how fragmentation of a short fibre progresses so that the critical fibre length can be defined. What is interesting is how long a fibre should be for a reinforcing efficiency close to a continuous fibre. This can be estimated by examining eq. (5.81). Table 5.4 provides estimates of η_l as a function of L/L_c.

The values of η_l in Table 5.4 show that a fibre needs to have a length of at least $100\,L_c$ to achieve a reinforcing efficiency close to that of a continuous fibre. The value of L_c is a function of the actual filament strength, so stronger fibres will have higher critical lengths, but it is not really a controllable variable. However, the shear stress that can be accommodated at the interface is most important. Thus, strongly bonded fibres will ensure that L_c is smaller. However, with strongly bonded fibres the shear strength of the matrix determines stress transfer mechanisms. For polymer matrices, the shear strengths are relatively low so that interphasal or interfacial failure will determine the maximum stress, which can be transferred to the fibres (Table 5.5).

5.6.2   Elastic Matrices

For elastic matrices we assume that the interfacial bond is perfect and that the yield stress of the matrix is in excess of the shear stress, which develops at a fibre-break.

Under these conditions Cox analysis (Section 5.4.2) applies. Equation (5.86) provides an estimate of the average stress on a fibre:

$$\sigma_{f(ave)} = E_f \varepsilon \left[1 - \frac{\tanh\left(\beta^l/_2\right)}{\left(\beta^l/_2\right)} \right]. \tag{5.86}$$

From eq. (5.80) we get:

$$\eta_l = \frac{\sigma_{f(ave)}}{\sigma_{f(max)}} = \frac{E_f \varepsilon}{E_f \varepsilon} \left[1 - \frac{\tanh\left(\beta^l/_2\right)}{\beta^l/_2} \right]. \tag{5.87}$$

With $\eta_\theta = 1$,

$$E_c = \left[1 - \frac{\tanh\left(\beta^l/_2\right)}{\beta^l/_2} \right] E_f V_f + E_m V_m. \tag{5.88}$$

β is a function of the modulus ratio, G_m/E_f, and therefore shear stresses develop in the matrix at a fibre-break. The magnitudes of these are given in Table 5.2, and inform that this analysis probably only applies at 'low' applied strains. Thus, to determine the strength of aligned discontinuous composites we use the constant shear model.

5.7 Strength of an Aligned Discontinuous Fibre Composite

Most applications of short-fibre composites utilize ductile polymers as matrices, so we will confine ourselves to ductile matrix composites of $V_f > V_{crit}$. In this case the model described by eq. (5.27) is appropriate:

$$\sigma_{1u} = \sigma_{fu} V_f + \sigma'_m (1 - V_f), \tag{5.89}$$

where σ'_m is the stress on the matrix when the fibres fracture, becoming

$$\sigma_{cu} = \eta_l \eta_\theta V_f \sigma_{fu} + (1 - V_f) \sigma'_m. \tag{5.90}$$

With $\eta_\theta = 1$ and η_l given by either eq. (5.81) or (5.84) we can obtain equations for the strength of a composite. For $L > L_c$,

$$\sigma_{cu} = V_f \sigma_{fu} \left[1 - \frac{L_c}{2L} \right] + (1 - V_f) \sigma'_m. \tag{5.91}$$

We conclude from this analysis that failure occurs by *fibre fracture*. The fracture toughness is low because the stress concentrations at fibre ends initiate matrix cracks, which propagate.

For $L < L_c$,

$$\sigma_{cu} = V_f \left[\frac{\tau_u L}{2r_f} \right] + (1 - V_f) \sigma'_m, \tag{5.92}$$

where $\sigma'_m = \varepsilon_{fu}E_m$. We conclude that failure occurs by fibre pull-out and matrix fracture.

Clearly, supercritical fibres will fracture, but the fragments will eventually become subcritical and pull out. Thus, the fracture surface will exhibit differing degrees of pull-out and, because fractured supercritical fibres are likely to initiate matrix fracture, the original fibre length can be inferred in microscopy studies.

From eq. (5.92) we see that the ratio L/d_f is the important factor in determining the failure mechanism of a composite. Most discontinuous fibre composites are referred to as reinforced plastics. In this context the analysis shows that fibre dispersion is important. Non-dispersed bundles behave as 'short, fat' fibres and pull out, but under impact loads they disintegrate, providing a mechanism for creating surface area and hence energy absorption. Therefore, fibre dispersion is a mechanism for controlling fracture toughness and impact strength.

5.7.1   Brittle Matrices

For *brittle matrices* the matrix term $[(1 - V_f)\sigma'_m]$ in eqs (5.91) and (5.92) is not included in the analogous analysis, and strength is given by:

$$\sigma_{cu} = \eta_l\eta_\theta V_f \sigma_{fu}. \tag{5.93}$$

Thus, supercritical fibres will ensure that matrix cracking is contained. Multiple cracking will occur and composite fracture will be determined by the statistics of fibre strength. With subcritical fibres, the interfacial bond will determine failure of the composite and the resistance to pull-out of the fibres is critical. The shear stresses, which develop where matrix cracks impinge the fibres, will lead to debonding so the frictional bond strength will be a controlling factor. With brittle matrices, therefore, it is best to employ longer, discontinuous fibres so that $L_{(ave)} \gg L_c$.

5.8   Performance of 'Real' Composites

As discussed above, short-fibre composites are perhaps best referred to as reinforced plastics, although short-fibre reinforcement is regularly used to introduce fracture toughness to brittle materials such as ceramics or highly crosslinked thermoset polymers. We will concentrate on thermoplastics, which have low modulus and require reinforcement, but can be moulded into 3D artefacts often in rapid processes such as injection moulding. For the purpose of this analysis we will assume that the fibres have a random distribution of lengths and are randomly oriented to the load direction. In an actual moulding, the fibres will have a preferred orientation at certain locations determined by the flow of the polymer melt or compound and the design of the tooling. In this case, some fibres will be subcritical and some supercritical; the model developed by Bowyer and Bader [19] sums the contributions of each distribution. To do this, eq. (5.46) is rewritten to define a critical aspect ratio, R_c:

$$R_c = \frac{L_c}{d_f} = \frac{\sigma_{fu}}{2\tau_i} = \frac{\varepsilon_c E_f}{2\tau_i}. \tag{5.94}$$

R_c is a function of composite strain, ε_c, so eq. (5.93) should be written as

$$R_{c(\varepsilon)} = \frac{\varepsilon_c E_f}{2\tau_i}. \tag{5.94}$$

At low applied strains most fibres are supercritical, but become subcritical as the strain increases even without fibre fracture. Fibre fracture adds to the complexity of the model. Thus, there is a y-distribution of *subcritical* fibres with $R_y < R_c$:

$$\sigma_{fy(ave)} = R_y \tau_i. \tag{5.96}$$

There is a z-distribution of *supercritical* fibres with $R_z > R_c$:

$$\sigma_{fz(ave)} = \varepsilon_c E_f \left(1 - \frac{\varepsilon_c E_f}{4 R_z \tau_i}\right). \tag{5.97}$$

Summation of the contributions provides the equation for the stress–strain curve of a discontinuous fibre composite:

$$\sigma_c = \eta_\theta \left[\sum V_y R_y \tau_i + \sum \varepsilon_c E_f V_z \left(1 - \frac{\varepsilon_c E_f}{4 R_Z \tau_i}\right)\right] + \varepsilon_c E_m (1 - V_f), \tag{5.98}$$

where $R_Z > R_{c(\varepsilon)} > R_y$ and η_θ is the orientation parameter, which is a measure of the average orientation angle of the fibres.

η_θ and τ_i are usually obtained from fitting the stress–strain curve to the model. The model predicts the quasi-elastic properties, but cannot predict strength. Bader and Collins [20] experimentally verified the equation and showed that eq. (5.97) could be used to predict the stress–strain curves for glass-fibre polyamide-6 (Figure 5.30). η_θ and τ_i were typically in the range 0.55–0.64 and 35–48 MPa, respectively. In order to predict strength, a failure criterion is required. Thomason [21] has studied the prediction of performance of short glass-fibre polyamide-66 in detail and found that this approach could be used. He obtained microscopic measurements of η_θ. Values interpolated from the modulus were in good agreement and were typically 0.6–0.68. τ_i was found to be 30–36 MPa.

5.8.1 Properties of Short-Fibre Reinforced Thermoplastics

The properties of injection moulded reinforced plastics are determined by a number of factors, including (1) notched sensitivity of the polymer matrix and (2) the mould flow-induced fibre orientation. The former is briefly discussed here, whereas the latter is considered in Chapter 4. Figure 5.31 shows how the properties of short carbon and glass fibres are affected by fibre volume fraction. As the theoretical analysis shows, carbon fibres have a higher reinforcing efficiency arising from their higher modulus.

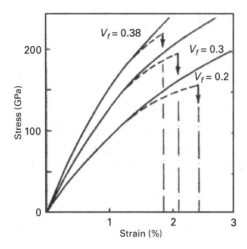

Figure 5.30 Stress–strain curves for glass-fibre polyamide-6 injection moulded composites with differing fibre volume fractions (V_f): 0.2, 0.3, 0.38. Dashed line = experimental; prediction of eq. (5.97) [20,22].

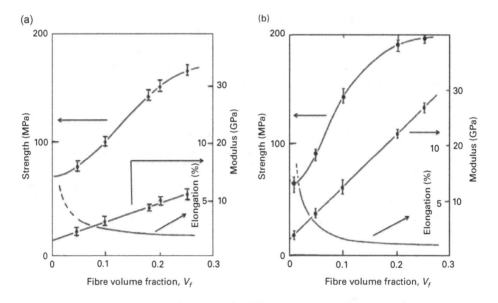

Figure 5.31 Mechanical properties of short-fibre injection moulded polyamide-66 with differing volume fraction (V_f): (a) glass; (b) carbon [22,23].

The following conclusions were drawn from this study:

1. Low strain modulus increases linearly.
2. σ_{cu} increases non-linearly.
3. ε_{cu} drops significantly.

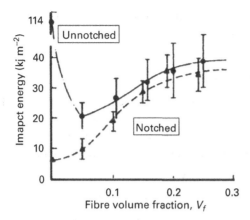

Fibre 5.32 Impact strength of a range of polyamide-66 mouldings containing a range of volume fractions of glass fibres. The initial fibre length was 500 μm.

The decrease in composite failure strain results from the fact that many polymers, such as polyamides, but it results in reduced impact strength. The impact strength of the analogous glass-fibre composites was also measured in the same research. In order to limit the effect of fibre length, Bader and coworkers [22] used a novel technique to ensure that initial fibre length was constant. This research led to the development of long-fibre reinforced thermoplastics. Figure 5.32 shows the impact strength of a range of glass-fibre reinforced polyamide-66 mouldings. Addition of the glass fibre significantly reduces impact strength of the polymer. The fibre ends initiate brittle crack growth in the polymer. This is confirmed by the comparison of notched and unnotched impact strength measurements on the unreinforced polymer.

Above V_f of 5–10% the absorbed impact energy increases with V_f when the mechanism of energy absorption changes from being dominated by matrix fracture to interfacial debonding. There is also a significant contribution from poorly dispersed bundles, which behave as short-fat 'fibres' with small L/d ratio and can disintegrate under impact load. This mechanism is supported by the reduced reinforcing efficiency as indicated by the reduction in modulus (E_c) shown in Figure 5.31a.

By recognizing that the fracture toughness of the matrix is a major limitation, a solution can be identified. Thus, short-fibre reinforced thermoplastics need to be impact modified by rubber toughening. For example, carboxyl-terminated ethylene propylene rubber.

5.8.1.1   Fibre Orientation within Mouldings

The flow of a plastics melt into a mould is subject to flow under shear, as discussed in Chapter 4. We recognize that this can lead to regions of differing fibre orientation. These heterogeneous fibre arrangements can lead to regions of differing strengths and hence defects that can initiate premature failure. The other consequence is the generation of thermal strains resulting from differential shrinkage. These effects can be mitigated by careful design of the flow paths in the tooling. In some cases, multi-live-feed techniques

[24,25] can be used to either randomize fibre orientations or provide fibre alignment to strengthen potential weak areas of a moulding.

5.8.2   Long Discontinuous Fibre Composites

Early 'glass fibre' used chop strand mat (CSM), which is an example of a 2D in-plane reinforced material. In this case, $\eta_l = 1$ because the average length of the fibres is larger than L_c; therefore, we need to consider the choice of a value for η_θ.

Krenchel [26] showed that an empirical prediction could be employed:

$$E_c \approx \frac{5}{8}E_t + \frac{3}{8}E_l,\tag{5.99}$$

where E_t is the modulus of a continuous aligned fibre composite in the transverse direction, and E_l is the modulus in the longitudinal direction at the equivalent V_f. E_t is given by the Halpin–Tsai equation (eq. 5.14) and E_l by the law of mixtures (eq. 5.6).

5.9   Characterization of Fibre–Matrix Interfaces

The interfacial strength between a fibre and matrix within a composite can be characterized by *direct* and *indirect* methods.

Indirect methods are mainly tests on composite coupons in which the material is placed under shear. Figure 5.33 describes two commonly used test methods: the *short beam test* for the measurement of interlaminar shear strength (ILSS) and the *Iosipescu* test, which employs a loading jig and specimen dimensions that is designed to subject the coupon to a pure shear stress. The *short beam test* has to be conducted on coupons with a minimum load span to thickness of 5:1 to ensure the specimen fails in shear.

The maximum shear stress occurs at the neutral axis in a loaded flexural beam so that failure also involves matrix fracture, so that an absolute value of the interface shear strength cannot be obtained. Thus, the maximum interlaminar shear strength will be that of the shear strength of the matrix and not the interface. For a typical epoxy resin-based composite, the interlaminar shear strength will reach about 80 MPa when

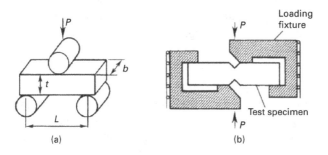

(a)   (b)

Figure 5.33 Indirect methods for interfacial characterization: (a) short beam shear test; (b) Iosicpescu test [28].

the interfacial bond is perfect. Lower values can be attributed to interfacial failure. To ensure that the interface dominates the failure, the fibre volume fraction of the coupons needs to be about 50–60%. Plotting the interlaminar shear strength against fibre volume fraction is one way to ensure that a limiting value has been achieved. The *Ioscipescu test* requires a more complex notched specimen, together with a loading fixture that concentrates the shear stress between the areas of the two notches. However, the shear stress in this region is not necessarily constant and, therefore, this method may not be one of the most appropriate for interfacial characterization.

The off-axis tensile strength is one alternative technique. With a fibre angle of $\simeq 5°$ the resolved shear stress at the interface usually is sufficient to cause matrix or interfacial shear failure before tensile fracture of the fibres, providing comparative interfacial properties.

Analogously, the tensile transverse strength of a 90° composite can also provide the relative properties of the interface as long as it is recognized that fracture also includes a contribution from the matrix. The presence of flaws leads to a statistical range of values. Alternatively, the statistics of fracture of the transverse ply of a 0°/90°/0° laminate was used by Peters and Anderson [27] to determine the transverse strength of a defect-free 90° ply, which was related to interfacial shear strength.

The disadvantage of the indirect methods is that they do not measure a pure interface characteristic. Four standard tests for *direct* measurement are given in Figure 5.34 [28,29]. These are single filament experiments, and only one can be applied to high V_f materials. This is the *push-in* test, in which a load is applied to an individual filament in a composite sample, which is pushed into the matrix until the interface fails under shear. For this test, a well-polished perpendicular section of a composite is required. While the major advantage is that it can be conducted on real specimens, the technique has serious experimental difficulties. It is difficult to avoid

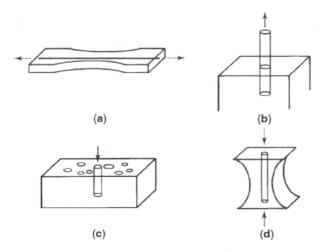

(a)

(b)

(c)

(d)

Figure 5.34 Direct methods for interfacial characterization: (a) Fragmentation test; (b) pull-out test; (c) push-in (or micro-indentation) test; (d) transverse debond test [28,29].

matrix deformation, as with many polymers, prior to interfacial failure, so the technique is more applicable to ceramic matrix composites. Furthermore, with highly oriented fibres, fibrillation damage can also be a problem. For PMCs the data is of limited value because it is experimentally difficult to avoid interfacial damage during specimen preparation and matrix deformation during the test. The interfacial shear strength can be obtained using eq. (5.100):

$$L_d = \frac{2P_f^2}{\pi^2 d_f^3 \tau_i E_f} - \frac{2\gamma}{\tau_i}, \tag{5.100}$$

where L_d is the displacement of the fibre under load P_f, and γ is the fracture surface energy.

The other three techniques are the fragmentation test, the pull-out test, and the transverse debond test of a single embedded filament.

5.9.1   The Fragmentation Test

In the fragmentation test, a continuous fibre is embedded in a coupon of resin and loaded axially (Figure 5.34a). A tensile load is applied to a specimen until the embedded fibre ceases to fragment further, when the interface is fully debonded. Stress transfer occurs through frictional stress transfer, and the Kelly–Tyson model is applicable (eqs 5.42–5.46). Thus, the average fragment length is a measure of the critical length of the embedded fibre in *that resin system*. Equation (5.45) can be redrafted as:

$$\tau_u = \frac{\sigma_{fu}(L_c) r_f}{L_c}, \tag{5.101}$$

where $\sigma_{fu}(L_c)$ is the fibre strength at length L_c. Since a fragment of length L_c can fracture, the fragment length will range from $\frac{1}{2}L_c$ to L_c. Thus, $L_c = \frac{4L_f}{3}$ where L_f is the average fragment length. $\sigma_{fu}(L_c)$ is determined experimentally from the Weibull statistics of fibre strengths. The strength of fibres at a fixed length needs to be measured for a minimum statistical number of specimens, normally ~50, in addition to the fragmentation test.

This simple analysis ignores the contribution of the matrix and/or interphase to the stress transfer function and, in many composite systems, sufficient ductility within the matrix is a contributing factor. As a result, the fragmentation process can cease with only partial debonding of the fibre. In this case, several authors have combined the models of Cox (eq. 5.52) and Kelly–Tyson (eq. 5.46) for data analysis. Thus, it is necessary to measure the debonded and bonded lengths during fragmentation.

Tripathi et al. [30,31] proposed a method of including matrix properties and the degree of debonding into the analysis. They used the elastic analysis of Nairn [32] to calculate the shear stress profile over the transfer length prior to interphasal matrix yield. Thus, from the bonded and debonded lengths a stress transfer profile for each could be calculated. An average cumulative stress transfer function and stress transfer efficiency is calculated by summing across all fragments. The interphase properties can be included in the analysis [33] using the three-phase model of Wu et al. [34].

5.9.2 The Pull-Out Test

The pull-out test (Figure 5.34b) has a number of variants. A fibre is embedded in a block of resin, which may have a differing geometrical shape. The force required to pull the fibre from the block of resin is recorded. The test requires a very sensitive load cell and an accurate technique for measuring the embedded length of the fibre [35,36]. From eq. (5.41) we get

$$\tau_{iu} = P_f / \pi d_f L_e, \tag{5.102}$$

where L_e is the embedded length of the test fibre.

Pitkethly and Doble [37] applied the Greszczuk model [38] to extract a value for interfacial shear strength from the pull-out load, P_f. From an analogous equation to eq. (5.63) we get

$$\tau_{(max)} = \tau_{(ave)} \alpha L_e \coth(\alpha L_e), \tag{5.103}$$

where $\tau_{(ave)} = \frac{P_f}{2r_f \pi L_e}$ and $\alpha = \left(\frac{2G_i}{br_f E_f}\right)^{1/2}$. b is the effective width of the interface, G_i the shear modulus of the interface. $\tau_{(max)}$ is estimated graphically by plotting $\tau_{(ave)}$ against L_e and extrapolating to $L_e = 0$.

5.9.2.1 Microdebond or Microbond Test

This is a variant of the pull-out test in which a microdroplet is deposited onto a single fibre (Figure 5.35a) [39]. The droplet is loaded by a micro-vice until the interface fails in shear and the droplet moves. The loading geometry is critical and is selected to avoid droplet damage so that the maximum force can be attributed to interfacial failure.

A typical load–displacement curve for an individual single filament pull-out test is illustrated in Figure 5.35b. Calculation of interfacial bond strength uses the constant shear model (eq. 5.102). It is essential to establish microscopically that interfacial failure has occurred. As above, τ_{iu} should be plotted against L_e and extrapolated to $L_e = 0$. The crack can initiate in the polymer before propagating along the interface, in which case an energy balance approach should be used [40,41]. It is important to ensure that the droplet is representative of the matrix polymer. Liu and Thomason [42] have extensively studied these variables, especially molten droplets of thermoplastics.

5.9.3 Measurement of Fibre Stress Profiles

In the fragmentation test, interpretation benefited from calculating the stress profile along a fragment. Young et al. [43–45] developed laser Raman spectroscopy and related techniques to measure the fibre stress (σ_f)/length profile directly. The fibre stress and fibre strain are obtained from the shift in wavelength of a specific spectral

(a)   (b)

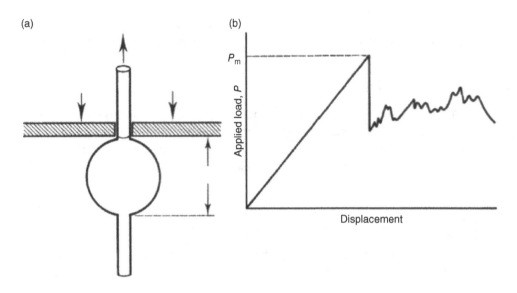

Figure 5.35 Schematic of the microbond test for interface characterization: (a) loading arrangement; (b) typical load–displacement curve showing the load drop occurring when the microdroplet debonds [39].

line. Figure 5.36 gives typical plots of fibre strain against length for an untreated high-modulus carbon fibre (T50) in an epoxy resin. This shows how debonding can be observed through changes in strain profile of a single filament, as the matrix strain increases. At a matrix strain of 0.7%, debonding is observed and debond lengths of ≈400 μm at each end of the fibre can be identified. At a strain of 0.9% the fibre broke. At higher strains fragmentation occurred. The fragments exhibited strain profiles typical of complete debonding. Full details are given elsewhere [44,45]. Laser Raman spectroscopy is ideally suited for the analysis of aramid and carbon fibres, but not glass fibres. A fluorescence peak associated with impurities in alumina fibres has also been found to be stress-dependent. The technique has been applied to fragmentation and pull-out of samarium-doped glass-fibre specimens [43]. The inter-facial shear stress profile can be calculated from the fibre stress profile using a force balance. The Cox analysis (eq. 5.51) is normally employed because the applied strain in the experiment is relatively low. By comparing these profiles, the integrity of the interface can be identified. The interfacial bond strength is normally equated with the recorded maximum shear stress. When debonding occurs, a Kelly–Tyson profile is observed and analysed accordingly. The major experimental difficulty concerns the 'transparency' of the matrix resin to the activating laser beam. Matrix fluorescence limits the range of interfaces that can be probed.

The interfacial shear stress profile along a short fibre or at a fibre-break can also be measured directly using phase-stepped photoelasticity [46,47]. In this technique, polarized light can be used to produce birefringent patterns, which can be employed to compute the principle stress difference $(\sigma_1 - \sigma_2)$ in the matrix. At the interface, the stress is dominated by the shear component and hence the interfacial shear stress can

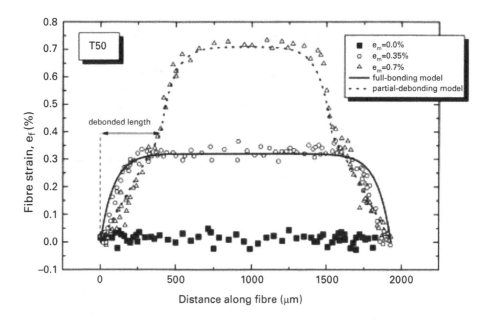

Figure 5.36 Fibre strain distribution obtained from strain-induced Raman band shifts at different levels of matrix strain for an untreated high-modulus T50 carbon fibre in an epoxy resin. Debonding of the interface is observed to have occurred at a matrix strain of 0.7% [45].

be measured directly. The major limitation of this technique is the magnification achieved using optical techniques so that carbon fibres are not readily interrogated. The transparency and the photoelastic constant of the matrix can also limit its application.

5.9.4 Transverse Debonding Test

This method makes use of the transverse Poisson expansion of a resin block under compression. Therefore, in the Broutman test (Figure 5.34d) an axial compressive load is applied to the embedded fibre to induce a radial tensile stress and interfacial debonding. This provides a direct measure of the interfacial bond strength under tension. This specimen is shaped to localize the debonding for observation.

To avoid the complications associated with compression tests, Koyanagi et al. [48,49] have developed a methodology for applying a tensile load to a 90° oriented fibre or shear load to an off-axis fibre to measure the debonding load.

5.10 Conclusions

We have discussed the fundamentals behind the mechanical properties of fibre composites. Thus, the laws of mixtures for calculating the moduli of unidirectional fibre composites have been presented together with a discussion of the validity of these equations. For the transverse modulus the Halpin–Tsai equation was developed. To understand the strength of a unidirectional composite, we demonstrated how multiple or single fracture arises. The differing cases for a variety of matrices and fibres are considered. To explain multiple fragmentation of fibres and matrices, we consider the main theories of stress transfer. The role of the interface is also discussed. More complex theories of stress transfer exist, but the objective is to present the basis of all the models. Clearly, more precise equations that describe the stress transfer profiles at fibre-breaks and matrix cracks are available, but the intention is to provide basic understanding. The concept of fracture statistics is also introduced, but the development of more sophisticated models for fragmentation or transverse cracking is beyond the scope of this book. The basic models are also applied to the performance of discontinuous fibre composites and reinforced plastics.

5.11   Discussion Points

1. Give the laws of mixtures for the prediction of the modulus of a unidirectional composite for fibres arranged at $0°$ and $90°$ to the applied load. Which equation gives the best prediction and what is the fibre orientation? Which equation gives a poor prediction? Give a reason for this observation.
2. Give the equation that addresses the deficiencies identified above?
3. Examine the models that provide the variation of strength with fibre volume fraction for $0°$ composites prepared from fibres and matrices. Discuss the concept of critical fibre volume fraction and describe the differing micromechanics that operate.
4. What are the assumptions behind the two stress transfer models, which operate at the interface between a short fibre and the matrix. Compare and contrast the fibre stress profiles observed. Which interface quantification model is employed for analysing the fragmentation test data? Explain why.
5. Discuss the concepts of subcritical and supercritical fibres.
6. What aspects of a moulded discontinuous fibre composite determine the modulus and strength of the material?
7. Briefly list the methods available for measuring the interfacial bond strength between a fibre and the matrix. Which methods are commonly employed?

References

1. J. G. Halpin and S. W. Tsai, *Environmental factors in composite design*, AFML Technical Report, TR67 423 (1967).
2. T. W. Clyne, *Cambridge composites lectures, C16*, University of Cambridge (2015).
3. D. Hull, *An introduction to composite materials* (Cambridge: Cambridge University Press, 1981)

4. J. E. Bailey and A. Parvizi, On fibre debonding effects and the mechanism of transverse ply failure in cross-ply laminates of glass fibre/thermoset composites. *J. Materials Sci.* **16** (1981), 649–659.

5. C. C. Chamis. Microchemical failure theory. In *Composite materials*, vol. 5, ed. L. J. Broutman and R. H. Krock (New York: Academic Press, 1974).

6. J. A. Kies, *Maximum strains in the resin of fiberglass composites*, US Naval Research Report 5752 (1962).

7. W. Weibull, A statistical distribution function of wide applicability. *J. Appl. Mech.* **18** (1951), 293–297.

8. N. G. McCrum, *Review of the science of fibre reinforced plastics* (London: HMSO, 1971).

9. B. D. Coleman, On the strength of classical fibres and fibre bundles. *J. Mech. Phy. Solids* **7** (1958), 60–70.

10. B. W. Rosen, Mechanics of composite strengthening. In *Fibre composites materials* (Metals Park, OH: ASM, 1965).

11. Z. Hashin and B. W. Rosen, The elastic moduli of fiber-reinforced materials. *J. Appl. Mech.* **31** (1964), 223–232.

12. J. P. Foreman, S. Behzadi, D. Porter, P. T. Curtis, and F. R. Jones, Hierarchical modelling of a polymer matrix composite. *J. Mater. Sci.* **43** (2008), 6642–6650.

13. D. V. Dunsford, J. Harvey, J. Hutchings, and C. H. Judd, *The effect of surface treatment of type 2 carbon fibres on CFRP properties*. (London: HMSO, 1981).

14. D. J. Hannant, *Fibre cements and fibre concretes*. (Chichester: Wiley, 1978).

15. A. Kelly and N. H. MacMillan, *Strong solids*, 3rd ed. (Oxford: Clarendon, 1990).

16. A. Kelly and W. R. Tyson, Tensile properties of fibre-reinforced metals: copper/tungsten and copper/molybdenum. *J. Mech. Phys. Solids* **13** (1965), 329–338.

17. H. L. Cox, The elasticity and strength of paper and other fibrous materials. *Br. J. Appl. Phys.* **3** (1952), 72.

18. J. E. Bailey, P. T. Curtis, and A. Parvizi, On the transverse cracking and longitudinal splitting behaviour of glass and carbon fibre reinforced epoxy cross ply laminates and the effect of poisson and thermally generated strain. *Proc. Roy. Soc. Lond. A* **366** (1979), 599–623.

19. W. Bowyer and M. G. Bader, On the re-inforcement of thermoplastics by imperfectly aligned discontinuous fibres. *J. Mater. Sci.* **7** (1972), 1315–1321.

20. M. G. Bader and J. F. Collins The effect of fibre-interface and processing variables on the mechanical properties glass-fibre filled nylon 6. *Fibre Sci. Technol.* **18** (1983), 217–231.

21. J. L. Thomason, Structure–property relationships in glass reinforced polyamide: 1) The effects of fibre content. *Polym. Compos.* **27** (2006), 552–562.

22. M. G. Bader, University of Surrey, UCLA Short Course Notes, vol.11 (1984).

23. P. T. Curtis, *The strength of fibre filled thermoplastics*, PhD Thesis, University of Surrey, 1976.

24. P. S. Allan and M. J. Bevis, Multiple live-feed injection molding. *Plast. Rubber Proc. Appl.* **7** (1987), 3–10.

25. P. S. Allan and M. J. Bevis, Injection moulding-fibre management by shear controlled orientation. In *Handbook of polymer-fibre composites*, ed. F. R. Jones (Harlow: Longman, 1994), pp. 171–176.

26. H. Krenschel, *Fibre reinforcement; theoretical and practical investigations of the elasticity and strength of fibre-reinforced materials*, Dissertation, Technical University of Denmark (1964).

27. P. W. M. Peters and S. I. Anderson, The influence of matrix fracture strain and interface strength on cross-ply cracking in CFRP in the temperature range of $-100°C$ to $+100°C$. *J. Compos. Mater.* **23** (1989), 944–960.

28. I. Verpoest, Interfacial bond strength determination: critique. In *Handbook of polymer-fibre composites*, ed. F. R. Jones (Harlow. Longman, 1994), pp. 230–235.

29. F. R. Jones, Interfacial analysis in fiber composite materials. In *Wiley encyclopaedia of composites*, ed. L. Nicolais and A. Borzacchiello, 2nd ed. (Chichester: Wiley, 2012).

30. D. Tripathi and F. R. Jones, Single filament fragmentation test for assessing adhesion in fibre reinforced composites: a review. *J. Mater. Sci.* **33** (1998), 1–16.

31. D. Tripathi, F. Chen, and F. R. Jones, A comprehensive model to predict the stress fields in a single fibre composite. *J. Compos. Mater.* **30** (1996), 1514–1538.

32. J. A. Nairn, A variational mechanics analysis of the stresses around breaks in embedded fibers. *Mech. Mater.* **12** (1992), 131–154.

33. A. C. Johnson, S. A. Hayes, and F. R. Jones, Data reduction methodologies for single fibre fragmentation test: role of the interface and interphase. *Composites A* **40** (2009), 449–454.

34. W. Wu, E. Jacobs, I. Verpoest, and J. Varna, Variational approach to the stress-transfer problem through partially debonded interfaces in a three-phase composite. *Comp. Sci. Technol.* **59** (1999), 519–535.

35. A. Hampe, G. Kalinka, S. Meretz, and E. Schulz, An advanced equipment for single-fibre pull-out test designed to monitor the fracture process. *Composites* **26** (1995), 40–46

36. E. Pisanova, S. Zhandarov, E. Mäder, I. Ahmad, and R. J. Young, Three techniques of interfacial bond strength estimation from direct observation of crack initiation and propagation in polymer–fibre systems. *Composites A* **32** (2001), 435–443.

37. M. Pitkethly and J. B. Doble, Characterizing the fibre/matrix interface of carbon fibre-reinforced composites using a single fibre pull-out test. *Composites* **21** (1990), 389–395.

38. L. B. Greszczuk. Theoretical studies of the mechanics of the fiber–matrix interface in composites. In *Interfaces in composites* (Philadelphia, PA: American Society for Testing and Materials, 1969) pp. 42–58.

39. L. S. Penn, Interfacial bond strength determination: microdebond pull-out test. In *Handbook of polymer-fibre composites*, ed. F. R. Jones (Harlow. Longman, 1994), pp. 238–242.

40. L. S. Penn and C. Chou, Identification of factors affecting single filament pull-out test results. *J. Compos. Technol. Res.* **12** (1990), 164–171.

41. P. S. Chua and M. R. Piggott, The glass fibre-polymer interface: II – work of fracture and shear stresses. *Comp. Sci. Technol.* **22** (1985), 107–119.

42. L. Yang and J. L. Thomason, Interface strength in glass fibre–polypropylene measured using the fibre pull-out and microbond methods. *Composites A* **41** (2010), 1077–1083.

43. M. Hejda, K. Kong, R. J. Young, and S. J. Eichhorn, Deformation micromechanics of model glass fibre composites. *Comp. Sci. Technol.* **68** (2008), 848–853.

44. M. A. Montes-Moran and R. J. Young, Raman spectroscopy study of HM carbon fibres: effect of plasma treatment on the interfacial properties of single fibre/epoxy composites part I: fibre characterization. *Carbon* **40** (2002), 845–855.

45. M. A. Montes-Moran and R. J. Young, Raman spectroscopy study of high-modulus carbon fibres: effect of plasma-treatment on the interfacial properties of single-fibre-epoxy composites part II: characterisation of the fibre–matrix interface. *Carbon* **40** (2002), 857–875.

46. F. M. Zhao, R. D. S. Martin, S. A. Hayes, et al., Photoelastic analysis of matrix stresses around a high modulus sapphire fibre by means of phase-stepping automated polariscope. *Composites A.* **36** (2005), 229–244

47. F. M. Zhao, S. A. Hayes, E. A. Patterson, and F. R. Jones, Phase stepping photoelasticity for the measurement of interfacial shear stress in single fibre composites. *Composites A* **37** (2006), 216–221.

48. S. Ogihara and J. Koyanagi, Investigation of combined stress state failure criterion for glass fiber/epoxy interface by the cruciform specimen method. *Comp. Sci. Technol.* **70** (2010), 143–150.

49. J. Koyanagia, S. Ogiharab, H. Nakatanic, T. Okabed, and S. Yoneyamae, Mechanical properties of fiber/matrix interface in polymer matrix composites. *Adv. Compos. Mater.* **23** (2014), 551–570.

6 Mechanical Properties of Laminates

In this chapter the micromechanics of unidirectional fibre composites (see Section 5.1) are extended to laminates, where strain transfer occurs at a matrix crack other than at a fibre-break. The stress distribution under load is also discussed to describe the accumulation of damage under differing loading conditions.

6.1 Laminate Mechanics

Chapter 5 considered the properties of fibre-reinforced materials with respect to aligned unidirectional composites. We see that these materials are highly anisotropic, with longitudinal modulus E_1 (or E_l) and strength σ_{1u} (or σ_{lu}) being significantly higher than the analogous properties in the transverse direction, E_2 (or E_t) and σ_{2u} (or σ_{tu}). Quasi-isotropy is achieved by laminating plies of unidirectional material at different angles. Thus, a typical lay-up might be $0_x°/\pm45_y°/90_z°/\pm45_y°/0_x°$. It is important that a symmetrical arrangement is employed, otherwise the cured plate will be warped because of differential shrinkage of the individual plies on cooling. In order to understand the micromechanics of failure, we develop the understanding of unidirectional laminates given in Chapter 5 to cross-ply laminates and angle-ply laminates here. We start by recognizing that a cross-ply or $0°/90°/0°$ laminate behaves analogously to a brittle matrix composite in that the $90°$ ply fails first and the crack is pinned by the $0°$ plies.

6.1.1 Cross-Ply $0°/90°/0°$ Laminates

Figure 6.1 shows the configuration of a balanced $0°/90°/0°$ laminate and an unbalanced $0°/90°$ laminate.

A $0_4°/90_4°/0_4°$ is illustrated. It can also be defined as $(0_4°/90_2°)_s$ symmetric (s) laminate. As shown in Figure 6.2, when loaded in the direction rotated by $90°$ the configuration is described as $90_4°/0_4°/90_4°$ or $(90_4°/0_2°)_s$ laminate. As shown in Figure 6.3, an unbalanced non-symmetric laminate such as a $90_4°/0_4°$ will become curved. The plate will be $90_4°/0_4°$ in one direction and $0_4°/90_4°$ in the perpendicular direction, and will tend to have a different curvature in each direction, and hence will take on the saddle shape shown in Figure 6.4.

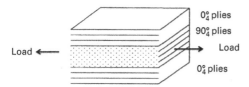

Figure 6.1 The definition of a $0_4°/90_4°/0_4°$ balanced laminate.

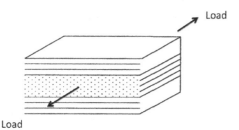

Figure 6.2 The laminate in Figure 6.1 loaded at 90°. This configuration is referred to as a $90_4°/0_4°/90_4°$ laminate.

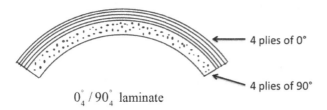

Figure 6.3 A $0_4°/90_4°$ unbalanced laminate takes on a curvature after cure and cooling. A thin strip is shown in cross-section.

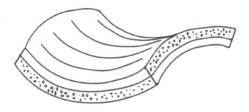

Figure 6.4 An unbalanced $0_x°/90_z°$ plate illustrating the curvature in both directions.

6.1.2　Failure Processes in Balanced Cross-Ply or 0°/90°/0° Laminates

Since the 90° ply failure strain is less than that for the 0° ply on tensile loading, the *first ply failure* is a transverse (also referred to as matrix) crack. As with 0° brittle matrix composites, the damage will progress by a multiplication of transverse cracks until the stress transferred back into the 90° ply by a shear lag process is insufficient

Table 6.1. Damage accumulation in a 0°/90°/0° cross-ply laminate

Failure sequence	Micromechanics	Ply location	Comment
1.	Transverse cracking	90°	Also referred to as matrix cracking
2.	Multiple transverse cracking	90°	
3.	Longitudinal splitting	0°	Not for carbon-fibre-reinforced polymers – ε_{fu} is too low and fibre fracture occurs first.
4.	Statistical fibre fracture	0°	

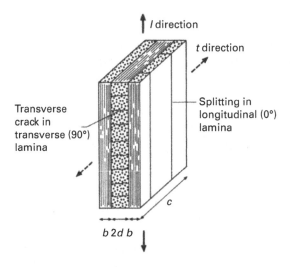

Figure 6.5 Schematic of a cross-ply laminate showing damage events under a uniaxial load in the longitudinal (*l*) or 1-direction [1].

for a new crack to form. The damage accumulation is summarized in Table 6.1. These events are illustrated in Figure 6.5.

    Figure 6.6 gives a typical stress–strain curve for a 0°/90°/0° laminate under a tensile load. This shows that the first ply crack forms at a strain (ε) of <0.4%, which corresponds with the 'knee' in the stress–strain curve. This is confirmed by acoustic emission [3,4]. Microscopic examination shows that this occurs when the first transverse or 90° ply crack forms.

6.1.2.1   Shear Lag Model of Transverse Cracking

To understand the accumulation of transverse cracks, we apply the shear lag model as illustrated in Figure 6.7. Garrett and Bailey [3] used a standard nomenclature to define the terms:

ε_{tl} = strain in the transverse (*t*) ply in the longitudinal (*l*) direction;

ε_{tlu} = the first transverse ply cracking strain, which is the ultimate (*u*) or failure strain of the transverse ply;

ε_{lt} = strain in the longitudinal ply in the transverse direction.

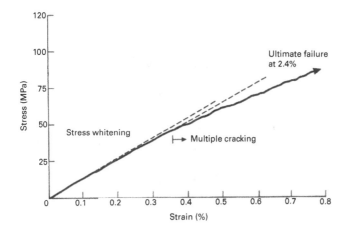

Figure 6.6 Typical stress–strain curve for a 0°/90°/0° cross-ply laminate loaded in the longitudinal (l) or 1-direction in uniaxial tension, showing the change in gradient with the onset of debonding (whitening) and transverse cracking [1,2].

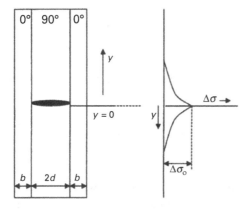

Figure 6.7 Schematic of a 0°/90°/0° laminate showing the reintroduction of the additional stress $\Delta\sigma_0$ into the 90° ply over distance y from the plane of the first transverse crack, by a shear lag mechanism.

The dimensions of the laminate are defined as: $b = 0°$ ply thickness, $d =$ semi 90° ply thickness (i.e. 90° ply thickness $= 2d$); and $c =$ coupon width.

The transverse crack forms and is pinned by the 0° plies, which remain bonded to the 90° ply. A shear stress will be introduced at this point, which decays over the adjacent material. Therefore, the shear lag model can be applied to the first transverse crack, as illustrated in Figure 6.7. When the first crack forms, an additional stress, $\Delta\sigma_0$, is thrown onto the 0° ply at a point defined by $y = 0$. $\Delta\sigma$ decreases over a distance y from the plane of the transverse crack as the stress is transferred back into the 90° ply. Since plies are well bonded, the rate of stress transfer occurs by a shear lag mechanism. A modified shear lag model was employed in the analysis.

Garret and Bailey [3,5,6] defined the rate of stress transfer in terms of the decay of $\Delta\sigma_0$ in the $0°$ ply:

$$\Delta\sigma = \Delta\sigma_o \exp\left(-\Phi^{1/2}y\right), \tag{6.1}$$

where $\Phi = \frac{E_c}{E_l}\frac{G_t}{E_t}\frac{(b+d)}{bd^2}$, E_c, E_l, E_t, and G_t are the Young's or elastic modulus of the composite (c), longitudinal (l) ply, transverse (t) ply, and shear modulus of the transverse ply in the y-direction.

The load, F, transferred back into the $90°$ ply is given by

$$F = 2bc\Delta\sigma_o[1 - \exp\left(-\Phi^{1/2}y\right)]. \tag{6.2}$$

A transverse crack forms when

$$F = 2\varepsilon_{tlu}E_t dc,$$

and occurs at $y = y_\infty$ when $\sigma_c = \varepsilon_{tlu}E_c$.

Therefore, $\Delta\sigma_0$ required to form a crack midway between two transverse cracks of spacing t is given by:

$$\Delta\sigma_o = \varepsilon_{tlu}E_t\frac{d}{b}\left[1 + \exp\left(-\Phi^{1/2}t\right) - 2\exp\left(-\Phi^{1/2}\frac{t}{2}\right)\right]^{-1}. \tag{6.3}$$

Equation (6.3) is a *stepped curve* indicating the minimum and maximum crack spacing. To place the curve, we need to estimate $\Delta\sigma_0$ when the first crack forms. Since $\sigma_0 = \varepsilon_{tlu}E_c$,

$$\Delta\sigma_0 = \varepsilon_{tlu}E_c\frac{(b+d)}{L} - \varepsilon_{tlu}E_l. \tag{6.4}$$

Figure 6.8 shows a plot of eq. (6.3), illustrating the prediction of crack spacing for a glass-fibre $0°/90°/0°$ epoxy laminate, while Figure 6.9 shows development of multiple transverse cracking in the $90°$ and longitudinal splitting in the $0°$ of the same. The average crack spacing, t_{ave}, is related to the lower bound value, t_{lb}, according to the statistical analysis of the car parking problem:

$$t_{ave} = 1.37t_{lb}. \tag{6.5}$$

The theory assumes that ε_{tlu} has a unique value, but in reality it will always have a statistical aspect. When the crack spacing, t, is greater than the transfer length for reintroduction of 99% of the stress, cracks will occur in the weakest regions. When t is less than the transfer length, cracking should theoretically cease, but cracks can still form in these regions if the stress and stress relaxation process has weakened the material.

Furthermore, ε_{tlu} is affected by local variations in V_f, and ply variables such as b and d. Local variations in thermal strains occur as a result of variations in ply parameters [7]. The prediction of transverse crack spacing has been subject to many studies, and refined theories are available [8–10].

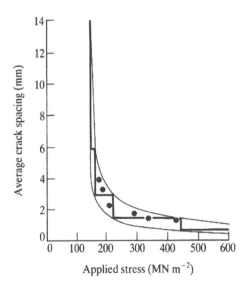

Figure 6.8 Predicted and experimental crack spacing for a glass-fibre 0°/90°/0° epoxy laminate of length $L = 190$ mm, 90° ply thickness $2d = 1.2$ mm. G_t was estimated from Halpin–Tsai as 5.1 GPa. The stepped curve is calculated from eq. (6.3), assuming that the first crack occurs in the middle of the laminate. • experimental data. The continuous curves represent the upper and lower bounds for crack spacing [5].

6.1.3 Residual Thermal Strains in Cross-Ply Laminates

Most laminates are cured and post-cured at elevated temperatures, so the effect of cooling to ambient temperature is critical to the performance of laminated structures. We saw in Figures 6.3 and 6.4 that an unbalanced laminate becomes curved because of the differential shrinkage of the plies. This can occur either by resin shrinkage during room temperature curing or thermal shrinkage on cooling. Mulheron et al. [11,12] showed that the thermal shrinkage dominates because the 'cure' residual strain is annealed out at elevated temperatures. This happens because post-curing normally occurs at temperatures above or within the glass transition region, where molecular relaxation can occur. Therefore, we will consider thermal strains in detail.

6.1.3.1 Linear Thermal Expansion Coefficient

The important point is that the linear thermal expansion coefficient of a polymer or cured thermoset resin, used as the matrix, is several orders of magnitude larger than that of the fibres, which are usually glass or ceramic. Typically α_m will have a value of ≈ 60–100×10^{-6} K^{-1} while α_f will be -0.1 to 5×10^{-6} K^{-1}. For example, for high-strength carbon fibre, $\alpha_{fl} = -0.26 \times 10^{-6}$ K^{-1} in the axial direction (subscript l) and $\alpha_{ft} = 26 \times 10^{-6}$ K^{-1} in the transverse or radial direction (subscript t), while E-glass fibres are considered isotropic and $\alpha_f = 5 \times 10^{-6}$ K^{-1}. High-modulus carbon fibres have values of $\alpha_{fl} \approx -1.6 \times 10^{-6}$ K^{-1} in the axial direction. These values are usually

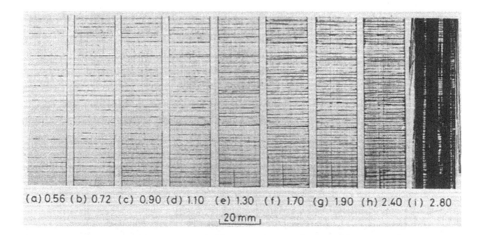

(a) 0.56 (b) 0.72 (c) 0.90 (d) 1.10 (e) 1.30 (f) 1.70 (g) 1.90 (h) 2.40 (i) 2.80

20 mm

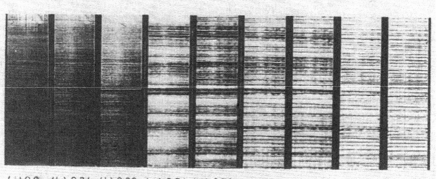

(j) 0.0 (k) 0.34 (l) 0.36 (m) 0.54 (n) 0.72 (o) 0.90 (p) 1.08 (q) 1.50 (r) 1.90

20 mm

Figure 6.9 Photographic record of the development of multiple transverse cracking in the 90° and longitudinal splitting in the 0° of an epoxy resin glass-fibre cross-ply 0°/90°/0° laminate taken in a *bright* field by illumination (a–i), *dark* field illumination (j–r). The applied strain (in %) is given. The first transverse crack appeared at 0.36% strain [2].

obtained from measurements of composites, especially in the transverse direction. Therefore, according to the Schapery equations, the linear thermal expansion coefficient of an aligned fibre composite in the transverse direction, α_t, is much larger than in the longitudinal direction, α_l [13]. The Schapery equations (eqs 6.6 and 6.7) can be used to calculate α_l and α_t.

$$\alpha_l = \frac{E_m \alpha_m V_m + E_f \alpha_f V_f}{E_m V_m + E_f V_f},\tag{6.6}$$

$$\alpha_t = \alpha_m V_m (1 + v_m) + \alpha_f V_f (1 + v_f) - \alpha_l v_c,\tag{6.7}$$

$$v_c = v_m V_m + v_f V_f.\tag{6.8}$$

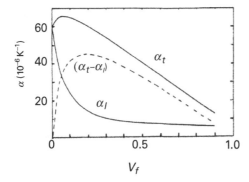

Figure 6.10 Theoretical variation of linear thermal expansion coefficients of a glass-fibre composite, α_t, α_l, and ($\alpha_t - \alpha_l$), at 20 °C with fibre volume fraction, V_f, calculated from the Schapery equations. $\alpha_m = 60 \times 10^{-6}$ K^{-1}, $\alpha_f = 5 \times 10^{-6}$ K^{-1}, $E_m = 4$ GPa, $E_f = 72$ GPa, $v_m = 0.4$, $v_f = 0.2$. E_1 and v_c are calculated from the law of mixtures [12].

These equations can be used to examine the effect of the volume fraction of fibres on the linear thermal expansion coefficients of unidirectional composites. Figure 6.10 shows how the important parameter ($\alpha_t - \alpha_l$) varies with V_f. We will see later that this variation in expansion coefficients is critical to the microcracking of laminates.

6.1.3.2 Thermal Residual Strains in Cross-Ply Laminates
Figure 6.10 shows that

$$\alpha_t \gg \alpha_l, \tag{6.9}$$

so on cooling from the cure or post-cure temperature the transverse ply will shrink more than the longitudinal ply, and because $E_l \gg E_t$ in a cross-ply laminate the shrinkage of the transverse ply will be constrained. Therefore, in the longitudinal direction the ply will go into compression, while in the transverse direction the ply will be put into tension. Figure 6.10 illustrates the generation of thermal strains in a cross-ply laminate. We refer to the thermal residual strains, given in Figure 6.11 using the standard nomenclature: ε_{tl}^{th} is the thermal strain in the transverse ply (t) in the longitudinal (l) direction; ε_{lt}^{th} is the thermal strain in the longitudinal ply (l) in the transverse (t) direction; ε_{ll}^{th} is the thermal strain in the longitudinal ply (l) in the longitudinal (l) direction; and ε_{tt}^{th} is the thermal strain in the transverse ply (t) in the transverse direction (t).

As a result of this analysis, we see that the transverse cracking of the 90° ply will occur at a lower applied tensile strain:

$$\varepsilon_{tlu} = \varepsilon_{tu} - \varepsilon_{tl}^{th}, \tag{6.10}$$

where ε_{tlu} is the measured first transverse cracking strain and ε_{tu} is the transverse ply failure strain (failure strain of the isolated 90° ply).

For glass fibres, a thermal strain of $\varepsilon_{tl}^{th} \approx 0.45\%$ is typical, which reduces ε_{tu} from $\approx 0.6\%$ to $\varepsilon_{tlu} \approx 0.15\%$.

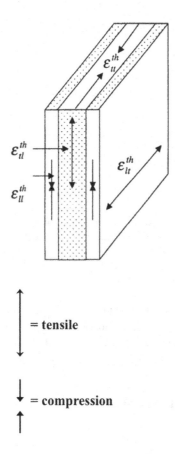

= tensile

= compression

Figure 6.11 Schematic illustrating the nature of the thermal strains induced in the plies of a 0°/90°/0° cross-ply laminate, on cooling after thermal excursion (post-cure). In the fibre direction the plies are in compression and in the transverse direction the plies are in tension.

For carbon fibres, a thermal strain of $\varepsilon_{tl}^{th} \approx 1.0\%$ is typical, which may cause transverse cracking of laminate on cooling. This observation led to the development of suitable matrix resins for carbon-fibre-reinforced polymers and optimization of interfacial performance.

6.1.3.3 Estimation of Thermal Strains

The thermal strains in laminates on cooling from T_1 to T_2 can be calculated from the following equations, which can be obtained from a consideration of balance of forces and strain compatibility:

$$\varepsilon_{tl}^{th} = \frac{E_l b (\alpha_t - \alpha_l)(T_1 - T_2)}{(E_l b + E_t d)}, \tag{6.11}$$

$$\varepsilon_{lt}^{th} = \frac{E_l d (\alpha_t - \alpha_l)(T_1 - T_2)}{(E_l b + E_t d)}, \tag{6.12}$$

where T_1 is the strain-free temperature and T_2 is the service temperature (usually ambient).

The strain-free temperature is the temperature at which the thermal strain is first induced on cooling. This could be related to either the cure or post-cure or glass transition temperature, T_g, of the cured resin. In the case of the latter, it will be close to the value defined by the temperature at which the storage modulus decreases in a thermal analysis test, as opposed to the maximum in tanδ, which is the standard definition of T_g.

6.1.3.4 Measurement of Thermal Strain

We see in Figure 6.3 that an unbalanced laminate in the form of a thin strip takes on a curvature related to the strain induced in the 90° ply of the beam if it was constrained to be flat. Therefore, we can obtain a measure of $(\alpha_t - \alpha_l)(T_1 - T_2)$ from the curvature of the beam.

In analogy to a bimetal strip, $(\alpha_t - \alpha_l)(T_1 - T_2)$ is related to the degree of curvature of an unbalanced beam:

$$(\alpha_t - \alpha_l)(T_1 - T_2) = \frac{(b+d)}{2\rho} + \left(\frac{(E_l b^3 + E_t d^3)}{6\rho(b+d)}\right)\left[\frac{1}{E_t b} + \frac{1}{E_t d}\right], \qquad (6.13)$$

where ρ is the radius of curvature of the 0°/90° beam.

The simplest technique for determining ρ is to match the curvature of the beam to circles of known radius. Independent experimental measurement of E_l and E_t and the laminae thicknesses enables $(\alpha_t - \alpha_l)(T_1 - T_2)$ to be estimated. Equations (6.11) and (6.12) can be used to calculate ε_{tl}^{th} and ε_{lt}^{th}.

The measurement of ρ is not straightforward, and the above technique is preferred. However, we can measure the stress-free temperature, T_1, by observing the point at which the beam just becomes flat ($\delta = 0$) on increasing temperature. Similarly, the variation in thermal strain with temperature on cooling is also useful. Thus, a direct measurement of the curvature of the beam is required. This can be done by recording x in Figure 6.12 with temperature. The following analysis can be used:

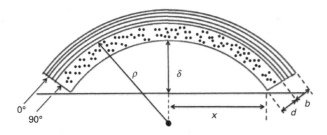

Figure 6.12 Illustration of an unbalanced 0°/90° beam defining the parameters required for the estimation of εth. ρ is the radius of curvature, δ is the displacement from flat, and x is the semi-chord length.

$$\rho = \frac{(\delta^2 + x^2)}{2\delta}, \tag{6.14}$$

$$\frac{1}{\rho} = \frac{2}{L} \cos^{-1}\left(1 - \frac{\delta}{\rho}\right), \tag{6.15}$$

$$L = \frac{\pi(x^2 + \delta^2)}{180\delta} \sin^{-1}\left[\frac{2x\delta}{(x^2 + \delta^2)}\right]. \tag{6.16}$$

Equations (6.14) and (6.15) can be solved iteratively from the length of the beam, L, which can be determined by eq. (6.16) from the values of δ and x at room temperature. Equation (6.11) is often presented in a simple form:

$$\varepsilon_{tl}^{th} \infty \Delta a \Delta T. \tag{6.17}$$

Table 6.2 gives some typical values of thermal strain in transverse directions of the 90° and 0° plies. Since the axial thermal expansion coefficient of carbon fibres is slightly negative, significantly higher tensile thermal strains are induced in comparison to glass fibres. The consequences of this on transverse cracking can be explained by examining eq. (6.10), where it shown that first ply cracking will occur earlier.

We are also interested in the magnitude of the compressive thermal strains in the fibre direction. These are given by a consideration of strain compatibility:

$$\varepsilon_{ll}^{th} = -\frac{E_t}{E_l}\varepsilon_{tl}^{th}, \tag{6.18}$$

$$\varepsilon_{tt}^{th} = -\frac{E_t}{E_l}\varepsilon_{lt}^{th}. \tag{6.19}$$

Table 6.2. Typical values of thermal strain (in %) for epoxy resin 0°/90°/0° composites prepared from high-strength carbon and glass fibres as a function of varying semi-inner ply thickness, d (mm) and $b = 0.5$ mm

		Thermal strains in the longitudinal direction of the transverse or 90° ply (ε_{tl}^{th}) and in transverse direction of longitudinal ply (ε_{lt}^{th})			
Semi-inner ply	Inner ply thickness	Carbon fibres		Glass fibres	
d (mm)	$2d$ (mm)	$\varepsilon_{tl}^{th}(\%)$	$\varepsilon_{lt}^{th}(\%)$	$\varepsilon_{tl}^{th}(\%)$	$\varepsilon_{lt}^{th}(\%)$
2.0	4.0	–	–	0.053	0.115
1.5	3.0	–	–	0.062	0.112
1.0	2.0	–	–	0.075	0.107
0.5	1.0	0.332	0.322	0.094	0.094
0.25	0.5	0.332	0.303	0.107	0.075
0.15	0.3	–	–	0.113	0.095
0.125	0.25	0.337	0.271	–	–
0.0625	0.125	0.340	0.224	–	–

6.1.4 Effect of Ply Thickness on Transverse Cracking: Constraint Cracking Theory

In the previous discussion we examined the situation in which the transverse crack instantaneously runs across the 90° ply of a 0°/90°/0° laminate under an applied tensile load. Observations show that the thickness of the 90° ply, $2d$, is greater than 0.4 mm. In this section we will examine the transverse cracking of plies where $2d < 0.4$ mm. In this case, the cracks initiate at the edges and propagate slowly across the ply. Parvizi and Bailey [4] hypothesized that for cracks to grow at a constant load, thermodynamics principles apply. Therefore, a crack cannot form unless the work done exceeds the stored energy:

$$\Delta w - \Delta u > 4\gamma_t dc, \tag{6.20}$$

where Δw is the work done in the formation of a crack under an applied stress, σ; Δu is the increase in stored elastic energy; γ_t is the fracture surface energy per unit area.

$$\Delta w = 2\sigma_c \Delta L(b+d)c, \tag{6.21}$$

where ΔL is the extension of the composite as a result of crack formation.

Garrett et al. [5] showed that eq. (6.21) could be expanded to give eq. (6.22):

$$\Delta w = \frac{4d(b+d)cE_t E_c(\varepsilon_{tlu} + \varepsilon_{tl}^{th})\varepsilon_{tlu}}{bE_l \Phi^{1/2}}, \tag{6.22}$$

where E_c is the modulus of the 0°/90°/0° composite and

$$\Phi = \frac{E_c G_t(b+d)}{E_t E_1 b^2 d}.$$

Δu has several components, as given by eq. (6.23):

$$\Delta u = \Delta u_{ll} + \Delta u_{tl} + \Delta u_{ts}, \tag{6.23}$$

where Δu_{ll} is the energy stored in the 0° plies during crack formation, Δu_{tl} is the energy in the 90° ply during crack formation, and Δu_{ts} is the energy gained by the build-up of shear stresses in the transverse ply.

The components of Δu were estimated from a modified shear lag analysis including thermal strains [6]:

$$\Delta u_{ll} = \frac{cd^2 E_t^2}{bE_l \Phi^{1/2}}(\varepsilon_{tlu}^i + \varepsilon_{tl}^{th})^2 + \frac{4cdE_t}{\Phi^{1/2}}\left(\varepsilon_{tlu}^t - \varepsilon_{tl}^{th}\frac{dE_t}{bE_l}\right)(\varepsilon_{tlu}^i + \varepsilon_{tl}^{th}),$$

$$\Delta u_{tl} = \frac{-3cdE_t}{\Phi^{1/2}}(\varepsilon_{tlu}^t - \varepsilon_{tl}^{th})^2,$$

$$\Delta u_{ts} = \frac{cd(b+d)E_t E_{cl}(\varepsilon_{tlu}^i + \varepsilon_{tl}^{th})^2}{bE_l \Phi^{1/2}},$$

Since $E_{cl} = \left[\dfrac{E_1 b}{(b+d)} + \dfrac{E_t d}{(b+d)}\right].$

Thus, Δu becomes:

$$\Delta u = \frac{2d(b+d)cE_tE_{cl}}{bE_l\Phi^{1/2}}[(\varepsilon_{tlu}^i)^2 - (\varepsilon_{tl}^{th})^2], \qquad (6.24)$$

where ε_{tlu}^i is the transverse cracking strain of the inner (i) or 90° ply.

Combining eqs (6.20), (6.22), and (6.24), a minimum cracking strain can be identified:

$$\varepsilon_{tlu}^{min} = \left[\frac{2\gamma_t bE_l\Phi^{1/2}}{E_tE_c(b+d)}\right] - \varepsilon_{tl}^{th}, \qquad (6.25)$$

where ε_{tlu}^{min} is the minimum transverse cracking strain of a 0°/90°/0° laminate. Figure 6.13 is a plot of ε_{tlu}^{min} as a function 90° ply thickness, which shows that the experimental data for a glass-fibre epoxy laminate can be predicted. This illustrates that at small inner ply thicknesses the cracking process is constrained, while at thicknesses >0.5 mm transverse cracking is instantaneous because the stored energy in the ply exceeds that released by crack formation.

Therefore, we see that ply thickness is a potentially critical parameter for the design of laminates since we can inhibit transverse cracking by dispersing the plies through-out the thickness of the composite. The limitation is the fibre tex, which determines the thickness of the prepreg. The development of fibre-spreading technology to exploit this phenomenon is in progress.

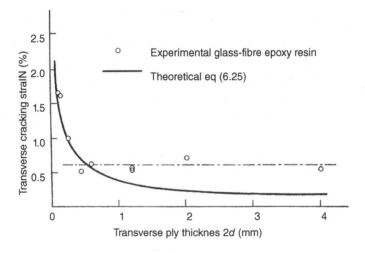

Figure 6.13 Plot of ε_{tlu}^{min} (eq. 6.25) against 90° ply thickness, $2d$, of glass-fibre epoxy 0°/90°/0° laminate compared to experimental data (○). The horizontal line represents the limiting value of ε_{tlu} for large ply thicknesses [4].

6.1.5 Transverse Cracking (Splitting) of 0° Plies of 0°/90°/0°

Equation (6.10) shows how the induction of thermal strains strongly influences the development of microcracking damage in laminates. In analogy,

$$\varepsilon_{lts} = \varepsilon_{ltu} = \varepsilon_{tu} - \varepsilon_{lt}^{th}, \qquad (6.26)$$

where ε_{ltu} is the transverse cracking strain of the longitudinal or 0° ply, often referred to as splitting strain,

$$\varepsilon_{lls} = \left(\frac{1}{\nu_1}\right)\left[(E_1 d + E_t b)\frac{\varepsilon_{lts}^p}{E_1 d} + \nu_t \varepsilon_{tlu}^i\right],$$

where ε_{lt}^{th} is given by eq. (6.12).

In order to understand this observation (see Figure 6.9g,h), we examine how the transverse stress in the 0° ply develops. Essentially, under a tensile load the 0° ply tries to shrink in the transverse direction according to its Poisson ratio, but this is restrained by the high modulus of the fibres in 90° ply. Therefore, a Poisson strain will develop, which is given by balancing forces and transverse strain compatibility:

$$\varepsilon_{lt}^p = \frac{E_1 d \varepsilon_{ll}(\nu_l - \nu_t)}{(E_1 d + E_t b)} \qquad (6.27)$$

Equation (6.27) represents the transverse Poisson strain prior to the formation of transverse cracks in the 90° ply. At higher applied stress, in the presence of transverse cracks, ε_{lt}^p will be given by:

$$\varepsilon_{lt}^p = \left[\frac{E_1 d(\varepsilon_{ll}\nu_l - \varepsilon_{tlu}\nu_t)}{(E_1 d + E_t b)}\right]. \qquad (6.28)$$

In glass-fibre composites, values of ε_{lt}^p can reach 0.5%, and in carbon-fibre composites 0.3% at the composite fracture strain. Therefore, splitting arises when

$$\varepsilon_{lt}^p + \varepsilon_{lt}^{th} \geq \varepsilon_{lts}. \qquad (6.29)$$

In an analogous analysis to that for transverse cracking, a minimum composite strain for splitting of the 0° plies of a 0°/90°/0° is given by

$$\varepsilon_{lls}^{min} = \frac{1}{\nu_l}\left\{\frac{E_{ct}(b+d)}{E_1 d}\left[\left[\frac{(\varepsilon_{lt}^{th})^2}{16} + \frac{\gamma_t d E_l \Phi^{1/2}}{(b+d)E_{ct}E_t}\right]^{1/2} - \frac{3}{4}\varepsilon_{lt}^{th}\right] + \nu_t \varepsilon_{tlu}^i\right\}. \qquad (6.30)$$

For a thin-ply glass-fibre 0°/90°/0° laminate, minimum splitting strains as high as 2% have been observed, and are predicted by eq. (6.30). Therefore, to induce sufficient Poisson strain in the transverse direction of the 0° ply to exceed the splitting strain is more likely with fibres of 'high' failure strain. Thus, splitting is unlikely with carbon fibres since the Poisson induced strains are insufficient to cause splitting prior to fibre fracture (i.e. composite fracture). However, we should not forget that thermal strains are higher with carbon-fibre-reinforced polymers.

6.2   Angle-Ply Laminates

So far we have discussed the micromechanics of unidirectional and simple cross-ply laminates. However, to achieve isotropy or to tailor the properties of a structure to meet the stress state, we need to understand the interactions between plies. To do this we need to consider the basics of laminate theory, which is used to model the stresses within $\pm\theta°$ laminates. It is essential to understand this in order to employ available computer codes effectively.

6.2.1   Laminate Theory

6.2.1.1   Stresses at Point

The stresses acting at a point in a solid can be represented by the stresses acting on the surfaces of a cube, as shown in Figure 6.14. There are three normal stresses, σ_{11}, σ_{22}, σ_{33}, and three shear stresses, τ_{23}, τ_{31}, τ_{12}. The first number refers to the direction normal to the plane.

These stresses have corresponding strains ε_{11}, ε_{22}, ε_{33}, γ_{23}, γ_{31}, γ_{12}. By convention we treat tensile stresses as positive and compressive stresses as negative.

It is also convention to use a contracted notation, where $\sigma_{11} = \sigma_1$, $\sigma_{22} = \sigma_2$, $\sigma_{33} = \sigma_3$. Therefore, the state of stress at a point needs nine stress components: σ_{11}, σ_{22}, σ_{33}, τ_{12}, τ_{23}, τ_{31}, τ_{32}, τ_{13}, τ_{21}.

However, $\tau_{23} = \tau_{32}$, $\tau_{31} = \tau_{13}$, and $\tau_{12} = \tau_{21}$, which reduces the number of components to six. These are referred to as σ_1, σ_2, σ_3, σ_4, σ_5, σ_6, where $\sigma_4 = \tau_{23}$, $\sigma_5 = \tau_{31}$, $\sigma_6 = \tau_{12}$.

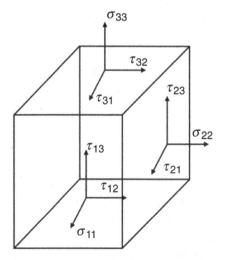

Figure 6.14 The stresses acting at a point in a solid defined by three normal stresses, σ_{11}, σ_{22}, σ_{33}, and three shear stresses, τ_{23}, τ_{31}, τ_{12}, acting on a cube.

A unidirectional tensile stress acting on a solid is defined by Hooke's law, where E is Young's modulus and ε is the elastic strain:

$$\sigma = E\varepsilon. \tag{6.31}$$

The normal strain transverse to the applied stress is given by $-\varepsilon v$, where v is Poisson's ratio. For an isotropic material the shear modulus, G, is given by

$$\tau = G\gamma, \tag{6.32}$$

where γ is the engineering shear strain and

$$G = \frac{E}{2(1+v)}. \tag{6.33}$$

The generalized form of Hooke's law is given by

$$\sigma_i = \sum_{j=1}^{6} C_{ij}\varepsilon_j, \tag{6.34}$$

where the σ_i is given in contracted notation form $\sigma_1 \ldots \sigma_6$. C_{ij} is the stiffness matrix, and $C_{ij} - C_{ji}$ and eq. (6.31) in expanded form become

$$
\begin{bmatrix} \sigma_1 \\ \sigma_2 \\ \sigma_3 \\ \tau_{23} \\ \tau_{31} \\ \tau_{12} \end{bmatrix}
=
\begin{bmatrix}
C_{11} & C_{12} & C_{13} & C_{14} & C_{15} & C_{16} \\
C_{12} & C_{22} & C_{23} & C_{24} & C_{25} & C_{26} \\
C_{13} & C_{23} & C_{33} & C_{34} & C_{35} & C_{36} \\
C_{14} & C_{24} & C_{34} & C_{44} & C_{45} & C_{46} \\
C_{15} & C_{25} & C_{35} & C_{45} & C_{55} & C_{56} \\
C_{16} & C_{26} & C_{36} & C_{46} & C_{56} & C_{66}
\end{bmatrix}
\begin{bmatrix} \varepsilon_1 \\ \varepsilon_2 \\ \varepsilon_3 \\ \gamma_{23} \\ \gamma_{31} \\ \gamma_{12} \end{bmatrix}. \tag{6.35}
$$

This matrix notation represents six individual equations, which relate stress to strain:

$$\sigma_1 = C_{11}\varepsilon_1 + C_{12}\varepsilon_2 + C_{13}\varepsilon_3 + C_{14}\gamma_{23} + C_{15}\gamma_{31} + C_{16}\gamma_{12}$$
$$\sigma_2 = C_{12}\varepsilon_1 + C_{22}\varepsilon_2 + C_{23}\varepsilon_3 + C_{24}\gamma_{23} + C_{25}\gamma_{31} + C_{26}\gamma_{12}$$
$$\sigma_3 = C_{13}\varepsilon_1 + C_{23}\varepsilon_2 + C_{33}\varepsilon_3 + C_{34}\gamma_{23} + C_{35}\gamma_{31} + C_{36}\gamma_{12}$$
$$\tau_{23} = C_{14}\varepsilon_1 + C_{24}\varepsilon_2 + C_{34}\varepsilon_3 + C_{44}\gamma_{23} + C_{45}\gamma_{31} + C_{46}\gamma_{12}$$
$$\tau_{31} = C_{15}\varepsilon_1 + C_{25}\varepsilon_2 + C_{35}\varepsilon_3 + C_{45}\gamma_{23} + C_{55}\gamma_{31} + C_{56}\gamma_{12}$$
$$\tau_{12} = C_{16}\varepsilon_1 + C_{26}\varepsilon_2 + C_{36}\varepsilon_3 + C_{46}\gamma_{23} + C_{56}\gamma_{31} + C_{66}\gamma_{12}.$$

For isotropic material, the stiffness matrix is much simpler because the elastic properties are the same in all directions and eq. (6.32) reduces to

$$
\begin{bmatrix} \sigma_1 \\ \sigma_2 \\ \sigma_3 \\ \tau_{23} \\ \tau_{31} \\ \tau_{12} \end{bmatrix}
=
\begin{bmatrix}
C_{11} & C_{12} & C_{12} & 0 & 0 & 0 \\
C_{12} & C_{11} & C_{12} & 0 & 0 & 0 \\
C_{12} & C_{12} & C_{11} & 0 & 0 & 0 \\
0 & 0 & 0 & (C_{11}-C_{12}) & 0 & 0 \\
0 & 0 & 0 & 0 & (C_{11}-C_{12}) & 0 \\
0 & 0 & 0 & 0 & 0 & (C_{11}-C_{12})
\end{bmatrix}
\begin{bmatrix} \varepsilon_1 \\ \varepsilon_2 \\ \varepsilon_3 \\ \gamma_{23} \\ \gamma_{31} \\ \gamma_{12} \end{bmatrix}. \tag{6.36}
$$

There are a set of corresponding equations that relate strain to stress:

$$\varepsilon_i = \sum_{j=1}^{6} S_{ij}\varepsilon_j, \tag{6.37}$$

where S_{ij} is the compliance matrix, which for an isotropic material is represented by

$$\begin{bmatrix} \varepsilon_1 \\ \varepsilon_2 \\ \varepsilon_3 \\ \gamma_{23} \\ \gamma_{31} \\ \gamma_{12} \end{bmatrix} = \begin{bmatrix} S_{11} & S_{12} & S_{12} & 0 & 0 & 0 \\ S_{12} & S_{11} & S_{12} & 0 & 0 & 0 \\ S_{12} & S_{12} & S_{11} & 0 & 0 & 0 \\ 0 & 0 & 0 & 2(S_{11}-S_{12}) & 0 & 0 \\ 0 & 0 & 0 & 0 & 2(S_{11}-S_{12}) & 0 \\ 0 & 0 & 0 & 0 & 0 & 2(S_{11}-S_{12}) \end{bmatrix} \begin{bmatrix} \sigma_1 \\ \sigma_2 \\ \sigma_3 \\ \tau_{23} \\ \tau_{31} \\ \tau_{12} \end{bmatrix}. \tag{6.38}$$

Since $S_{11} = 1/E$, $S_{12} = -v/E$, $2(S_{11} - S_{12}) = 1/G$, we can write eq. (6.35) as

$$\begin{bmatrix} \varepsilon_1 \\ \varepsilon_2 \\ \varepsilon_3 \\ \gamma_{23} \\ \gamma_{31} \\ \gamma_{12} \end{bmatrix} = \begin{bmatrix} 1/E & -v/E & -v/E & 0 & 0 & 0 \\ -v/E & 1/E & -v/E & 0 & 0 & 0 \\ -v/E & -v/E & 1/E & 0 & 0 & 0 \\ 0 & 0 & 0 & 1/G & 0 & 0 \\ 0 & 0 & 0 & 0 & 1/G & 0 \\ 0 & 0 & 0 & 0 & 0 & 1/G \end{bmatrix} \begin{bmatrix} \sigma_1 \\ \sigma_2 \\ \sigma_3 \\ \tau_{23} \\ \tau_{31} \\ \tau_{12} \end{bmatrix}. \tag{6.39}$$

These equations have only two independent elastic constants: the compliances, S_{11} and S_{12}, or the engineering constants, E and v (which also defines G in eq. 6.33). In a simple uniaxial tensile test, $\sigma_1 = \sigma$ and $\sigma_2 = \sigma_3 = \tau_{23} = \tau_{31} = \tau_{12} = 0$. Therefore eq. (6.39) becomes

$$\varepsilon_1 = (1/E)\sigma$$

and

$$\varepsilon_2 = \varepsilon_3 = -(v/E)\sigma.$$

For laminae we assume they are sufficiently thin for through-thickness stresses to be zero under plane stress, so that $\sigma_3 = \tau_{23} = \tau_{31} = 0$. Then, eq. (6.39) for isotropic materials becomes

$$\begin{bmatrix} \varepsilon_1 \\ \varepsilon_2 \\ \gamma_{12} \end{bmatrix} = \begin{bmatrix} 1/E & -v/E & 0 \\ -v/E & 1/E & 0 \\ 0 & 0 & 1/G \end{bmatrix} \begin{bmatrix} \sigma_1 \\ \sigma_2 \\ \tau_{12} \end{bmatrix}. \tag{6.40}$$

Equation (6.40) can be recast in terms of the more conventional stress–strain relation:

$$\begin{bmatrix} \sigma_1 \\ \sigma_2 \\ \tau_{12} \end{bmatrix} = \begin{bmatrix} E/(1-v^2) & vE/(1-v^2) & 0 \\ vE/(1-v^2) & E/(1-v^2) & 0 \\ 0 & 0 & G \end{bmatrix} \begin{bmatrix} \varepsilon_1 \\ \varepsilon_2 \\ \gamma_{12} \end{bmatrix}. \tag{6.41}$$

6.2.1.2 Orthotropic Lamina

We saw in Chapter 5 that continuous fibre composites are anisotropic. The simplest non-isotropic body utilized in laminate construction is an orthotropic one. An orthotropic lamina has three mutually perpendicular planes of material symmetry where the properties at any point are different in three mutually perpendicular directions. A unidirectional lamina is shown in Figure 6.15 to be orthotropic because it has three perpendicular planes of symmetry. Lamina from woven rovings or chopped strand mat can also be assumed to be isotropic within the plane of the lamina.

Since the material properties in the plane normal to the 1-direction in Figure 6.15 are assumed to be isotropic, plane stress conditions operate so that the strain–stress relation of the lamina is given by eq. (6.42). Therefore, in a unidirectional lamina there are two orthotropic planes of symmetry and $\sigma_3 = 0$, $\tau_{23} = 0$, and $\tau_{31} = 0$:

$$\begin{bmatrix} \varepsilon_1 \\ \varepsilon_2 \\ \gamma_{12} \end{bmatrix} = \begin{bmatrix} S_{11} & S_{12} & 0 \\ S_{12} & S_{22} & 0 \\ 0 & 0 & S_{66} \end{bmatrix} \begin{bmatrix} \sigma_1 \\ \sigma_2 \\ \tau_{12} \end{bmatrix}, \tag{6.42}$$

where $S_{11} = 1/E_1$, $S_{22} = 1/E_2$, $S_{66} = 1/G_{12}$, $S_{12} = -v_{12}/E_1 = -v_{21}/E_2$. v_{12} is Poisson's ratio, which refers to the strain produced in the 2-direction under a stress in the 1-direction, $-\varepsilon_2/\varepsilon_1$. Similarly, $v_{21} = -\varepsilon_1/\varepsilon_2$.

The stress–strain response of a *unidirectional lamina* is given by eq. (6.43):

$$\begin{bmatrix} \sigma_1 \\ \sigma_2 \\ \tau_{12} \end{bmatrix} = \begin{bmatrix} Q_{11} & Q_{12} & 0 \\ Q_{12} & Q_{22} & 0 \\ 0 & 0 & Q_{66} \end{bmatrix} \begin{bmatrix} \varepsilon_1 \\ \varepsilon_2 \\ \gamma_{12} \end{bmatrix}, \tag{6.43}$$

where Q_{11}, Q_{12}, and Q_{66} are the reduced stiffnesses. These constants are defined by the engineering constants, as shown in eq. (6.44):

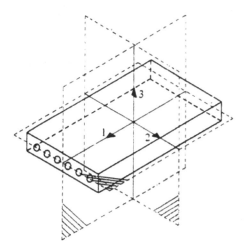

Figure 6.15 A unidirectional lamina showing the three mutually perpendicular planes of material symmetry required to be orthotropic [14].

$$Q_{11} = C_{11} = \frac{E_1}{(1 - v_{12}v_{21})}$$

$$Q_{12} = C_{12} = \frac{v_{12}E_2}{(1 - v_{12}v_{21})} = \frac{v_{21}E_1}{(1 - v_{12}v_{21})}$$

$$Q_{22} = C_{22} = \frac{E_2}{(1 - v_{12}v_{21})}$$

$$Q_{66} = \frac{1}{2}(C_{11} - C_{12}) = G_{12}.$$

(6.44)

Equation 6.43 shows that orthotropic materials tested in tension or compression in the principal directions do not exhibit shear strains in these directions. Further, the deformation is independent of the shear modulus G_{12}. Also, a shear stress, τ_{12}, produces shear strains, which are independent of E, v_{12}, and v_{21}. Therefore, there is no coupling between tensile and shear strains.

However, when tested at angles other than the principal directions, the above does not apply. Therefore, we consider a lamina tested with a different coordinate system, as shown in Figure 6.16. The stress–strain response is given by

$$\begin{bmatrix} \sigma_x \\ \sigma_y \\ \tau_{xy} \end{bmatrix} = \begin{bmatrix} \bar{Q}_{11} & \bar{Q}_{12} & \bar{Q}_{16} \\ \bar{Q}_{12} & \bar{Q}_{22} & \bar{Q}_{26} \\ \bar{Q}_{16} & \bar{Q}_{26} & \bar{Q}_{66} \end{bmatrix} \begin{bmatrix} \varepsilon_x \\ \varepsilon_y \\ \gamma_{xy} \end{bmatrix},$$

(6.45)

where $\bar{Q}_{ij}$ is the transformed reduced-stiffness matrix.

$$\bar{Q}_{11} = Q_{11}c^4 + 2(Q_{12} + 2Q_{66})s^2c^2 + Q_{22}s^4$$

$$\bar{Q}_{12} = (Q_{11} + Q_{22} - 4Q_{66})s^2c^2 + Q_{12}(s^4 + c^4)$$

$$\bar{Q}_{22} = Q_{11}s^4 + 2(Q_{12} + 2Q_{66})s^2c^2 + Q_{22}c^4$$

$$\bar{Q}_{16} = (Q_{11} - Q_{12} - 2Q_{66})sc^3 + (Q_{12} - Q_{22} + 2Q_{66})s^3c$$

$$\bar{Q}_{26} = (Q_{11} - Q_{12} - 2Q_{66})s^3c + (Q_{12} - Q_{22} + 2Q_{66})sc^3$$

$$\bar{Q}_{66} = (Q_{11} + Q_{22} - 2Q_{12} - 2Q_{66})s^2c^2 + Q_{66}(s^4 + c^4),$$

(6.46)

where $s = \sin\theta$ and $c = \cos\theta$.

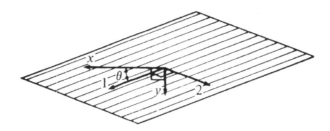

Figure 6.16 Schematic showing the rotation of axes from a 1-,2- to an x,y coordinate system [14].

The strain–stress response is provided by a corresponding equation:

$$\begin{bmatrix} \varepsilon_x \\ \varepsilon_y \\ \gamma_{xy} \end{bmatrix} = \begin{bmatrix} \overline{S}_{11} & \overline{S}_{12} & \overline{S}_{16} \\ \overline{S}_{12} & \overline{S}_{22} & \overline{S}_{26} \\ \overline{S}_{16} & \overline{S}_{26} & \overline{S}_{66} \end{bmatrix} \begin{bmatrix} \sigma_x \\ \sigma_y \\ \tau_{xy} \end{bmatrix}. \tag{6.47}$$

The $\overline{S}_{ij}$ matrix can be represented by a similar set of equations to eq. (6.46). Since $\overline{S}_{16}$ and $\overline{S}_{26}$ are not zero, a unidirectional stress, σ_x, will create normal and shear strains. Using this approach, a series of equations for the elastic properties E_x, E_y, G_{xy}, and ν_{xy} on the x,y coordinate system are given in by eq. (6.48):

$$\frac{1}{E_x} = \frac{1}{E_1}c^4 + \left(\frac{1}{G_{12}} - \frac{2\nu_{12}}{E_1}\right)s^2c^2 + \frac{1}{E_2}s^4$$

$$\frac{1}{E_y} = \frac{1}{E_1}s^4 + \left(\frac{1}{G_{12}} - \frac{2\nu_{12}}{E_1}\right)s^2c^2 + \frac{1}{E_2}c^4$$

$$\frac{1}{G_{xy}} = 2\left(\frac{2}{E_1} + \frac{2}{E_2} + \frac{4\nu_{12}}{E_1} - \frac{1}{G_{12}}\right)s^2c^2 + \frac{1}{G_{12}}(s^4 + c^4)$$

$$\nu_{xy} = E_x\left[\frac{\nu_{12}}{E_1}(s^4 + c^4) - \left(\frac{1}{E_1} + \frac{1}{E_2} - \frac{1}{G_{12}}\right)s^2c^2\right]. \tag{6.48}$$

Figure 6.17 shows the plot of E_x against the loading angle θ for a unidirectional carbon-fibre-reinforced polymer (type 1 carbon fibre) epoxy composite. The predictions from eq. (6.48) are compared to the experimental data [15]. The modulus is seen to reduce rapidly with angle.

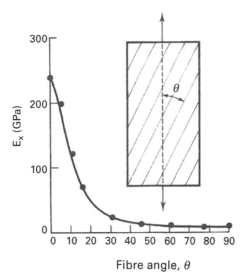

Figure 6.17 The angular dependence of modulus of a unidirectional high-modulus carbon-fibre-reinforced polymer lamina of $V_f = 0.5$. The continuous line is the prediction of eq. (6.48). • are the experimental data of Sinclair and Chamis [14,15].

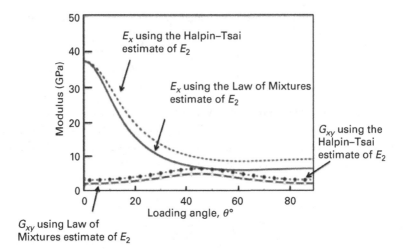

Figure 6.18 Variation of (a) Young's modulus, E_x, and shear modulus, G_{xy}, with loading angle ($\phi°$) for epoxy glass-fibre reinforced polymer lamina ($V_f = 50\%$) [16].

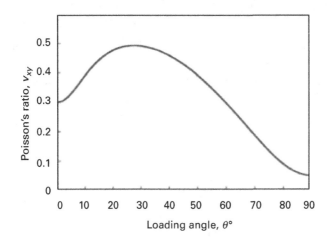

Figure 6.19 Variation of Poisson's ratio, v_{xy}, with loading angle ($\phi°$) for epoxy glass-fibre reinforced polymer lamina ($V_f = 50\%$) [16].

Figure 6.18 shows the angular dependence of the predicted Young's modulus and shear modulus. Note that the shear modulus is at a maximum at 45°, so we can expect higher shear stresses to develop in 45° plies of a laminate. The angular dependence of Poisson's ratio is shown in Figure 6.19 and is at a maximum at 30°.

6.2.2   Multi-Ply Laminates

As a consequence of the anisotropy of a unidirectional lamina as shown in Figures 6.17 and 6.18, laminates are prepared by stacking plies of differing orientation

Table 6.3. Definition of symmetric and non-symmetric laminate constructions.

Symmetric laminates	Non-symmetric
$+\theta/-\theta/-\theta/+\theta$ $(\pm\theta)_s$	$+\theta/-\theta/+\theta/-\theta$
$+\theta/-\theta/+\theta/+\theta/-\theta/+\theta$ $(+\theta/-\theta/+\theta)_s$	$+\theta/-\theta/-\theta/+\theta_2$

in order to cater for multidirectional stresses. Generally, the stacks should be symmetric so that the laminates are balanced. Figure 6.4 illustrates the shapes that unbalanced laminates can take on. Examples of unbalanced and balanced laminates are given in Table 6.3.

The elastic properties of a laminate depend on the properties of the individual laminae. Therefore, to use laminate theory we assume that (1) the plies are perfectly bonded so there is no slippage; (2) the interface between plies is infinitely thin; and (3) a laminate has the properties of a thin sheet. Therefore, a laminate can be treated as a thin elastic plate and the classical analysis of Kirchhoff can be used to derive the strain distribution throughout the plate under external forces. The laminate is made from laminae of differing orientations, but having the stress–strain response described by eq. (6.45). The stress–strain relationship of the kth layer of the laminate can be expressed according to eq. (6.49):

$$\begin{bmatrix} \sigma_x \\ \sigma_y \\ \tau_{xy} \end{bmatrix}_k = \begin{bmatrix} \overline{Q}_{11} & \overline{Q}_{12} & \overline{Q}_{16} \\ \overline{Q}_{12} & \overline{Q}_{22} & \overline{Q}_{26} \\ \overline{Q}_{16} & \overline{Q}_{26} & \overline{Q}_{66} \end{bmatrix}_k \begin{bmatrix} \varepsilon_x \\ \varepsilon_y \\ \gamma_{xy} \end{bmatrix}_k . \tag{6.49}$$

Thus $\overline{Q}_{ij}$ is evaluated for each layer and classical lamination theory is a method of calculating resultant forces and moments by integrating the stresses on each lamina through the thickness of the laminate. The ratio of thicknesses is related to the volume fraction of the plies; therefore, the reduced-stiffness matrix for the angle-ply laminate, Q_{aij}, where the subscript a refers to a balanced angle-ply laminate, is given by

$$Q_{aij} = \sum_{k=1}^{N} [Q_{ij}]_k V_k. \tag{6.50}$$

In balanced laminates, Q_{a16} and Q_{a26} cancel each other out because the total thickness of all the $-\theta$ layers equals that of the $+\theta$ layers. Therefore, eq. (6.44) becomes

$$Q_{a11} = \frac{E_{a1}}{1 - v_{a12}v_{a21}}$$

$$Q_{a12} = \frac{v_{a12}E_{a2}}{1 - v_{a12}v_{a21}} = \frac{v_{a21}E_{a1}}{1 - v_{a12}v_{a21}} \tag{6.51}$$

$$Q_{a22} = \frac{E_{a2}}{1 - v_{a12}v_{a21}}$$

$$Q_{a66} = G_{a12}.$$

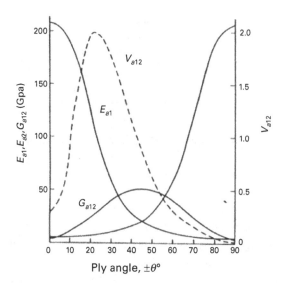

Figure 6.20 Elastic constants of $\pm\,\theta$ laminate type 1 high-modulus carbon-fibre epoxy composite [14].

Figure 6.20 shows plots of the elastic constants ($\nu_{a12}, E_{a1}, E_{a2}, G_{a12}$) for $\pm\theta$ laminates from type 1 high-modulus carbon-fibre epoxy. We see that the variation in moduli illustrates the need for the introduction of additional 90° plies for quasi-isotropy.

6.2.3   Failure of Angle-Ply Laminates

6.2.3.1   Stresses Acting on Individual Laminae

In order to predict the failure of a laminate, we need to be able to calculate the stresses in individual plies. We have discussed briefly the development of laminate theory for examining stresses in laminated fibre-reinforced composites. It is especially important to establish the resolved stresses acting parallel, perpendicular, and at an angle to the fibres. In this way we can identify failure mode by equating the predicted stress with known values of fracture stresses for the individual components. Thus, laminate theory can be used to calculate σ_0, σ_{90}, σ_θ directions to the fibres. Figure 6.21 gives the variation of lamina stresses with loading angle.

Failure of a lamina occurs when the resolved stresses in the principal material directions (parallel or perpendicular to the fibres) or shear stresses exceed their ultimate values. Thus, failure can result from (1) fibre fracture, (2) transverse cracking, and (3) shear fracture (Table 6.4). Laminate theory can be used to calculate the stress in the relevant directions $\sigma_{0°}$, $\sigma_{90°}$, and τ_{12}, and to compare them to the strength values. Figure 6.21 shows that with $\theta \leq$ 5–10° it is highly probable that the failure will occur through fibre fracture. When $\theta \geq$ 10° (5–20°) shear stresses are significant and failure will occur through shear, either at the fibre–matrix interface or by intralaminar shear. At $\theta \approx$ 45°, transverse stresses are significant and the shear stresses are

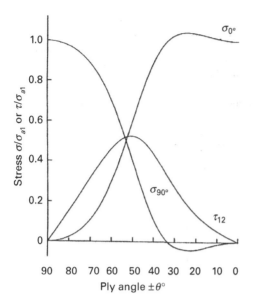

Figure 6.21 Relative stresses referred to σ_{a1} or $\sigma_{0°}$ in the direction parallel to the fibres, $\sigma_{90°}$ transversely, and τ_{12} in the $\theta°$ plane to the fibres in a symmetric angle-ply laminate under load at angle $\theta°$. Glass-fibre UPE laminate with $V_f = 0.5$ [14,17].

maximized, so it is possible that failure can occur by transverse or matrix fracture or through shear. Thus, we might expect a mixed-mode failure. With $\theta = 45–90°$, transverse stresses are at a maximum so matrix or transverse fracture will dominate.

6.2.3.2   Strength of an Off-Axis Unidirectional Composite

The simplest failure criterion is referred to as the **maximum stress theory**, in which fracture occurs when the stress in the principal directions reaches a critical value, as shown in Figure 6.22. Thus, the composite strength, σ_{cu}, could be a function of either longitudinal strength, σ_{lu}, or transverse strength, σ_{tu}, or in-plane shear strength, τ_{12u}, depending on which stress becomes critical first.

1. At $\theta = 0°$, the composite strength in the fibre direction is $\sigma_{cu} = \sigma_{lu}$.
2. At $\theta \approx 0–5°$ the angle between the stress and fibres is relatively small ($\theta < 5°$). The fibre tensile stress reaches the fibre fracture stress before any other failure event can occur. Resolving the applied force, F, over the area on which it acts, we get:

$$\sigma_1 = \frac{F \cos \theta}{A_1} = \frac{F \cos \theta \cos \theta}{A_o} = \sigma_\theta \cos^2 \theta.$$

$$\sigma_\theta = \sigma_1 \sec^2 \theta$$

(6.52)

Here, fibre fracture still dominates failure.

Table 6.4. Failure modes of an individual unidirectional lamina as a function of loading angle, θ

Load angle, θ (°)	Failure mode	Locus/ mechanism
0	Longitudinal tensile fracture	Fibre fracture
5–20	Intralaminar shear	Interfacial or matrix
20–45	Mixed mode	Intralaminar shear and transverse tensile failure
45–90	Transverse tensile failure	Matrix and fibre–matrix interface

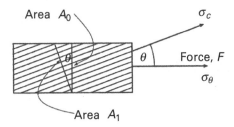

Figure 6.22 Resolution of forces, F, acting on a unidirectional angle-ply composite. σ_c is the stress in the fibre direction of the composite.

3. At larger values of $\theta(\approx 5\text{–}10°)$,

$$\sigma_\theta = \frac{\tau_{12}}{\sin\theta\cos\theta} = 2\tau_{12}\operatorname{cosec} 2\theta. \tag{6.53}$$

Here, there is shear failure – invariably interfacial debonding because the shear stress that develops will exceed the shear strength of the interface.

4. At much larger angles of θ,

$$\sigma_t = \sigma_\theta \sin^2\theta$$
$$\sigma_\theta = \sigma_t \operatorname{cosec}^2\theta. \tag{6.54}$$

The resolved stress perpendicular to the fibre–matrix interface causes failure at the interface and within the matrix, at relatively low stresses.

5. At $\theta° = 15\text{–}45°$ there is mixed-mode failure. Here, the properties of the interface and matrix significantly determine which failure mechanism operates. Since $\sigma_{\theta°} = \sigma_{lu}$, $\sigma_{90°} = \sigma_{tu}$, and $\tau_{12} = \tau_u$, these represent the failure criteria as a function of the *maximum stress* in the principal directions of the composites. This is formulated in Figure 6.23.

The equations are plotted in Figure 6.24, showing how the maximum stress theory has limitations in predicting the failure of a lamina across the whole range of angles. The most uncertainty arises over the mixed-mode region of $\theta = 15\text{–}45°$. Since laminates often include plies at $\pm 45°$, and sometimes other loading angles are utilized, alternative failure criteria needed. It is an ongoing issue of debate.

$$\sigma_\theta \geq \frac{\sigma_{lu}}{\cos^2\theta}$$

$$\sigma_\theta \geq \frac{\sigma_{tu}}{\sin^2\theta}$$

$$\sigma_\theta \geq \frac{\tau_u}{\sin\theta\cos\theta}$$

Figure 6.23 Maximum stress criteria for predicting the fracture mode of a unidirectional angle-ply lamina.

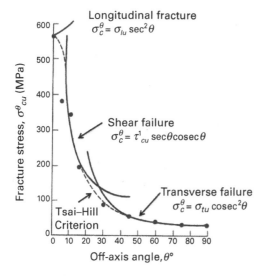

Figure 6.24 The angular dependence of fracture strength of a unidirectional carbon-fibre-reinforced polymer lamina with $V_f = 0.5$. • are the experimental data of Sinclair and Chamis [15]; - - is the Tsai–Hill criterion (eq. 6.56) (redrawn from Hull [14] and Jones [1]).

One alternative predictive model would be the analogous **maximum strain theory**, which should be more effective because we have seen above and in Chapter 5 that strain criteria are often more effective. However, the discrepancies between experiment and prediction are greater than the **maximum stress theory**.

The **Tsai–Hill criterion** has been developed from the equivalent criterion for failure of a metal. It is a development of the maximum work theory to provide a von Mises failure criterion for composite materials.

In plane stress, von Mises failure criterion for a metal is given by

$$\left(\frac{\sigma_1}{\sigma_y}\right)^2 + \left(\frac{\sigma_2}{\sigma_y}\right)^2 - \left(\frac{\sigma_1}{\sigma_y}\right)\left(\frac{\sigma_2}{\sigma_y}\right) = 1,$$

where σ_y is the yield stress of the metal, and σ_1 and σ_2 are the stresses in the 1- and 2-directions.

The von Mises equation can be adapted for composite lamina, as shown in eq. (6.55):

$$\left(\frac{\sigma_l}{\sigma_{lu}}\right)^2 + \left(\frac{\sigma_t}{\sigma_{tu}}\right)^2 - \left(\frac{\sigma_l\sigma_t}{\sigma_{lu}^2}\right) + \left(\frac{\tau_{12}}{\tau_u}\right)^2 = 1. \tag{6.55}$$

For most composite materials,

$$\sigma_{lu} > \sigma_{tu}$$

Therefore, the term $\left(\frac{\sigma_l\sigma_t}{\sigma_{lu}^2}\right)$ in eq. (6.55) is small and can be ignored.

Equation (6.55) becomes

$$\left(\frac{\sigma_l}{\sigma_{lu}}\right)^2 + \left(\frac{\sigma_t}{\sigma_{tu}}\right)^2 + \left(\frac{\tau_{12}}{\tau_u}\right)^2 = 1. \tag{6.56}$$

This equation is known as the Tsai–Hill failure criterion. It is plotted in Figure 6.24 and appears to provide a closer prediction of failure of a lamina at angles close to $\pm\theta$. The strength of a θ lamina is given by

$$\sigma_{\theta u} = \left[\frac{\cos^4\theta}{\sigma_{lu}^2} + \left(\frac{1}{\tau_u^2} - \frac{1}{\sigma_{lu}^2}\right)\sin^2\theta\cos^2\theta + \frac{\sin^4\theta}{\sigma_{tu}^2}\right]^{1/2}. \tag{6.57}$$

A detailed discussion of the failure criteria of composite laminates is beyond the scope of this book, and readers are referred to the text by Daniel and Ishai [18] and the worldwide failure exercise [19–21].

6.2.3.3 Strength of a Multi-Ply Laminate

We can see above that we can understand the failure of a laminate in terms of the sequential fracture of individual lamina in a laminate. Equations (6.49) and (6.50) provide the stress–strain response, so we can consider the stresses in the principal directions of individual plies using the maximum stress theory at low and high angles. The Tsai–Hill criterion can be used for plies of 20–45°.

The stresses acting on individual laminae of a cross-ply or angle-ply laminate are required to predict the failure mechanisms within each layer and hence the damage accumulation mechanism. Equation (6.45) can be applied to a cross-ply or angle-ply laminate so that stresses parallel to the fibres ($\sigma_{\parallel}$) or at 90° ($\sigma_{\perp}$) and the shear stress (τ) in individual plies can be calculated using laminate analysis. The strain–stress equation of the laminate is given by

$$\begin{bmatrix} \varepsilon_{a1} \\ \varepsilon_{a2} \\ \gamma_{a12} \end{bmatrix} = \begin{bmatrix} S_{a11} & S_{a12} & S_{a16} \\ S_{a12} & S_{a22} & S_{a26} \\ S_{a16} & S_{a26} & S_{a66} \end{bmatrix} \begin{bmatrix} \sigma_{a1} \\ \sigma_{a2} \\ \tau_{a12} \end{bmatrix}. \tag{6.58}$$

Cross-ply and angle-ply laminates with laminae of equal thickness are tested with principal stresses parallel to the directions of symmetry. Then, $\tau_{a12} = 0$, $S_{a16} = S_{a26} = 0$, and $\gamma_{a12} = 0$.

The strains ε_{a1}, ε_{a2}, and γ_{a12} are expressed in terms of strains parallel and transverse to the fibres according to the analysis given in Section 6.2.3.2 using stress and strain transformation equations:

$$
\begin{bmatrix} \sigma_x \\ \sigma_y \\ \tau_{xy} \end{bmatrix} = \begin{bmatrix} \cos^2\theta & \sin^2\theta & -2\sin\theta\cos\theta \\ \sin^2\theta & \cos^2\theta & +2\sin\theta\cos\theta \\ \sin\theta\cos\theta & -\sin\theta\cos\theta & \cos^2\theta - \sin^2\theta \end{bmatrix} \begin{bmatrix} \sigma'_x \\ \sigma'_y \\ \tau'_{xy} \end{bmatrix}, \tag{6.59}
$$

$$
\begin{bmatrix} \varepsilon_x \\ \varepsilon_y \\ \gamma_{xy} \end{bmatrix} = \begin{bmatrix} \cos^2\theta & \sin^2\theta & -2\sin\theta\cos\theta \\ \sin^2\theta & \cos^2\theta & +2\sin\theta\cos\theta \\ \sin\theta\cos\theta & -\sin\theta\cos\theta & \cos^2\theta - \sin^2\theta \end{bmatrix} \begin{bmatrix} \varepsilon'_x \\ \varepsilon'_y \\ \gamma'_{xy} \end{bmatrix}. \tag{6.60}
$$

For a cross-ply laminate, $\theta = 0°$ or $90°$ in alternate laminae, so that in laminae in which the 1-direction is parallel to the fibres:

$$
{}^1\varepsilon_{||} = \varepsilon_{a1}; \; {}^1\varepsilon_{||} = \varepsilon_{a1} \text{ and } {}^1\gamma = \gamma_{a12}. \tag{6.61}
$$

For laminae in which the 2-direction is parallel to the fibres:

$$
{}^2\varepsilon_{||} = \varepsilon_{a2}; \; {}^2\varepsilon_{\perp} = \varepsilon_{a1} \text{ and } {}^2\gamma = \gamma_{a12}. \tag{6.62}
$$

The stresses in these individual laminae can be obtained from the stress–strain equations for the laminae, eqs (6.43) and (6.44), in combination with the conditions given for the two cases in eqs (6.61) and (6.62).

Thus, individual laminae will fracture once the resolved stress in the relevant direction exceeds a critical value. Therefore, the off-axis plies will tend to fracture first and progressively develop further damage. Damage accumulation follows the sequence according to weakest ply consideration. A typical quasi-isotropic laminate $0°_2/90°_2/\pm45°/90°_2/0°_2$ will exhibit the following damage sequence:

1. transverse cracking of 90° and 45° plies;
2. interlaminar shear failure initiated by transverse cracks;
3. delamination;
4. fibre fracture in 0° plies;
5. composite failure.

6.3   Strength of a Multi-Ply Laminate under Complex Loads

Flexural loading of a beam leads to a distribution of stresses through the thickness. Figure 6.25 illustrates three-point loading showing that a neutral axis exists. A shear stress develops under load at the neutral axis. Often, four-point loading is employed in mechanical tests because the stress is more uniform through the thickness of the beam (Figure 6.26).

As the L/d ratio decreases, the shear stress becomes dominant. This test method with $L/d < 5$ is used to measure the interlaminar shear strength (ILSS) of a composite.

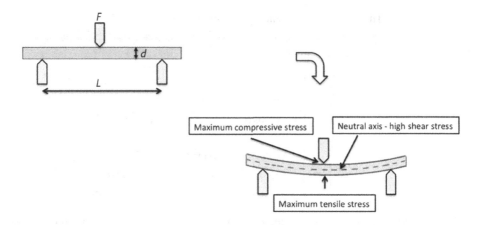

Figure 6.25 Flexural loading under a three-point configuration showing the high tensile and compressive stress regions. The midpoint line of the beam is at the neutral axis.

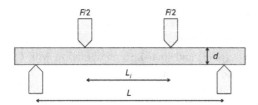

Figure 6.26 Flexural loading under a four-point configuration.

Therefore, under flexural loading the arguments given in Section 6.2.3.3 still apply, and the response of the laminate is determined by the stress operating in the individual plies. In other words, first ply failure will occur in the tensile face, usually by transverse cracking followed by an increased probability of shear failure initiated at the ply–ply interface, where the transverse cracks are pinned because shear stresses are generated. Figure 6.27 shows the damage cascade induced by a low-energy impact in which the material responds in flexure.

6.3.1   Impact Response

Impact response is a function of the rate of loading. Drop weight loading is used to simulate real impact events:

1. dropped tool – low-rate event;
2. bird strike – intermediate-rate event;
3. ballistic impact – very high-rate event.

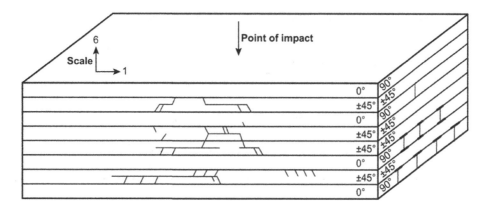

Figure 6.27 Typical damage cascade pattern of a $(0°/\pm 45°/0°/\pm 45°)_s$ laminate after low-energy impact.

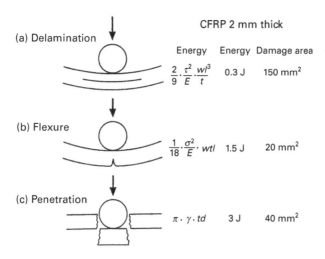

Figure 6.28 Principal failure modes of composite laminates under impact loading. Energies involved and damage area of 2 mm thick carbon-fibre-reinforced polymer. w, l, and t are the width, unsupported length, and thickness of the specimen; d is the diameter impactor; τ is the ILSS; σ is flexural strength; E is the flexural modulus; and γ is the fracture energy [22].

Dorey [22] has described the impact response at different rates and impact energies, as shown in Figure 6.28 (see also Table 6.5).

The damage, which develops at intermediate impact energies, results from transverse cracking in the tensile ply furthest from the impact, as shown in Figure 6.27. Transverse cracks form first, which initiate interply delamination, and the damage cascades into the material. This damage cannot usually be seen from the impacted surface and is referred to as barely visible impact damage

Table 6.5. Summary of damage resulting from impact events of differing energy

Impact energy/ velocity	Damage type	Location	Comment
Low	delamination	Interfacial or matrix shear	Material responds in flexure as in short beam test
Intermediate	Transverse cracking with initiated delamination	Tensile face opposite from impact	Barely Visible Impact Damage (BVID)
High	Fibre fracture	Fibre fracture in opposite face - cascade leads to puncture	Plug fracture

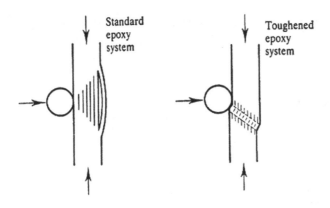

Figure 6.29 Failure modes associated with compression loading of impact-damaged carbon-fibre-reinforced polymer laminates [23].

(BVID). Much effort is expended to identify sensors for this damage (see Section 6.4).

6.3.2   Residual Strength after Impact

The most important property of a composite material is the compression strength after impact. This property is commonly used to assess the quality of materials subjected to impact. The residual compression strength is directly related to the delaminated area produced in the impact event. Compression strength is a strong function of the shear modulus of the matrix resin, since fibre buckling is the main failure mechanism in the 0° plies. Figure 6.29 gives a schematic of mechanisms of unidirectional plies using matrix resins with different fracture toughnesses.

The compression strength, σ_{lcu}, of a unidirectional composite is a function of the shear modulus of the matrix, as shown in eq. (6.63), because of the support, which prevents buckling. In terms of a laminate, localized damage will reduce the support provided by the stiffness of the matrix and adjacent fibrous plies. The critical nature of

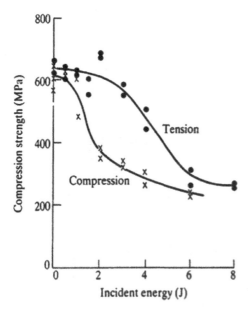

Figure 6.30 Residual tensile and compression strengths of a $(0°/90°/0°/\pm45°/0°)_s$ carbon-fibre epoxy laminate [23].

delamination damage is shown in Figure 6.30, where the residual compressive and tensile strengths after impact are given:

$$\sigma_{lcu} = \frac{G_m}{V_m} = \frac{G_m}{1 - V_f}. \tag{6.63}$$

6.4   Damage Detection

For completion, we list the methods of detecting damage in laminates either as manufactured or subjected to a loading event such as impact. We should differentiate between non-destructive testing (NDT) [24–25] and structural health monitoring (SHM) [26]. The NDT techniques are summarized in Table 6.6.

Non-destructive testing techniques generally are inspection approaches in which the structure is examined before service for manufacturing defects and post-service for damage formation. Table 6.6 lists possible techniques together with a brief description of the monitoring system and their applications. The advantages and disadvantages are also briefly described. On the other hand, SHM refers to sensors permanently placed or embedded in the composite. The sensors can be interrogated continuously during service, either in static or dynamic mode. They should be in unobtrusive locations and lightweight, and not impose a weight or cost penalty nor affect the strength of the composite material and structure. Table 6.7 describes the principal of SHM techniques. The use of carbon-fibre resistance to detect impact damage is reported [27].

Table 6.6. Techniques for non-destructive testing

NDT technique	Damage type	Material	Inspection	Comment
Visual	Voids, Resin-rich regions Cracks Debonds	Glass fibre	1. Manufacturing defects 2. Impact damage	BVID limits use
Tap test	Surface damage	–	–	Not recommended
Ultrasound, transmission	Delaminations	Carbon, aramid, and glass fibre	Manufacturing defects in laminates Impact damage	
Ultrasound, pulsed echo	Delaminations	1. Composite panels 2. Composite structures: carbon, aramid, glass fibres	1. Manufacturing defects in laminates 2. Impact damage	With scanning of a sample in a water bath or with water jet for large structures: 1. A scan: through thickness using simple back surface reflection 2. B Scan: in one direction of specimen 3. C scan: of both width and breadth directions
X-radiography	Voids and stress-induced defects	1. Glass fibre; carbon fibre needs a radio-opaque penetrant 2. CT-digital assembly provides location of damage within a volume	Manufacturing defects	1. Low-*energy X-rays, e.g. 5–50 keV in transmission 2. Computed tomography (CT) scans: uses a thin X-ray beam to obtain absorption profile of transmitted beam at intervals as the beam moves around to construct a cross-sectional image

Technique				
Thermography	Damage but location through thickness is difficult	1. Defects appear as hot spots in the thermal image. 2. Can be used to detect damage in a fatigue test	Manufacturing and load-induced damage	Measurement of thermal transients arising from a thermal pulsed thermal stimulation 1. single-sided 2. double-sided 3. vibro-thermography: detects hot spots where high-energy dissipation or damping occurs
Acoustic emission	Detects dynamic cracking; identifying the location of a crack is complex because of attenuation of the stress wave as it propagates through the material over a potential considerable distance	Composite structures	Proof testing of structures but employs a load to generate cracks	Detects elastic strain energy released as stress waves. These propagate through the material to the surface where a transducer converts them into an electrical signal
Acoustic, ultrasonics	Measures a stress wave factor, which is related to residual strength	Composite structures	Proof testing of structures	Employs pulsed ultrasound wave stimulation
Electromagnetic: eddy current measurement	Detects cracks in some components	1. MMC 2. Glass-fibre composite		Uses a changing magnetic field to create eddy currents in the substrate

Table 6.7. Techniques for structural health monitoring

Technique	Damage type	Inspection approach	Advantages	Disadvantage
Conventional strain gauges	Difficult to specify	Converts relative change in strain into electrical resistance Needs a precision instrument (Wheatstone Bridge)	Simply adhered to surface	Resistance may be affected by geometric factors and maintenance of adhesion
Fibre optic sensors	Matrix cracks, Debonding and other physical changes inducing local strains 1. Light intensity change, microbending, or fracture 2. Twisting 3. Mechanical strain or thermal expansion 4. Applied strain 5a. Strain from cavity between two optical fibres 5b. Etched grating (after removal of cladding) acts as a strain gauge.	Embedded optical fibre: consisting of 'doped' silica core, with a cladding (polymeric or low refractive index glass) and coating Modes: 1. Intensity modulation 2. Polarization modulation 3. Phase modulation 4. Scattering modulation 5. Spectral modulation: a. Fabry–Perot interferometer (FPI) b. Fibre Bragg grating (FBG)	1. Immune to electromagnetic interference 2. Direct embedment 3. Corrosion resistance 4. Multiplexing 5. Multimode fibres have a larger diameter core.	1. Optoelectronic equipment for complex signal processing 2. Optical fibres have typical diameters of 120–250 μm, significantly larger than reinforcing fibres of 5–20 μm. Embedment leads to an area of distorted reinforcement and compromises composite strength

Piezoelectric wafer active sensors (PWAS)	Lamb waves travel long distances over thin-wall structures, detecting structural anomalies, cracks, and delaminations 1. One PWAS generates waves for detection by a second PWAS 2. The same PWAS acts as a transmitter and receiver for echo from a crack or damage 3. PWAS receives elastic waves generated by AE event	Thin, inexpensive piezoceramic wafers poled in the thickness direction with electrodes are bonded to the surface. At ultrasonic frequencies they sense and excite Lamb waves. Modes: 1. Pitch-catch 2. Pulse-echo 3. Thickness impact with acoustic emission (AE)	Relatively simple method of inspecting thin-walled structures	Needs further development and validation
Electrical properties sensors: Resistance (conductance)	Self-sensing is possible by measuring resistance of reinforcing carbon fibres Detects fibre-breaks by increase in resistance Detects matrix cracks and delamination by decrease in resistance	Carbon fibres are electrical conducting Resins are non-conducting	When carbon fibres break, resistance increases When matrix cracks, carbon-fibre resistance decreases; release of compressive thermal strain and/or piezoelectric effect in carbon fibres is responsible	Further development and application of self-sensing techniques required
Dielectric Impedance		Glass fibres are insulators but GRP exhibits a change in dielectric properties with damage	Dielectric spectroscopy	

Smart systems, which employ techniques for self-sensing of damage and for activating a self-healing mechanism, are described in Chapter 9.

6.5   Conclusions

In this chapter we have reviewed the role of lamination in the creation of composite artefacts with good all-round properties and identified the mechanisms of failure. To achieve these aims the basics of 'laminate theory' are discussed. The reader is referred to texts such as Tsai's Composite Design series [28,29] and Daniel and Ishai [18] in order to use available laminate computer programs efficiently for specific design problems. The aim of this chapter was to provide the sufficient knowledge for careful consideration of the outcomes provided by theoretical analyses. Laminate theory-based computer programs are available, but without understanding of the principles poor design decisions could arise.

6.6   Discussion Points

1. What is understood by a laminate?
2. Explain the difference between anisotropic and isotropic laminates? Give an example of a balanced configuration of a typical isotropic laminate.
3. What is the origin of thermal strains in cross-ply laminates? How can the thermal strains be estimated?
4. What is the sequence of damage formation in a balanced $0_x°/90_y°/0_x°$ cross-ply laminate under load? How does the value of y influence the nature of the crack growth?
5. Briefly describe the principle of the shear lag analysis for describing the development of cross-ply cracking.
6. Examine the failure of a unidirectional lamina at different loading angles. Derive the failure criteria by resolving forces.
7. Use the failure criteria to predict the fracture behaviour of a unidirectional composite across the range of loading angles from 0° to 90°.
8. Apply the knowledge about tensile failure modes of unidirectional and cross-ply laminates to the failure modes of a balanced angle-ply laminate under a low-energy impact event.

References

1. F. R. Jones, Micromechanics and properties of fibre composites. In *Composite materials in aircraft structures*, ed. D.H. Middleton (Harlow: Longman, 1990), pp. 69–90.
2. P. W. Manders, T.-W. Chou, F. R. Jones, and J. W. Rock, Statistical analysis of multiple fracture in 0°/90°/0° glass fibre/epoxy resin laminates. *J. Mater. Sci.* **18** (1983), 2876–2889.

3. K. W. Garrett and J. E. Bailey, Multiple transverse fracture in 90° cross-ply laminates of a glass fibre-reinforced polyester. *J. Mater. Sci.* **12** (1977), 157–168.

4. A. Parvizi and J. E. Bailey, On multiple transverse cracking in glass fibre epoxy cross-ply laminates. *J. Mater. Sci.* **13** (1978), 2131–2136.

5. K. W. Garrett, A. Parvizi and J. E. Bailey, Constrained cracking in glass fibre-reinforced epoxy cross-ply laminates. *J. Mater. Sci.* **13** (1978), 195–201.

6. J. E. Bailey, P. T. Curtis, and A. Parvizi, On the transverse cracking and longitudinal splitting behaviour of glass and carbon fibre reinforced epoxy cross ply laminates and the effect of Poisson and thermally generated strain. *Proc. Roy. Soc. A.* **366** (1979), 599–623.

7. P. A. Sheard and F. R. Jones, Computer simulation of the transverse cracking process in glass fibre composites. In *Proceedings of the international conference on composite materials ICCM VI/ECCM 2*, vol. 3, ed. F. L. Matthews, N. C. R. Buskell, J. M. Hodgkinson, and J. Morton (London: Elsevier, 1987), pp. 123–135.

8. J. A. Nairn, S. Hu, and J. S. Bark, A critical evaluation of theories for predicting microcracking in composite laminates. *J. Mater. Sci.* **28** (1993), 5099–5111.

9. J. A. Nairn, Matrix cracking in composites. In *Polymer matrix composites*, ed. R. Talreja and J.-A. Manson (Oxford: Elsevier Science, 2000).

10. A. S. D. Wang, Fracture mechanics of sublaminate cracks. *Compos. Mater. Comp. Tech. Rev.* **6** (1984), 45–62.

11. F. R. Jones, M. Mulheron, and J. E. Bailey, Generation of thermal strains in GRP part 1: effect of water on the expansion behaviour of unidirectional glass fibre reinforced laminates. *J. Mater. Sci.* **18** (1983), 1522–1532.

12. F. R. Jones, M. Mulheron, and J. E. Bailey, Generation of thermal strains in GRP part 2: the origin of thermal strain in polyester crossply laminates. *J. Mater. Sci.* **18** (1983), 1533–1539.

13. R. A. Schapery, Thermal expansion coefficients of composite materials based on energy principles. *J. Comp. Mat.* **2** (1968), 380–404.

14. D. Hull, *An introduction to composite materials* (Cambridge: Cambridge University Press, 1981).

15. J. H. Sinclair and C. C. Chamis, Fracture modes in off-axis fiber composites. In *Proc. 34th SPI/RP annual technology conference* (New York: Society of the Plastics Industry, 1978), paper 22A.

16. T. W. Clyne, Cambridge composites lectures, C16. Cambridge University.

17. B. Spencer and D. Hull, Effect of winding angle on the failure of filament wound pipe. *Composites* **9** (1978), 263–271.

18. I. M. Daniel and O. Ishai, *Engineering mechanics of composite materials*, 2nd ed. (Oxford: Oxford University Press, 2006).

19. M. J. Hinton, A. S. Kaddour, and P. D. Soden. *Failure criteria in fibre reinforced polymer composites: the world-wide failure exercise* (Oxford: Elsevier, 2004).

20. A. S. Kaddour and M. J. Hinton, The background to the Second World-Wide Failure Exercise (WWFE-II). *J. Compos. Mater.* **46** (2012), 2283–2294.

21. A. S. Kaddour and M. J. Hinton, Maturity of 3D failure criteria for fibre-reinforced composites: comparison between theories and experiments: Part B of WWFE-II. *J. Compos. Mater.* **47** (2013), 925–966.

22. G. Dorey, Impact performance : CFRP laminates. In *Handbook of polymer-fibre composites*, ed. F. R. Jones (Harlow: Longman, 1994), pp. 327–330.

23. G. Dorey, Impact performance: residual compression strength. In *Handbook of polymer-fibre composites*, ed. F.R. Jones (Harlow: Longman, 1994), pp. 330–334.

24. F. L. Matthews and R. D. Rawlings, *Composite materials: engineering and science* (London: Chapman and Hall, 1994), pp. 415–447.

25. R. H. Bossi and V. Giurgiutiu, Nondestructive testing of damage in aerospace composites. In *Polymer composites in the aerospace Industry*, ed. P. E. Irving and C. Soutis (Cambridge: Woodhead, 2015), pp. 413–448.

26. V. Giurgiutiu, Structural health monitoring (SHM) of aerospace composites. In *Polymer composites in the aerospace Industry*, ed. P. E. Irving and C. Soutis (Cambridge: Woodhead, 2015), pp. 449–507.

27. T. J. Swait, F. R. Jones, and S. A. Hayes, A practical structural health monitoring system for carbon fibre reinforced composite based on electrical resistance. *Compos. Sci. Technol.* **72** (2012), 1515–1523.

28. S. W. Tsai, *Composites design*, vol. 1, 4th ed. (Palo Alto, CA: Think Composites, 1988).

29. S. W. Tsai, *Theory of composites design* (Palo Alto, CA: Think Composites, 1992).

7 Fatigue Loading of Laminates

In this chapter, the analysis in Chapter 6 is extended to dynamic loading. The main aim is to provide sufficient knowledge for predicting the life of a composite structure.

The term *fatigue* refers to time-dependent fracture behaviour. Both static and dynamic fatigue phenomena occur. The static fatigue of glass fibres is discussed in Chapter 2. With respect to composite materials, we are mainly concerned with dynamic loading of a structure and the prediction of service life.

Homogeneous materials such as a metal or ceramic fail by the propagation a single crack. In ceramics, the fracture toughness is relatively low and elastic fracture mechanics apply. With metallic materials, a plastic zone forms at the tip of the crack. Thus, *fracture mechanics* attempts to identify the criticality associated with the presence of a crack or pre-existing flaw. Thus, the following questions are posed:

1. What is the residual strength as a function of crack size?
2. What are the critical crack dimensions?
3. What is the possible service life before the crack size becomes critical?
4. What size of pre-existing flaw can be allowed?
5. How often should structural inspections for cracks take place?

There are three modes of loading that the material could experience:

1. mode 1, opening mode: the two crack surfaces move away symmetrically in the y-direction;
2. mode 2, sliding mode: the two crack surfaces slide away in the x-direction;
3. mode 3, tearing mode: the two crack surfaces move in the z-direction, introducing tearing in the x–y plane, with the surfaces remaining parallel to the crack front.

These modes are illustrated in Figure 7.1.

Here, we consider mode 1, linear fracture mechanics as illustrated in Figure 7.1a. In **brittle** materials, fracture is a low-energy process (low energy dissipation), in which catastrophic failure occurs without warning because crack velocity is generally high. Little or no plastic deformation occurs before fracture.

With **tough** materials, ductile fracture involves a high-energy process, which involves dissipation of a large amount of energy, which is associated with a large plastic deformation zone before crack instability occurs. Crack growth is relatively slow.

We can describe these processes using Griffith's criterion [1] and the extensions provided by Irwin [2].

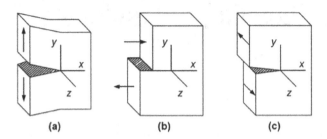

Figure 7.1 Schematic showing the fundamental fracture modes: (a) Mode 1 (opening), (b) Mode 2 (sliding), (c) Mode 3 (tearing).

7.1   Griffith's Criterion

The energy release rate for crack growth, or *strain energy release rate*, is obtained from the change in elastic strain energy per unit area of crack growth, given by:

$$G = \left[\frac{\delta U}{\delta a}\right]_p = -\left[\frac{\delta U}{\delta a}\right]_u, \tag{7.1}$$

where U is the elastic energy of the system and a is the crack length. The load, P, or the displacement, u, are kept constant.

Irwin showed that for a mode 1 crack (opening mode), the strain energy release rate and the stress intensity factor are related. Consider a thin rectangular plate with a crack perpendicular to the load:

$$G = \frac{\pi\sigma^2 a}{E}, \tag{7.2}$$

where G is the strain energy release rate, σ is the applied stress, a is half the crack length, and E is Young's modulus, which for the case of plane strain should be divided by the plate stiffness factor $(1 - v^2)$. v is Poisson's ratio.

$$\text{For plane stress:} \qquad G = G_1 = \frac{K_1^2}{E}. \tag{7.3}$$

$$\text{For plane strain:} \qquad G = G_1 = \frac{(1 - v^2)K_1^2}{E}. \tag{7.4}$$

The strain energy release rate can be better understood as *the rate at which energy is absorbed by growth of the crack*.

However, we also have a critical value for crack growth, G_c:

$$G_c = \frac{\pi\sigma_u^2 a}{E}, \tag{7.5}$$

where σ_u is the ultimate or failure stress. The criterion for when the crack will begin to propagate is that $G_1 \geq G_c$.

Irwin extended Griffith's theory, where the stress intensity factor, K_1, replaces the strain energy release rate. The critical value K_{1c} is often referred to as fracture toughness. Both of these terms are simply related to the energy terms that Griffith used:

$$K_1 = \sqrt{\pi a}. \tag{7.6}$$

For plane stress: $\quad K_{1c} = \sqrt{EG_c}, \tag{7.7}$

For plane strain: $\quad K_{1c} = \dfrac{\sqrt{EG_c}}{(1 - v^2)}, \tag{7.8}$

where K_1 is the stress intensity factor, K_{1c} is the fracture toughness, and v is Poisson's ratio. It is important to recognize the fact that the fracture parameter K_{1c} has different values when measured under plane stress and plane strain conditions.

For plane strain deformation in mode 1, fracture occurs when $K_1 \geq K_{1c}$ and is considered a material property. The material can be loaded in different ways, as shown in Figure 7.1, to enable a crack to propagate. Tests for modes 1–3 are available.

For geometries other than the centre-cracked infinite plate, it is necessary to introduce a dimensionless correction factor, Y. We thus have:

$$K_1 = Y\sigma\sqrt{\pi a}, \tag{7.9}$$

where Y is a function of the crack length and width of sheet: $Y\left(\frac{a}{w}\right)$.

For example, in a sheet of finite width w containing a through-thickness edge-crack of length, a, Y is given by

$$Y\left(\frac{a}{w}\right) = 1.12 - \frac{0.41}{\sqrt{\pi}}\left(\frac{a}{w}\right) + \frac{18.7}{\sqrt{\pi}}\left(\frac{a}{w}\right)^2 - \cdots. \tag{7.10}$$

Fatigue testing can either involve (1) a constant load, where the time to failure is measured (static fatigue) or (2) the life of a structure under a dynamic load.

7.2 Fatigue Processes in Composite Materials

Fatigue damage is the progressive failure that occurs when a material is subjected to cyclic loads. The maximum stress is less than the static stress required to fail the material in a monotonic test.

1. Prediction of fatigue performance is essential to ensure material selection and performance matches service conditions.
2. Design of a structure requires the prediction of durability in fatigue from knowledge of material performance and service conditions.

In contrast to single-phase materials, which usually fail by the propagation of a single crack, fibre-reinforced materials exhibit diffuse damage formation. As discussed in Chapter 5, composite materials normally develop damage in the form of interfacial debonds, transverse cracks, delaminations, and eventually fibre fracture. These events manifest themselves in the following effects with the increasing number of load cycles:

1. visual damage;
2. progressive reduction in strength;
3. progressive reduction in stiffness; and
4. final failure, when the residual strength has reduced to the maximum applied stress.

7.2.1   Cyclic Loading

Loading regimes in a fatigue test are illustrated in Figure 7.2. The load is cycled between maximum and minimum values so that three differing regimes can be employed: tension–tension, tension–compression, and compression–compression. We define these according to the stress or strain range R, which is the ratio of peak stress or strain to minimum stress or strain:

$$R = \frac{\sigma_{min}}{\sigma_{max}}. \tag{7.11}$$

As shown in Figure 7.2,

R has a positive value for tension–tension and compression–compression tests; and R is negative for tension–compression tests.

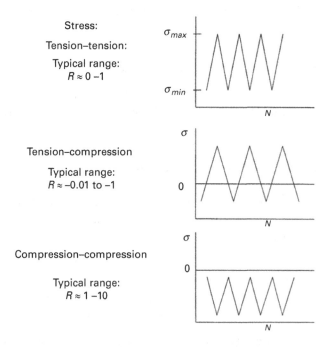

Figure 7.2 Schematic of the differing modes of fatigue testing and the definitions of stress or strain range, R, and typical values. N is the number of cycles.

Further details of the experimental design are given in Figures 7.3 and 7.4, where the terms are used to define the cycling of stress or strain in the fatigue experiment. Figure 7.3 describes the sine waveform of a fully reversed stress cycle, which applies to a tension–compression fatigue test. The terms mean stress, alternating stress, or stress amplitude are defined below:

$$\text{Mean stress:} \quad \sigma_m = \frac{\sigma_{max} + \sigma_{min}}{2}, \tag{7.12}$$

$$\text{Alternating stress:} \quad \sigma_a = \frac{\sigma_{max} - \sigma_{min}}{2}, \tag{7.13}$$

$$\text{Stress range:} \quad \Delta\sigma = \sigma_{max} - \sigma_{min}, \tag{7.14}$$

$$\text{Stress amplitude:} \quad A_s = \frac{\sigma_a}{\sigma_m}. \tag{7.15}$$

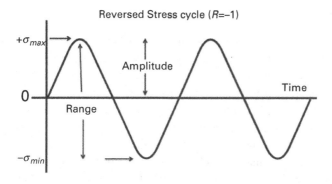

Figure 7.3 Sinusoidal waveform employed in a tension–compression fatigue experiment. Maximum (σ_{max}) and minimum (σ_{min}) stresses are defined.

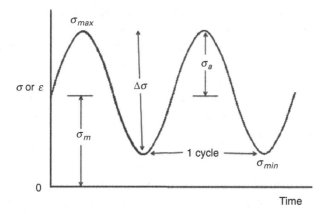

Figure 7.4 Sinusoidal waveform employed in a tension–tension fatigue experiment. Alternating (σ_a) and mean (σ_m) stresses are defined. $\Delta\sigma$ is the stress range.

Figure 7.4 illustrates the more typical tension–tension fatigue stress or strain regime, which applies to many scenarios experienced by a composite structure. The terms used to describe the design of the test are also shown.

7.3   Nature of the Damage Accumulation in Composite Materials

7.3.1   Unidirectional Laminates

The basic failure modes of a unidirectional laminate under monotonic loading, which are described in Chapter 5, also occur under cyclic loading. These are:

1. matrix cracking;
2. fibre fracture;
3. fibre–matrix debonding; and
4. delamination

Thus, 0° laminates will fail by either the accumulation of fibre-breaks or matrix cracks. The former occurs in ductile polymer matrix composites (PMCs) and continues until an i-plet of critical dimensions forms, which leads to rapid fracture. The process involves the statistics of strength of the embedded fibres. As we discussed in Chapter 5, the stress transfer at the broken fibre ends leads to high shear stresses and potential for debonding. The ineffective length of the fragmented fibres will determine the extent of multiple fracture of the fibres and the eventual failure.

With brittle matrices, matrix cracking dominates. In real PMCs it is the relative failure strains of the fibres and matrix that are critical. Thus, differing behaviours of a glass fibre ($\varepsilon_{fu} \approx 2$–3%), carbon fibre ($\varepsilon_{fu} \approx 0.6$–1.5%), and aramid fibre ($\varepsilon_{fu} \approx 2$–4%) will lead to differing failure mechanisms. Aramid fibres have poor radial strength, so fibrillation is an additional mechanism, especially under compression when buckling can initiate fibrillation.

Talreja [3] first recognized the differing performances of a composite material under fatigue conditions. Figure 7.5 shows the strain–life curves for a composite. Since $\varepsilon_c = \varepsilon_f = \varepsilon_m$ in the 0° load-bearing plies, it is easier to use strain than stress to present the data. The following micromechanics dominate the damage accumulation with time at the given strain range. Time is represented by the number of cycles, N.

Figure 7.5a illustrates the critical strains defining the micromechanical events, which occur depending on failure strains of the component matrix and fibres. Thus, the component with the lowest failure strain, which is usually the transverse cracking strain of the composite, often referred to as matrix cracking, is responsible.

With glass-fibre (and other high strain-to-failure fibres) composites, where the strain range $\Delta\varepsilon_c > \Delta\varepsilon_{mu}$, damage accumulation involves a mixed mode of mainly matrix cracking and interfacial failure. As shown in Figure 7.5b, this results in decay in properties with the number of cycles and a significant curvature to the strain–life plot.

Figure 7.5c shows that with carbon fibres where $\Delta\varepsilon_c > \Delta\varepsilon_{mu}$, a relatively flat strain–life curve, which is dominated by fibre fracture, is observed.

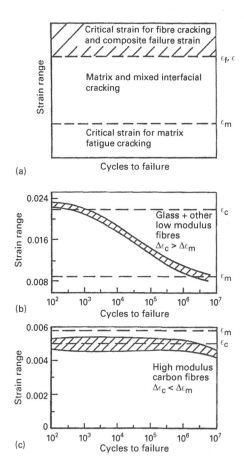

Figure 7.5 Strain–life plots for a cross-ply composite laminate showing the different regions of damage accumulation micromechanics mechanisms. ε_f, ε_m, ε_c are the failure strains of fibres, matrices, and composites. (a) Fibre fracture; (b) mixed-mode matrix cracking and interfacial failure; (c) matrix cracking [3,4].

We need to remember that E_f(glass) $< E_f$(carbon) and ε_{fu} (glass) is approximately 4 times ε_{fu} (carbon), so that to achieve a similar stress on glass-reinforced polymers, the composite will be under a higher strain. This is illustrated in Figure 7.6, where the fatigue strength–life performance curves of unidirectional carbon-, glass-, and aramid-reinforced polymers are compared.

7.3.2   Matrix Effects

When fibre failure dominates the fracture of unidirectional composites, the stress released by the fibre-break is placed on the matrix, which may lead to the immediate generation of a crack in the adjacent matrix or to an interfacial debond. If the matrix

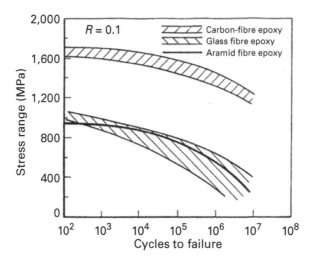

Figure 7.6 Fatigue strength–life curves for unidirectional carbon, glass, and aramid fibre; $R = 0.1$ [4,5].

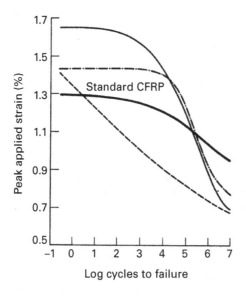

Figure 7.7 Tensile fatigue of 0° carbon-fibre composite from different matrix polymers but similar fibres: standard epoxy (thick black line); tough BMI blend (thinner black line); tough epoxy (dot-dash line); thermoplastic (dashed line) [6,7].

can respond by deformation, then stress transfer is spread over a larger volume. Also, the nature of the interface determines the volume of the affected matrix, which carries the additional load. Thus, the stress transfer process depends on the properties of the matrix. As a result, the fatigue behaviour of composites made from differing types of matrix differs, as shown in Figure 7.7.

7.3.3   Cross-Ply Laminates

To cater for multi-axial stresses, fibre-reinforced composites often include off-axis plies in the lay-up of a laminate. We saw in Chapter 5 that 90° plies will exhibit transverse cracking under load; therefore, we should expect the strength–life curves of angle-ply laminates to exhibit additional curvature. With increasing damage in off-axis plies, the stress concentrations on the load-bearing 0° fibres arising from the transverse cracks reduce the failure load of the 0° plies. Curtis [6] studied the effect of including 90° plies on the strength–life curves of unidirectional carbon-reinforced polymers, as shown in Figure 7.8.

The damage accumulation follows the micromechanics of a cross-ply laminate, which was discussed in Chapter 5. Thus, the failure of 0°/90°/0° laminate involves the following damage sequence:

1. early generation of transverse cracks, often at coupon edges and extending across the width and thickness of the 90° ply or plies;
2. increase in the density of transverse cracks until saturation occurs, resulting in a reduction in laminate stiffness;
3. nucleation and growth of longitudinal cracks (splits) in the 0° plies;
4. delamination initiated by the shear stresses present where transverse cracks impinge at the 0° ply interfaces;
5. delamination initiated at the ply interfaces by longitudinal cracks impinges on the 90° ply interfaces;

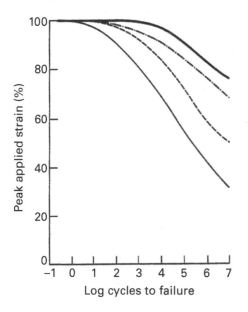

Figure 7.8 Normalized strain–life curves for laminates with varying fraction of 0° plies. 100% 0° (thick black line); 50% 0° (dot-dash line); 25% 0° (dashed line); 0% 0° (thin black line) [6].

6. fibre fracture in the 0° plies resulting from the stress concentrations at the tips of the 90° ply cracks;

7. final failure when sufficient fibre fractures mean that the remaining fibres cannot carry the load.

7.3.4   Angle-Ply Laminates

Most composite laminates include plies at angles other than 0° and 90°. For quasi-isotropy, ±45° plies are included in a balanced configuration. We saw in Chapter 5 that the micromechanics of failure of 45° involves transverse cracking, delamination, and debonding under the shear component of the stress. As shown in Figure 7.9, the introduction of 45° plies reduces the peak stress available but the rate of degradation from the formation of matrix cracks and associated damage is lower than the unidirectional (0°) and cross-ply (0°/90°) laminates.

Figure 7.9 also demonstrates that the variety of glass-fibre composites exhibit differing performances, which reflects the gradual deterioration of the load-bearing capability of the fibres [9,10]. These observations are consistent with the simple data presentation given by eq. (7.16).

7.3.5   Improving Fatigue Performance of Cross-Ply and Angle-Ply Laminates

It is apparent that controlling transverse cracking can offer a major improvement in fatigue performance. This can be achieved in three ways:

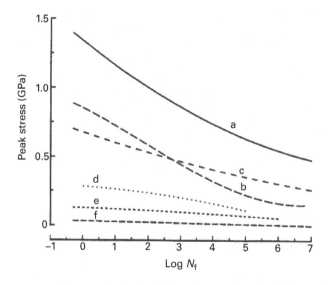

Figure 7.9 Stress–life curves for various GRP materials: 0° (a); [(0°/90°)₂]ₛ (b); [(±45°/0°₂)₂]ₛ (c); woven cloth laminate (d); CSM from polyester (e); DMC (f) [8].

1. Toughening of the matrix increases the first ply failure strain and hence delays the damage accumulation mechanisms. Epoxy resins often employ thermoplastic modification at a concentration that provides a co-continuous phase-separated morphology and enhanced fracture toughness.
2. Toughening of ply interfaces by incorporating polymeric particulates in this region of prepreg materials.
3. Reducing the off-axis ply thickness, as shown in Chapter 6, to increase the first ply failure strain by ensuring that the stored energy is insufficient for the transverse cracks to propagate. In this case the transverse cracks will propagate at an observable rate so that the damage accumulation process occurs over a longer period during the fatigue experiment.

7.4 Life Prediction

A number of techniques are employed to analyse the stress–life (or strain–life) curves. We provide a brief review of these methodologies.

7.4.1 Wear-Out

This methodology is illustrated in Figure 7.10. Defining the point of failure is difficult because there is not a critical flaw that causes fracture. We see that damage accumulates with a loss of stiffness and /or residual strength, which is referred to as 'wear-out'. Thus, we can define failure as the point at which the residual strength has fallen to the peak cyclic stress.

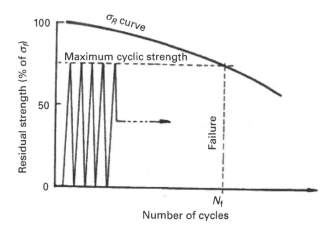

Figure 7.10 Degradation by 'wear-out' occurs until the residual strength declines to the level of the maximum in the cyclic stress, which is defined as failure.

7.4.2   Empirical Stress: Life Curves

For typical '*flat*' stress–life (*S–N*) curves typical of *unidirectional composites*, an empirical log life (log N_f) curve can be used for predictive purposes:

$$\sigma_{max} = \sigma_{cu} - B\log N_f, \qquad (7.16)$$

where σ_{max} is the peak tensile stress in the cycle; σ_{cu} is the monotonic tensile strength of the composite; *B* is a constant; and N_f is the number of cycles to failure.

This approach works reasonably well for many composites, but the presence of matrix cracking in unidirectional composites and transverse cracking in laminated materials containing off-axis plies leads to highly curved *S–N* plots.

Thus, *cross-ply laminates* and related materials require a different predictive equation, such as a power law in residual strength.

Figure 7.11 gives the peak stress during tensile fatigue loading on the residual strength of a 0°/90° glass-fibre epoxy laminate. Various cumulative damage models are proposed for the fatigue processes in glass-reinforced polymers. In Figure 7.12, a global power law is used to normalize the data into a single curve:

$$t^a + S^b = 1. \qquad (7.17)$$

S is a residual strength function, $(\sigma_R - \sigma_{max})/(\sigma_{cu} - \sigma_{max})$, where σ_R is the residual strength of the laminate after a given number of fatigue cycles at the peak stress of σ_{max}. σ_{cu} is the normal dynamic tensile strength obtained at the same loading rate as used in the fatigue experiment.

t is a function of the number of cycles, $(\log n - \alpha)/(\log N_f - \alpha)$, where *n* is the number of cycles sustained at the given maximum stress, σ_{max}, for which the expected fatigue life is N_f. $\alpha(= \log 0.5 = -0.3)$ accounts for the reason that normal strength corresponds to the lower limit cycles scale at 1/2 cycle [8,9,11].

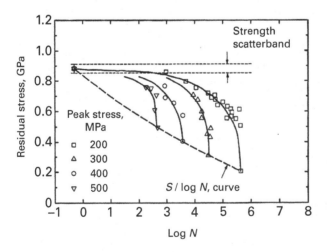

Figure 7.11 Changes in the residual strength during cycling of 0°/90°/0° glass-fibre epoxy at differing maximum stress levels, σ_{max}, with $R = 0.1$, at RH = 65%. $\sigma_{max} = 200$ (□), 300 (Δ),400 (O),500 (▽) MPa [9,11].

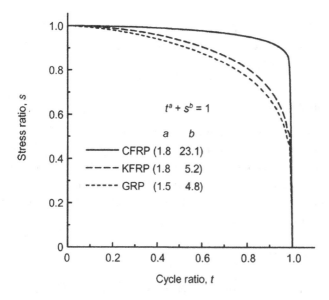

Figure 7.12 Normalized residual strength curves for 0°/90°/0° laminates prepared from HTS carbon-fibre, aramid (Kevlar 49), and E-glass fibres in a standard epoxy resin. The values of *a* and *b* in eq. (7.18) are also given for each curve [9].

The exponents, *a* and *b*, are material- and environment-dependent parameters usually obtained by curve fitting. Thus, $(\sigma_{cu} - \sigma_R)$ represents the loss in strength from the cycling of applied stress and is a damage function.

Figure 7.12 shows how this model has been used to provide the potential to predict the residual strength or the remaining life, once the material parameters *a* and *b* are known.

According to Harris [9], eq. (7.17) can be rearranged to provide a prediction for residual strength:

$$\sigma_R = (\sigma_{cu} - \sigma_{max})(1 - t^a)^{1/b} + \sigma_{max}. \tag{7.18}$$

Figure 7.12 shows how this analysis can be applied to differing composites, providing a curve fit with differing values of *a* and *b*. In this way, eq. (7.18) can be used to estimate the residual strength at any fatigue stress. For these materials, a single damage mechanism probably applies, namely transverse cracking. However, it is unlikely that a single micromechanical mechanism operates, so more complex models are required. Figure 7.5 shows that differing mechanisms operate at different strain levels.

7.4.3 Residual Strength Degradation

Residual strength degradation theories are dependent on a correlation with the damage state, but unfortunately most non-destructive techniques are not sufficiently sensitive to provide this, so empirical laws of this form are employed:

$$\sigma_R = \sigma_a[1 + f(N_f - 1)]^S, \tag{7.19}$$

where σ_R is the residual strength and f and s are constants. In some models f and s are functions of $R(= \sigma_{min}/\sigma_{max})$.

7.4.4   Stiffness Degradation

Alternative empirical models have examined the degradation of laminate stiffness that arises from matrix cracking, delamination, and eventually fibre fracture. The extent of damage growth can be monitored by a reduction in the 'modulus' of the laminate, referred to as stiffness. Failure occurs when the accumulated damage becomes critical. The rate of damage, D, and accumulation, dD/dN, with the number of cycles is given by Poursatip and Beaumont [12] as

$$\frac{dD}{dN} = A\left(\frac{1}{E_0}\frac{dE}{dN}\right), \tag{7.20}$$

where dD/dN is the rate of damage (D) accumulation of all types, dE/dN is the rate of stiffness ('modulus') degradation, E_o is the original stiffness, and A is a constant. Typically for $[+45°/90°/-45°/0°]_s$ carbon-fibre laminate, A was found experimentally to be 2.857 [12].

Integration gives an expression for the life of the laminate:

$$N_f = A\left(\frac{\sigma_a}{\sigma_{cu}}\right)^B\left[C\left(\frac{1-R}{1+R}\right)\right]^P\left[1 - \frac{\sigma_a}{(1-R)\sigma_{cu}}\right], \tag{7.21}$$

where $A = 3.108 \times 10^4$; $B = -6.393$, $C = 1.22$, $P = 1.6$, and $\sigma_{cu} = 586$ MPa. R is the stress or strain ratio.

7.4.5   Laminate-Related Mechanisms

In the previous discussion we assumed that damage arising from fatigue loading is associated with the micromechanics of composite fracture and involves matrix cracking and eventual fibre fracture in 0–5° plies. More complex theories are discussed by Chou et al. [13] and Reifsnider [14].

In real structures, edge effects play a big role. Thus, edges, holes, bonded joints, and ply drop-offs can act as stress concentrations that cause delamination. Impact damage can also promote delamination.

The strain energy release rate (G) has been used to study the effect of cyclic loading. The critical value, G_c, is reported to be 20% of the static value. Sendeckyi [15] has reported on edge delamination subjected to tensile loading, and found that

$$G_c = \frac{\varepsilon_c^2 t_k}{2}(E_c - E_c^*), \tag{7.22}$$

where ε_c is the applied composite strain, t_k is the laminate thickness, E_c is the original laminate stiffness, and E_c^* is the delaminated laminate stiffness.

ΔG has been related to delamination growth rate in fatigue in analogy to crack growth in metals in the form of ΔK. The extent of delamination will depend on the nature of the loading; for example, in-plane compression leads to rapid delamination growth. Delamination occurs when G exceeds ΔG_c:

$$\frac{da}{dN} = A(\Delta G)^n, \tag{7.23}$$

where A and n are constants.

7.4.6 Miner's Linear Cumulative Damage Rule

This approach was developed in 1945 [16] and was used extensively and successfully for predicting the fatigue of metals where the stress or strain ratio, R, is a significant variable. Since the nature of the failure process of a composite is more complex, Miner's rule can only be used qualitatively and can have significant errors. Miner's rule is:

$$\sum_{m}^{t=n} \frac{n_i}{N_f} = 1, \tag{7.24}$$

where n_i is the number of cycles to failure under a given stress range, N_f is the number of cycles to failure at that given stress level, and m is the number of stress levels.

A simple version for linear strength reduction is given by:

$$\sigma_R = \sigma_{cu} - (\sigma_{cu} - \sigma_{max}) \frac{N}{N_f}, \tag{7.25}$$

where $\frac{N}{N_f}$ is the fractional fatigue life at a peak stress of σ_{max}.

As already discussed, a linear strength reduction curve is unrealistic for the majority of composites. Empirical laws are strongly dependent on experimental data and their value as predictive tools over a range of conditions is limited.

7.4.7 Residual Strength and Damage Accumulation over a Range of R Values

Progressive damage of the nature described above occurs both in tension–tension fatigue and compression–compression fatigue. To understand the contributions of these stress regimes, Harris et al. [17] have described the use of constant-life or Goodman diagrams where the R ratio can be illustrated in normalized constant-life diagrams using the function

$$a = f(1 - m)^u (c + m)^v, \tag{7.26}$$

where $a = \sigma_{alt}/\sigma_t$, $m = \sigma_m/\sigma_t$, and $c = \sigma_{uc}/\sigma_{ut}$. σ_{alt} is the alternating component of stress given by half the stress range $[= (\sigma_{max} - \sigma_{min})]$, σ_m is the mean stress $[= (\sigma_{max} + \sigma_{min})]$, σ_{uc} is the monotonic compressive strength, and σ_{ut} is the monotonic tensile strength. f is the stress function, which is material-sensitive, and the values σ_{uc}/σ_{ut}, u, and v are functions of $\log N_f$.

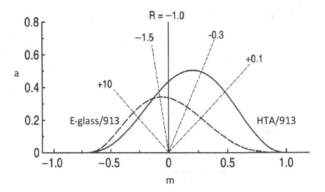

Figure 7.13 Constant-life plots for HTA/913 CFRP and E-glass/913 GRP $[(\pm45/0_2)_2]_s$ laminates with a life (N_f) of 10^5 cycles [18].

The extreme values of m for $a = 0$ are 1 on the tensile side and $-c$ on the compression side, so that the mean stress is $(1 + c)$. Figure 7.13 gives the curves at a constant life of $N_f = 10^5$ cycles for carbon-fibre- and glass-reinforced polymers (CFRP/GRP) manufactured from the same resin system (913 and with an identical lay-up). The curve for glass is more asymmetric than for carbon-fibre and has a maximum at a lower stress function (a).

The differences confirm that the glass-fibre composites exhibit more complex micromechanics because matrix cracking occurs in the unidirectional plies, as shown in Figure 7.5. This illustrates the benefits of constant-life diagrams.

7.5   Ceramic Matrix Composites

There are differences between PMCs and ceramic matrix composites (CMCs), which can be briefly summarized as follows:

1. *The matrix failure strain* is significantly lower than that of the fibres, so multiple matrix cracking dominates the failure process. Thus, composites technology is used principally to 'toughen' the ceramic.
2. *The fabrication process* involves either gaseous (chemical vapour deposition, or CVD) or liquid (slurry) impregnation, or a combination of both. However, high temperatures are required to consolidate the composite. Thus, the fibres need to resist the process temperatures and matrix environment. For example, Nicalon SiC fibres can have a reactive SiO_2 surface and are highly reactive with typical matrices. To protect the fibres during fabrication, graphite-coated fibres are often employed.
3. *Young's modulus* of the matrix will be similar to that of the fibres, so the conventional principles of reinforcement are generally not the explanation for the incorporation of fibres or fibrous fillers.

4. *Toughening of a ceramic* employs discontinuous fibres and a variety of micromechanical mechanisms will operate. Thus, matrix cracks will be pinned at fibre interfaces enabling an extended stress–deformation curve. Degrees of fibre pull-out will further contribute to the energy absorbed. Cycling of mechanical stress will mainly occur below the cracking strain, so this type of fatigue is of limited concern.

5. *Thermal fatigue* is potentially the most important. An example is carbon–carbon composites, where a CF textile preform is infiltrated with a carbon source. Commonly a mesophase pitch could be used for liquid infusion into the fibres, but the consolidation shrinkage leads to an imperfect impregnation, which needs to be followed by further infiltration using gaseous methods. Thus, CVD is used to complete the impregnation, usually in multiple stages.

   Alternatively, multiple CVD stages are used for many structures (mostly high-performance applications). Thermal fatigue of these structures is mainly dominated by 'chemical' degradation of the carbon matrix rather than micromechanical effects. Such structures should be designed to minimize microcracking, but if cracks do form then chemical degradation such as oxidation would be promoted.

6. *Models for the fatigue* of continuous fibre composites need to include the various mechanisms responsible for the degradation in strength. These follow the expected micromechanics of a brittle matrix composite:

   Unidirectional CMC:
   a. matrix cracking, which affects the fibre–matrix interface;
   b. either the crack propagates through a well-bonded fibre or initiates interfacial debonding through the shear stress that develops;
   c. interfacial debonding and shear cracks within the interphasal region, resulting from the presence of a fibre coating or fibre–matrix reaction.
   d. continuous cyclic loading causes a sliding stress at the fibre, provides a frictional mechanism of wear at the debonded interface.
   e. the release of thermal strains.
   f. environment gases such as oxygen can penetrate through the matrix cracks to the fibres, leading to additional interface and fibre corrosion mechanisms.

   Cross-ply and angle-ply CMC:
   a. Additional transverse cracking occurs at a lower stress.

These mechanisms have been reviewed in detail elsewhere [19–21]. Figure 7.14 illustrates the mechanisms operating at the fibre–matrix interface in materials containing matrix cracks.

7.6   Metal Matrix Composites

Metal matrix composites (MMCs) differ from PMC in that the ceramic fibres (or high-temperature wires such as tungsten or boron) will behave elastically even at high service temperatures. Under cyclic loads in the fatigue experiment, crack initiation

Table 7.1. Fatigue data showing improved endurance limit of MMCs [22]

Metal	Approximate composition	Reinforcement	Type	V_f (%)	Increase in endurance at 10^7 cycles (%)
Aluminium alloy	86Al/12Si/1Cu/1Ni	Alumina		20	30
Aluminium alloy 6061-T6	97Al/1Mg/0.6Si/0.2Cr/ 0.2Cu/Mn/Traces	Silicon carbide	Whisker	20	91
Magnesium alloy AZ91	90Mg/ 9Al/0.61Zn/0.18Mn	Alumina	Saffil fibre	16	46
Magnesium alloy AZ91	90Mg/ 9Al/0.61Zn/0.18Mn	Alumina	Saffil fibre	20	106

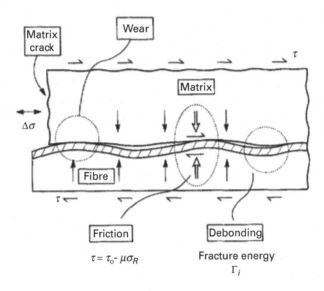

Figure 7.14 Schematic of the micromechanics operating during the fatigue of continuous fibre CMCs as included in the 'interface model' [19].

sites are often at internal defects such as poorly bonded clusters of reinforcement (particles or discontinuous fibres) or brittle intermetallic compounds. Despite these issues, the reinforcement of the metal can provide an improvement to fatigue resistance. Table 7.1 provides some examples of improved fatigue performance.

Full details of fatigue processes in MMCs are discussed by Johnson [23]. Particulate-reinforced MMCs are reviewed in detail by Ibrahim et al. [24]. As shown in Table 7.1, SiC whisker reinforcement is seen to be effective; however, concerns over toxicity akin to asbestos fibres has limited this usage.

More recently the use of carbon nanotube (CNT) fillers has become a practical option because of their high aspect ratio (length to diameter ratio), which provides a more appropriate reinforcement for metals. The use of CNTs in MMCs has been reviewed recently by Bakshi et al. [25].

7.7 Conclusions

In contrast to conventional materials, where crack propagation under fatigue is critical, composite materials fail progressively. The fatigue of a composite material involves complex mechanisms such as accumulation of fibre-breaks and/or matrix cracks, depending on the failure strains of the components. When these cracks impinge on the interface between the elements, such as fibres or plies, debonding or delamination add to the development of damage. The models available for predicting the fatigue life of a composite are reviewed here. A detailed description of these models is beyond the scope of this chapter and the reader is referred to the texts by Harris [26] and Reifsnider [27].

7.8 Discussion Points

1. Revise your knowledge of the development of damage in laminated composites and explain the phenomena identified in Figure 7.5.
2. Which model provides a reasonable prediction of the life of a unidirectional composite where fibre fracture is the major micromechanic?
3. How does the inclusion of angle plies affect the strain–life (S–N) plot?
4. Briefly describe the approaches employed to interpret the observed stress– or strain–life curves.
5. Describe the mechanisms operating in the mechanical fatigue of CMCs at different loads.
6. Is reinforcement beneficial to the performance of MMCs?

References

1. A. A. Griffith, The phenomena of rupture and flow in solids. *Phil. Trans. Roy. Soc. Lond. A* **221** (1920), 163.
2. A. Kelly and N. H. MacMillan, *Strong solids*, 3rd ed. (Oxford: Clarendon, 1990).
3. R. Talreja, Fatigue of composite materials: damage mechanisms and fatigue–life diagrams. *Proc. Roy. Soc. Lond. A* **378** (1981), 461–475.
4. P. E. Irving, Fatigue–life prediction: role of materials and structure. In *Handbook of polymer-fibre composites*, ed. F. R. Jones (Harlow: Longman, 1994), pp. 317–322.
5. P. T. Curtis, The fatigue of organic matrix composite materials. In *Advanced composites*, ed. I. K. Partridge (Amsterdam: Elsevier, 1989), pp. 331–368.
6. P. T. Curtis, Fatigue: carbon fibre composites. In *Handbook of polymer-fibre composites*, ed. F. R. Jones. (Harlow: Longman, 1994), pp. 305–309.
7. P. T. Curtis, *DSTL Technical Reports: TR82031, TR86021, TR87031* (London: Ministry of Defence, 1982–1987).
8. B. Harris, Fatigue: glass fibre reinforced plastics. In *Handbook of polymer-fibre composites*, ed. F. R. Jones. (Harlow: Longman, 1994), pp. 309–316.
9. B. Harris, *Engineering composite materials*, 2nd ed. (London: Institute of Materials, 1999).

10. J. F. Mandell, Fatigue behaviour of fibre–resin composites. In *Developments in reinforced plastics 2*, ed. G. Pritchard (London: Applied Science, 1982), pp. 67–108.

11. T. Adam, R. F. Dickson, C. J. Jones, H. Reiter, and B. Harris, A power law fatigue damage model for fibre-reinforced plastic laminates. *Proc. Inst. Mech. Eng. Sci. C* 200 (1986), 135–166.

12. A. Poursatip and P. W. R. Beaumont, Fatigue mechanics of a carbon fibre composite laminate: part 2. *Composites Sci. Tech.* 25 (1986), 283–299.

13. P. C. Chou, A. S. D. Wang, and H. Millar, *Cumulative damage models for advanced composites*. Air Force Wright Aeronautical Laboratory (AFWAL) Report TR-82-4083 (1982).

14. K. Reifsnider, Damage and damage mechanics. In *Fatigue of composite materials*, ed. K. L. Reifsnider (Amsterdam: Elsevier, 1991), pp. 1–77.

15. G. P. Sendeckyi, Life prediction for resin–matrix composite materials. In *Fatigue of composite materials*, ed. K. L. Reifsnider (Amsterdam: Elsevier, 1991), pp. 431–483.

16. M. A. Miner, Cumulative damage in fatigue. *J. Appl. Mech.* 12 (1945), A159–64.

17. B. Harris, N. Gathercole, J. A. Lee, H. Reiter, and T. Adam, Life-prediction for constant-stress fatigue in carbon-fibre composites. *Phil. Trans. Roy. Soc. London, A* 355 (1997), 1259–1294.

18. M. H. Beheshty, B. Harris, and T. Adam, An empirical fatigue–life model for high-performance fibre composites with and without impact damage. *Composites A* 30 (1999), 971–987.

19. A. G. Evans and F. W. Zok, The physics and mechanics of fibre-reinforced brittle matrix composites. *J. Mater. Sci.* 29 (1994), 3857–3896.

20. V. Birman and L. W. Byrd, Fracture and fatigue in ceramic matrix composites. *Appl. Mech. Rev.* 53 (2000), 147–174.

21. P. Reynard, Cyclic fatigue of ceramic-matrix composites at ambient and elevated temperatures. *Composite Sci. Tech.* 56 (1996), 809–814.

22. F. L. Matthews and R. D. Rawlings, *Composite materials: engineering science* (Oxford: Chapman and Hall, 1994).

23. W. S. Johnson, Fatigue of metal matrix composites. In *Fatigue of composite materials*, ed. K. L. Reifsnider (Amsterdam: Elsevier, 1991), pp. 199–229.

24. I. A. Ibrahim, F. A. Mohamed and E. J. Lavernia, Particulate reinforced metal matrix composites: a review. *J. Mater. Sci.* 26 (1991), 137–1156.

25. S. R. Bakshi, D. Lahiri, and A. Agarwal, Carbon nanotube reinforced metal matrix composites: a review. *Int. Mater. Rev.* 55 (2010), 41–64.

26. B. Harris, ed., *Fatigue in composites* (Cambridge: Woodhead, 2003).

27. K. L. Reifsnider, ed, *Fatigue of composite materials* (Amsterdam: Elsevier 1991).

8   Environmental Effects

Polymeric matrices absorb moisture, so here we examine how this affects the performance of a composite material. For an aerospace artefact, absorption and desorption is an important issue. For example, on the tarmac the relative humidity (RH) is high, whereas in flight the RH is low. Also, the ambient temperature can vary significantly, whereas the skin of a military aircraft may reach temperatures of 120 °C in flight. Therefore, we consider the effects of RH, temperature, and thermal excursions on moisture absorption and how they influence the micromechanics. Initially we can assume that the fibres are insensitive to water, which is realistic for most common reinforcements apart from aramid fibres.

8.1   Moisture Absorption

In humid environments, moisture diffuses into polymer matrix composites (PMCs) to differing degrees. The extent of moisture absorption depends on the polarity of the polymer molecules. Absorbed water reduces the glass transition temperature, T_g, and hence the service temperature of a composite. In addition, hydrolysis can degrade the thermomechanical properties further.

Unidirectional composites are highly anisotropic, so we need to recognize that transverse properties are more influenced by the matrix and are therefore more sensitive to moisture absorption.

The thermodynamic properties of the resin and environment will determine the maximum extent of moisture absorption, whereas diffusion is a kinetic property that is temperature-dependent. Normally, the latter is considered to be Fickian in nature [1–3]. However, highly polar polymers require more appropriate diffusion models [4,5]. For most structural composite materials, Fickian laws are applicable.

8.1.1   Fickian Diffusion

Fickian laws were developed for thermal equilibration, but thermal diffusion is significantly faster than moisture diffusion by a factor of $\approx 10^6$. For a 12-mm thick composite it can take 13 years for moisture to reach equilibrium at 350 K, but only 15 s for thermal equilibration. Thus, a typical aerospace structure will be in service for many years before it becomes saturated with water.

Fick's second law provides the analysis for one-dimensional diffusion through thickness, which is required for the prediction of moisture content of a structure:

$$\frac{dc}{dt} = D\frac{d^2c}{dx^2},$$ (8.1)

where D is the diffusion coefficient, c is the concentration, t is time, and x is the length scale.

In terms of moisture content, M, eq. (8.1) becomes

$$\frac{dM}{dt} = D\frac{d^2c}{dx^2},$$ (8.2)

where M is the concentration of moisture in the composite at distance x. The distribution of water throughout the composite will arise from an averaging of a series of moisture profiles over time.

Equation (8.2) can be solved by a finite difference technique to give

$$D = \frac{\pi d^2}{16M_\infty^2}\left(\frac{M_2}{\sqrt{t_2}} - \frac{M_1}{\sqrt{t_2}}\right)^2,$$ (8.3)

where d is the thickness of the specimen. M_1 and M_2 are the measured moisture contents at times t_1 and t_2. Usually, M is determined by measurement of weight at specified times. M_∞ is the equilibrium moisture content or maximum weight gain.

Figure 8.1 shows how the experimental measurements are analysed. Equation (8.3) requires the maximum moisture concentration (M_∞) to be determined.

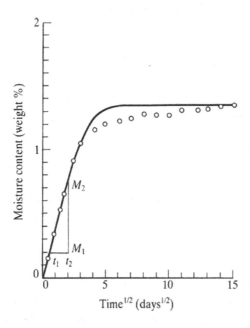

Figure 8.1 Comparison of the Fickian predicted (solid line) and measured (o) moisture diffusion concentrations in a carbon-fibre composite [6].

The experimental measurement of D uses coupons in which diffusion through all the faces occurs, so that an edge correction is required for the calculation of a true one-dimensional diffusion constant. For the prediction of the time required for moisture equilibration of a structure, the diffusion constant, D, for each material is required, but their measurement is time-consuming because the equilibrium moisture content (M_∞) has to be determined in long-term tests. Therefore, accelerated tests are used. As with other kinetic processes, D is related to temperature (T) according to the Arrhenius equation:

$$D = A \exp\left(-\frac{E_a}{RT}\right), \tag{8.4}$$

where E_a is the activation energy for moisture transport and A is the pre-exponential factor.

Composite materials are typically anisotropic, so the diffusion constants in the longitudinal and transverse directions can differ. These are related to each other through

$$D = D_x\left[1 + \frac{d}{l}\left(\frac{D_y}{D_x}\right)^{1/2} + \frac{d}{b}\left(\frac{D_z}{D_x}\right)^{1/2}\right]^2, \tag{8.5}$$

where D_x, D_y, and D_z are diffusion constants through the thickness, d, along the length, l, and across the breadth, b.

For a unidirectionally reinforced $0°$ ply, D_z and D_x are equivalent, both occurring at $90°$ to the fibres. Therefore, eq. (8.5) becomes

$$D = D_x\left[1 + \frac{d}{b} + \frac{d}{l}\left(\frac{D_y}{D_z}\right)^{1/2}\right]^2. \tag{8.6}$$

For perfectly bonded fibres, D_x will be significantly larger than D_y because of the relatively higher surface area of resin in the $90°$ direction. Typical values of the diffusion constant are 10^{-6} mm^2 s^{-1} for a resin and 10^{-7} mm^2 s^{-1} for a composite. Thus, a correction factor for D is given according to Shen and Springer [7], which defines the one-dimensional coefficient, D_∞:

$$D_\infty = D\left[1 + \left(\frac{d}{b}\right) + \frac{d}{l}\right]^{-2}. \tag{8.7}$$

It is possible to measure the influence of the fibres on the diffusion because, with a poor interfacial bond between the fibre and the resin, rapid transport will take place at the interface and can be differentiated from the resin-dominated diffusion at $90°$ to the fibres. If D_y is greater than D_x, then capillary diffusion at a poor interfacial bond must be occurring. Therefore, a plot of eq. (8.6), rearranged as eq. (8.8), enables D_x and D_y to be estimated from the diffusion constants for coupons of differing geometries. In this way, the degradation of the interface under aggressive environments in accelerated tests can be identified:

$$D^{\frac{1}{2}} = D\left(1 + \frac{d}{b}\right)D_x^{\frac{1}{2}} + \left(\frac{d}{l}\right)D_y^{\frac{1}{2}}. \tag{8.8}$$

8.1.2   Prediction of Moisture Content and Time Dependence

Figure 8.1 compares the Fickian prediction of moisture content with time of a typical carbon-fibre epoxy composite, which illustrates the use of eq. (8.2). It is seen that the Fickian analysis does not precisely predict water absorption into a composite. The main reason for this is that the diffusion model does not apply to the resin. Many resin systems exhibit Fickian diffusion, whereas others show non-Fickian behaviour. This usually occurs in highly polar resins in which moisture absorption is comparatively high, leading to significant swelling, which perturbs the diffusion mechanism.

In the case of matrix resins such as epoxies, the curing process may lead to a non-equilibrium resin structure. As a result, relaxation of the network structure occurs in the presence of absorbed moisture. This is confirmed by the observation that after desorption, reabsorption mostly exhibits Fickian behaviour [8].

M_∞ is needed for the calculation of D from eq. (8.3), and as shown in Figure 8.1 this is experimentally time-consuming. As a result, there have been several reports on accelerating moisture absorption process. This mostly involves higher conditioning temperatures and high RH atmospheres. Preconditioning is therefore often used [9] to speed up the determination. Alternatively, Ellis and Found [10] reported that after M_∞ had been measured, M_t/M_∞ can be used in conjunction with the diffusion half-life in an attempt to speed up the measurements on specimens of differing geometry.

8.1.3   Moisture Distribution in a Laminate

Figure 8.2 shows the through-thickness moisture profile in a carbon-fibre laminate. We see that the exact Fickian solution does not predict precisely the distribution of the moisture through the thickness of the laminate. However, Fickian analysis is adequate for most practical purposes.

Figure 8.2 provides a description of a one-sided diffusion profile. Note that diffusion from two opposite sides of a coupon will exhibit a symmetrical distribution, as indicated by adding a mirror image of Figure 8.2. Furthermore, a material with such a moisture distribution will be sensitive to a thermal event since, assuming no loss through evaporation, moisture will diffuse into the centre of the material or structure to equilibrate the moisture uniformly throughout the thickness.

8.2   Moisture Sensitivity of Matrix Resins

Moisture diffuses into polymers to differing degrees, which is a function of the morphology and molecular structure. The following aspects are critical:

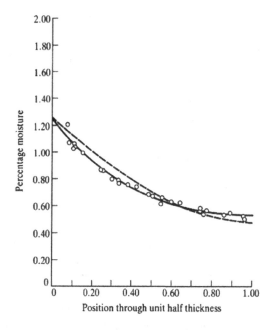

Figure 8.2 Through-thickness distribution of moisture in a carbon-fibre composite. Measured values (o) with best fit (continuous line), and Fickian prediction from a finite difference solution of Fickian second law (broken line). [7]

1. the polarity of the molecular structure;
2. the degree of crosslinking;
3. the degree of crystallinity in the case of a thermoplastic matrix; and
4. the presence of residuals in the material.

The residuals mentioned in item 4 are unreacted hardeners and other impurities which influence moisture absorption. For example, the sodium and potassium content of glass fibres has been steadily reduced over recent years because they were widely acknowledged to be responsible for the water sensitivity of glass-fibre composites in aqueous environments.

An important aspect of a composite material is that the matrix resin is often formed at the same time as the component. Therefore, the precise chemistry of the resin can vary. Table 8.1 gives typical values of the diffusion constant for a series of thermosetting matrices in which the effect of polarity of the resin on the equilibrium moisture concentration is illustrated.

8.2.1 Epoxy Resins

Epoxy resins are the network products of the reaction of a multifunctional epoxide monomer with a hardener. The final structure can be uncertain, and is a function of the nature of the epoxide, the hardener, and/or the catalyst, which determine the curing mechanism and hence the chemical structure with differing polarity. For example, an

Table 8.1. Typical moisture diffusion coefficients and maximum moisture concentrations of aerospace epoxy matrix resins [11,12]

Resin	Type	RH (%)	Temperature (°C)	M_∞ (%)	Diffusion const. (D)/ 10^{-7} mm^2 s^{-1}
Epoxy: general purpose	DGEBA	96	50	6.9	2.8
Epoxy: novolak	DEN 431	11	50	0.48	5.99
Epoxy: aerospace	Narmco 5245	96	45	1.96	11.0
		75	45	1.38	14.7
		46	45	0.8	19.4
		31	45	0.54	17.2
Epoxy: cyanate ester	Primaset PT30/DEN 431	96	50	2.96	3.62
		75	50	2.19	4.85
		46	50	1.44	6.34
		31	50	1.04	6.12
		16.5	50	0.70	5.46
		11	50	0.48	5.99

anhydride hardener (with tertiary amine accelerator) will produce a 'polyester'-type structure of low polarity, whereas an amine hardener will form a β-hydroxyl amine network of high polarity. Hydrogen bonding of water molecules at polar sites provides a mechanism for increasing the concentration of moisture, which can be absorbed. Catalytic curing involving chain polymerization through the epoxy ring yields a polyether chain structure of low polarity. Since mixed curing systems are often used for composite matrices, their polarity will be strongly dependent on the degree of incorporation of the individual components and the chemical mechanisms employed. In Table 8.1 it is clearly shown how the choice of resin and curing mechanism strongly influences the moisture sensitivity.

8.2.2   Advanced Matrix Resins

Table 8.2 gives typical values for M_∞ of composite materials from a variety of matrices containing thermoplastic and thermosetting modifiers. Thermoplastic matrices such as polyether ether ketone (PEEK) absorb much less moisture than the advanced epoxies. PEEK is an example of a partially crystalline linear polymer with very low moisture absorption. Polyether sulphone (PES) behaves similarly since addition of ≈20% to the matrix is shown to reduce the M_∞ of the composite proportionally. Thermoplastics, which are relatively non-polar and absorb lower concentrations of moisture, are also used as 'flow control' additives and/or toughening agents in advanced epoxy resins, and help to provide a reduction in moisture sensitivity.

In recent times, commodity thermoplastics have become significant contenders as matrices for glass-fibre composites where environmental resistance is sought; for example, polypropylene is now available in a prepreg form for fusion bonding. Since it has low polarity and is also partially crystalline, moisture absorption is very low.

Table 8.2. Moisture contents of carbon-fibre composites from thermoset (TS) and thermoplastic (TP) matrices, showing the benefit of modification [8]

Resin	Type	Designation	Modifier	RH (%)	Temperature (°C)	M_∞ (%)
Epoxy	TS	Fibredux 924E[a]	None	96	50	2.44
Epoxy	TS	924C	PES	96	50	1.72
Epoxy	TS	927C	Cyanate ester/ polyimide	96	50	0.98
Epoxy	TS	Narmco Rigidite 5255C	Bismaleimide	96	50	0.82
Bismaleimide	TS	PMR 15	None	96	50	0.32
PEEK	TP	APC 2	None	50	23 (350 h)	0.02
PEEK	TP	APC 2	None	100 (immersion)	100 (360 h)	0.23

[a] Calculated from data for cast resin (6.9%).

8.3 Matrix Glass Transition Temperature

When a polymer that contains a distribution of long-chain molecules is cooled from its liquid state, the molecules have insufficient time to organize into a regular arrangement in the form of a crystal. Therefore, the rate of cooling will determine the temperature at which the material freezes into a glass, which is a non-equilibrium state. As a result, the glass transition temperature, T_g, is a function of the molecular structure, as well as its thermal history. Many thermoplastics have a regular molecular structure but only exhibit partial crystallinity within the amorphous glass regions. Most thermosetting matrix resins exhibit a glassy structure. The value of T_g, determines the maximum service temperature of a composite. However, its magnitude will depend on the thermal history associated with the manufacturing process. Figure 8.3 shows how the specific volume of the polymer changes with cooling rate. With 100% crystallinity the close-packed molecules dictate the volume of the material and maximum density. Hence, the additional volume associated with a glass is referred to as *free volume*. Free volume is the unoccupied volume associated with that required for segments to cross over each other in the array of molecular random coils. Therefore, there will be a potential for relaxation of the glassy structure over time with a reduction in T_g, known as *ageing* [13].

Thus, water molecules can diffuse into a glassy polymer and occupy the free volume, but with a polar interaction the chains will be pushed apart, leading to an increase in volume (referred to as swelling).

Kelley and Bueche [14] used this concept to provide the model for predicting the change in glass transition temperature with a diluent, in this case water absorption:

$$T_g = \frac{\Delta\alpha_p V_p T_{g_p} + \Delta\alpha_d V_d T_{g_d}}{\Delta\alpha_p V_p + \Delta\alpha_d V_d},$$
(8.9)

where $\Delta\alpha$ is the change in linear expansion coefficient at the glass–rubber transition and V is the volume fraction of polymer (p) and diluent (d). T_{g_p} and T_{g_d} are the glass transition temperatures of the polymer and diluent (water).

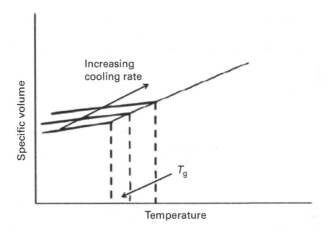

Figure 8.3 The effect of cooling rate on the glass transition temperature and density of a polymer.

While this equation is commonly used to predict the effect of absorbed water on T_g of the matrix resin, the empirical Fox equation [15] has also been used:

$$\frac{1}{T_g} = \frac{w_p}{T_{g_p}} + \frac{w_d}{T_{g_d}},$$ (8.10)

where w_p and w_d are weight fractions of polymer and diluent (in this case water).

Porter [16] developed the group interaction model (GIM), which provides for the role of absorbed water and its interaction with the polymer through direct incorporation of polar components and recognizing changes in the degrees of freedom of the system [17]. The Lennard–Jones potential describes the decrease in energy resulting from attraction forces as molecules, or in the case of a polymer the segments, are brought together. At a separation distance r_o, the segments repel each other and the energy increases. The shape of the Lennard–Jones potential is described by eq. (8.11):

$$\phi = \phi_o \left[\left(\frac{r_o}{r}\right)^{12} - 2\left(\frac{r_o}{r}\right)^{6} \right],$$ (8.11)

where ϕ is the interaction energy, $-\phi_o$ is the depth of the potential energy-well at a separation distance r_o, which is the equilibrium position of a pair of interacting molecules or segments at absolute zero. ϕ_o is therefore the energy of interaction for the molecular conformation of lowest energy.

The interaction energy, ϕ, has a number of components:

$$\phi = -\phi_o + H_c + H_T + H_M,$$ (8.12)

where H_c is the configuration energy, H_T is the thermal energy, and H_M is the mechanical energy. ϕ_o represents the energy that keeps the molecules together, and against which mechanical and thermal energies act. Therefore, ϕ_o is related to E_{coh}, the cohesive energy, which can be estimated from a measurement of the solubility parameter (δ_o):

$$\delta_o = \left(\frac{E_{coh}}{V_o}\right)^{1/2},$$ (8.13)

where V_o is the molar volume of a 'mer' unit in the polymer chain.

The GIM has a hexagonal cell of central polymer molecules surrounded by six equidistant polymer chains, as shown in Figure 8.4. Since the energy of interaction between near neighbours, ϕ, involves two 'mer' units, the total energy of interaction is 3ϕ. Therefore, eq. (8.11) can be rewritten in terms of cohesive energy and volume, as in eq. (8.14), plotted in Figure 8.5:

$$E = E_{coh}\left[\left(\frac{V_o}{V}\right)^6 - 2\left(\frac{V_o}{V}\right)^3\right],$$ (8.14)

$$E = -0.89E_{coh} + H_T.$$ (8.15)

r

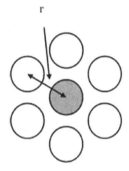

Figure 8.4 The basis of the GIM for prediction of polymer properties [16].

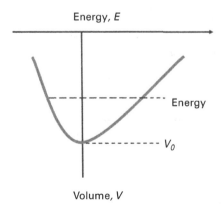

Figure 8.5 Plot of interaction energy, E, versus volume, V, according to eq. (8.14). The influence of thermal energy is also included.

In a polymer, the chain length is significantly larger than r and is therefore assumed to be invariant. E_{coh} refers to the zero-point cohesive energy, and V_o is the zero-point volume.

The potential function shown in eq. (8.14) is used as the basis for the prediction of the thermomechanical properties of a polymer as a function of temperature, strain, strain rate, and pressure. Since the change from glassy to rubbery properties occurs at T_g, when the attractive forces between adjacent polymers are zero and large-scale translational motion becomes possible, a predictive equation for T_g can be derived:

$$T_g = 0.224\theta_1 + 0.0513\frac{E_{coh}}{N},\qquad(8.16)$$

where N is the degree of freedom and θ_1 is the Debye temperature.

The Debye temperature is 550 K for phenyl-containing polymers (as with most composite matrices), while E_{coh} is a balance between the interaction energy and thermal energy, which acts against the attraction forces. The latter is governed by the degrees of freedom, N, in the molecular structure. Thus, for a high T_g, E_{coh} should be increased while ensuring N remains unchanged.

E_{coh} and N can be obtained by examination of the basic components of the molecular structure of the polymer. Table 8.3 gives the parameters used in the GIM model, while Table 8.4 shows how these quantities can be obtained from examination of the molecular structure of the resin matrix. Table 8.5 shows how crosslinking during curing influences the degrees of freedom of typical components of an aerospace composite epoxy resin. As can be seen, a composite matrix resin often contains mixed hardeners and epoxy monomers.

Equation (8.16) can be written in a more detailed form to calculate the T_g of a thermosetting epoxy resin using a blend of hardeners:

$$T_g = 0.224\theta_1 + 0.0513\left[\frac{E_{coh(mon)} + x_{st(H1)}E_{coh(H1)} + x_{st(H2)}E_{coh(H2)} - \Delta E_{et} - w}{(N_{mon} - fN_L) + x_{st(H1)}N_{H1} + x_{st(H2)}N_{H2}}\right],$$
$$(8.17)$$

where x_{st} are stoichiometric ratios of the hardeners, H1 and H2. The subscripts 'mon' refer to the monomeric resin component, H1 and H2 to hardeners 1 and 2, and et to ether links formed from secondary reactions. ω is the loss in cohesive energy per epoxy group during the formation of a crosslink [17,18].

Table 8.3. Parameters used in the GIM [16,17].

Parameter	Definition	Units
M	Molecular weight of 'mer' unit	g mol^{-1}
V_w	van der Waals volume of 'mer' unit	cm^3 mol^{-1} or m^3/(mer unit)
E_{coh}	Cohesive energy	J mol^{-1}
N	Degrees of freedom: skeletal modes of vibration of the polymer	(mer unit^{-1})
θ_1	Reference temperature of skeletal modes	K
L	Length of a 'mer' unit in the chain axis	nm

Table 8.4. Group contributions of the molecular elements in a polymeric matrix employed in the GIM model [17]

Group: balance	M	V_w (cc mol^{-1})	E_{coh}(J mol^{-1})	N
–CH$_n$–	14	10.23	4,500	2
–O–	16	5	6,300	2
–CO	28	11.7	17,500	2
–CO–O–(ester)	44	17	20,000	4
–O–CO–O–	60	22	19,500	6
–CH–(OH)–	30	14.8	(17,500)	(4)
–CO–NH–(amide)	43	19	40,000	4
–S–	32	10.8	8,800	2
	76	43.3	25,000	3 or 5
	76	43.3	25,000	4 or 6
	126	(65)	47,000	(6)
	90	54.5	29,500	4
	104	65.5	34,000	4
	194	119	63,500	8

Table 8.6 gives the experimental (exp.) and prediction (pred.) of T_g for the cured epoxy resins based on the components in Table 8.5.

Foreman and coworkers [19–21] have extended the model to the prediction of the mechanical properties of epoxy resins and composites. This is beyond the scope of this chapter, since we concentrate here on the effects of moisture absorption.

8.3.1 Effect of Water Absorption on T_g

Moisture diffuses into epoxy resins, causing a reduction in T_g. Typically, T_g is reduced by $\approx$–20 K for each 1% absorbed water present in the system [22]. It is important to have further understanding so that the durability of the mechanical properties can be

Table 8.5. Calculated values of degrees of freedom (GIM) of base epoxy resins and hardeners with respect to their extent of reaction [17].

Chemical name	Chemical formula	Degrees of freedom (N)		
		Unreacted N_{un}	Single crosslink trifunctional N_f	Two crosslinks tetrafunctional
Diamino diphenyl sulphone (DDS)	$H_2N-\bigcirc-\overset{\overset{O}{\|\|}}{\underset{\underset{O}{\|\|}}{S}}-\bigcirc-NH_2$	18	9	6
Dicyandiamide (DICY)	$H_2N-C=N-C\equiv N$ $\quad\quad\quad \|$ $\quad\quad\quad NH_2$	9	4	3
Tetrafunctional glycidyl amine (TGDDM or M720)	(tetrafunctional glycidyl amine structure)	46	28	22
Trifunctional glycidyl amine (TGAP or MY0510)	(trifunctional glycidyl amine structure)	34	14	–

Table 8.6. GIM predictions of the T_g for a range of base epoxy resins in comparison to the prediction of a commercial polymer blend [17]

Epoxy resin	Functionality	Hardener	Concentration (%)	T_g (exp.) (°C)	T_g (pred.) (°C)
MY0510	3	DDS	36	285	283
MY0510	3	DDS	45	276	268
MY0510	3	DICY	19	218	213
MY721	4	DDS	26	288	281
MY721	4	DICY	14	258	249
Epoxy 924	3.5	DDS/DICY	22/8	234	241

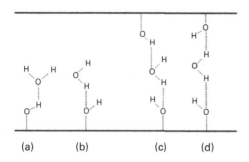

Figure 8.6 Hydrogen bonding of absorbed water and polymer backbone through hydroxyl groups (and also amino groups, not shown). (a) Singly bound water as acceptor; (b) singly bound water as donor; (c) doubly bound water as donor and acceptor; (d) doubly bound water as donor [8].

identified. Water adds significantly to the degrees of freedom of the system, but only causes a modest increase in the cohesive energy through hydrogen bonding.

Moisture absorption can be modelled by considering that the water molecule is a penetrant adding cohesive energy and degrees of freedom. Thus, eq. (8.17) becomes

$$T_g = 0.2240\theta_1$$
$$+ 0.0513 \left[\frac{E_{coh(mon)} + x_{st(H1)}E_{coh(H1)} + x_{st(H2)}E_{coh(H2)} - \Delta E_{et} - w + n_{H_2O}E_{coh(H_2O)}}{(N_{mon} - fN_L) + x_{st(H1)}N_{H1} + x_{st(H2)}N_{H2} + n_{H_2O}N_{H_2O}} \right],$$

(8.18)

where n_{H_2O} is the number of moles of water hydrogen-bonded to the repeat unit of the polymer (from the weight fraction). $E_{coh(H_2O)}$ is the cohesive energy of water (60 kJ mol^{-1}) and N_{H_2O} is the number of degrees of freedom (18).

The absorbed water molecules can be hydrogen-bonded to the polymer through hydroxyl or amine groups, either *singly* or *doubly*, as shown in Figure 8.6.

For the singly bonded water molecules, E_{coh} increases while N_{H_2O} decreases; thus, as shown by eq. (8.18), T_g is less affected. For doubly bonded water molecules the number of degrees of freedom is increased, leading to a further decrease in T_g. This mechanism can explain the formation of multiple relaxation peaks in the dynamic thermal analysis (DMTA) spectrum. Clustering of water molecules can occur.

8.3.2 Location of Water Molecules

Figure 8.7 shows an example of the cured structure of a thermoset resin using two different epoxide monomers with two different amine hardeners. The epoxies are triglycidyl p-aminophenol (TGAP) and tetraglycidyl 4,4-diaminodiphenylmethane (TGDDM), which are tri- and tetrafunctional, respectively ($n = 3$ and $n = 4$, respectively); the two hardeners are 4,4-DDS and dicyandiamide (DICY).

We see in Figure 8.7 that a cured structural epoxy resin can have various sites for hydrogen bonding of water molecules. Thus, the absorption of water can occur at a

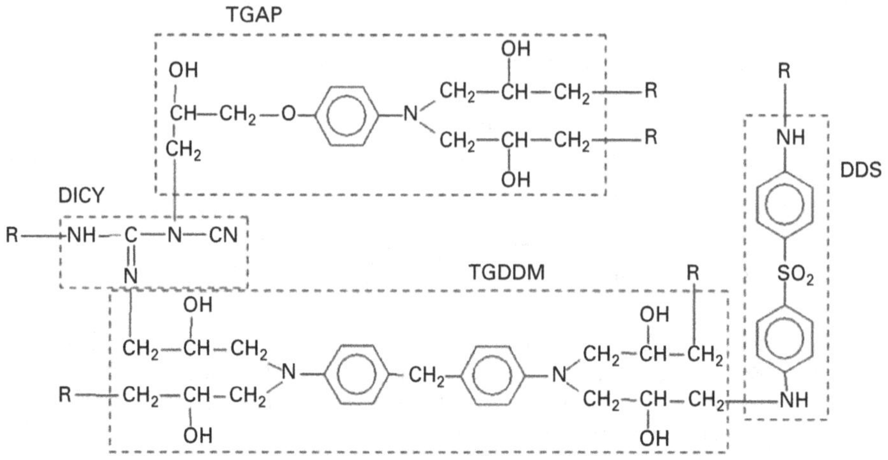

Figure 8.7 Schematic of the cured structure of an aerospace epoxy resin based on mixed hardeners DDS and DICY and mixed resins TGAP and TGDDM [8].

range of sites within the matrix polymer. This is often reflected in the number of peaks in the DMTA spectrum of the wet resin [23].

We saw in Figure 8.3 that T_g of a polymer is a function of the cooling rate. On rapid cooling, the glass forms at a higher temperature with a larger volume and lower density. Thus, the volume occupied by the molecules is a higher fraction of the material than for a glass formed slowly, and the unoccupied volume consists of so-called nano-voids and free volume. The latter is of molecular dimensions arising from the space required for segments within a random coil to cross over each other.

In a thermoset matrix resin, crosslinking contributes to the point at which the glass forms, and therefore the magnitude of the unoccupied volume, which is in the form of nano-voids, will be higher. An as-cured resin will initially be in a non-equilibrium state and have the potential to relax under a thermal stimulus. Relaxation of the resin structure can occur as a result of the presence of dissolved water, specifically on cooling from elevated temperatures.

Water molecules can therefore be located in the free volume or nano-voids. In the latter, clusters of water molecules can form without a significant impact, whereas water molecules within the free volume will be hydrogen-bonded and will lead to swelling. During a thermal event the water is redistributed as the network relaxes. One consequence is the effect of *thermal spiking* on moisture absorption.

8.3.3   Thermomechanical Response

From the discussion above we can understand the observed development of multiple peaks in the dynamic mechanical spectrum. Figure 8.8 shows the effect of absorbed water on the DMTA of a base epoxy resin used as a matrix. This is the epoxy blend used as the basis of Fibredux 924. The commercial system is normally toughened

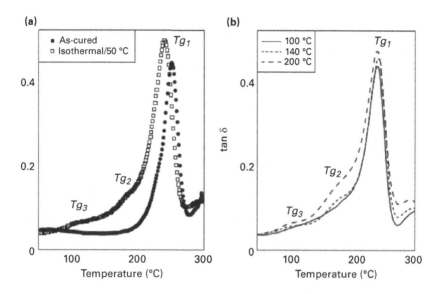

Figure 8.8 Tan δ spectra (DMTA) of a base epoxy resin blend (Fibredux 924E) showing the reduction in T_g and the development of secondary peaks (T_{g2}, T_{g3}) on water absorption at 50 °C and 96% RH (a) isothermally (b) after regular thermal excursions (thermal spiking) to 100 °C, 140 °C, and 200 °C [8,24].

with a thermoplastic. The main peak is associated with the translational ability of the whole chain, or network chain in the case of a thermoset resin, which defines T_g. Also shown is the development of secondary peaks associated with cooperative segmental crankshaft motions within the glass in the presence of absorbed water, assigned as T_{g2} and T_{g3}.

Examination of Figure 8.8 shows that after water absorption relaxation events associated with T_{g3} (and T_{g2}) will lead to a mechanism of 'yield' at $\approx$100 °C and a measurable fall in the storage modulus after isothermal absorption of 7% water. Table 8.7 shows that the modulus of the glass, E_g' falls on moisture absorption.

Conventionally we define T_g as the maximum in the tanδ, but composites engineers often use the temperature at which the storage modulus falls, $T_{g(E')}$, because it defines the onset of stiffness reduction. This is a measure of the heat distortion temperature and represents the maximum service temperature. Table 8.7 shows that T_g and $T_{g(E')}$ can differ by 25 °C, but water absorption increases the difference to 80 °C. Also, the breadth of the transition increases in the presence of absorbed water by the development of secondary transitions. We also see in Table 8.7 that the glassy storage moduli, E_g', in the presence of water is constant within experimental error. E_g' is less sensitive to network degradation, whereas the modulus in the rubbery region (i.e. above T_g), E_r' is strongly dependent on the crosslink density. If there is scission of crosslinks E_r' will be reduced. E_r' measured in flexure by DMTA is not precise, so the trend in values in Table 8.7 can be interpreted as showing that the epoxy resin is *not* hydrolytically degraded.

Table 8.7. The effect of water absorption at 96% RH on $T_{g(E')}$ and storage moduli, E' (DMTA), of a base epoxy resin (Fibredux 924E). E_g' is the value for the glass and E_r' (above T_g) for the rubbery region [24]

Condition	Temperature (°C)	M_∞ (wt%)	$T_{g\,(E')}$ (°C)	T_{g1} (°C)	T_{g2} (°C)	E_g' (GPa)	E_r' (MPa)
Dry/as-cured	–	0	225	251	–	1.9	50
Wet/isothermal	50	6.95	160	240	189	1.3	33
Wet/thermally spiked	100	7.57	163	240	189	1.4	43
	120	9.2	149	238	183	1.3	18
	140	9.66	150	241	187	1.3	31
	160	9.06	164	244	186	1.2	32
	180	8.33	143	240	188	1.3	46
	200	7.27	153	239	186	1.4	45

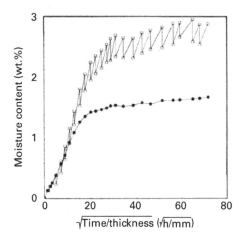

Figure 8.9 Typical moisture absorption curves for Fibredux 924C unidirectional laminate in a 96% RH environment at 50 °C: ● Isothermal, ○ with thermal spiking at 140 °C, where the individual points represent the individual weights immediately before and after a thermal spike [8,24,25].

8.3.4   Thermal Spiking

Aerospace composite structures are often subjected to cycles of RH and temperature in use. For example, military aircraft are likely to be exposed to high RH for days out of service followed by supersonic flight, where RH is low and the aircraft skin temperature can reach 120 °C. Experimental simulation has showed that moisture absorption is enhanced by brief excursions to elevated temperatures. This is referred to as thermal spiking.

Thermal spikes as short as one minute can cause a significant increase in moisture content. Figure 8.9 provides a typical moisture absorption curve for a carbon-fibre composite measured in the presence of a series of thermal spikes.

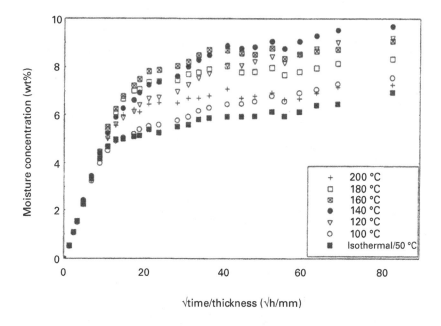

Figure 8.10 Moisture absorption curves for the base epoxy blend for the commercial Fiberdux 924 resin system conditioned over 5,000 hours at 96% RH at 50 °C and subjected to 22 short thermal cycles (spikes) to elevated temperatures of 100–200 °C [24].

Figure 8.10 shows how the spike temperature influences moisture absorption. Close inspection reveals that above 4 wt%, the water isothermal absorption diffusion curve deviates from Fickian behaviour and continues to absorb moisture. It is noticeable that with a thermal spike the absorbed moisture concentration is increased. The *maximum absorbed moisture* occurs with thermal spikes to 140 °C. At higher spike temperatures the moisture content decreases. The clue to the phenomenon comes from examining Table 8.7, where the next higher spike temperature of 160 °C reaches $T_{g(E')}$ and relaxation of the network can occur. The non-equilibrium state of the resin will then move closer to equilibrium, with the loss of unoccupied volume reducing the concentration of nano-voids. Since T_{g1} and $T_{g(E')}$ of the wet resin are effectively constant while M_{∞} increased from 6.95 to 9.66 wt%, we can conclude that water is present as clusters located in the nano-voids. The clustering of water molecules in cyanate ester resins as a function of RH is discussed elsewhere [26].

Epoxy resins used as matrices need to be modified with thermoplastic or a linear polymer, in this case PES, which acts as a toughening agent. The effect of this additive is explored in Figure 8.11, where the maximum moisture concentrations of the toughened and reinforced composite and base epoxy resin are compared. The addition of the PES toughening agent reduces the maximum moisture content that is absorbed at 96% RH. Furthermore, the addition of carbon fibres reduces moisture content proportionally to the reduced volume fraction of polymer. The so-called thermal spiking effect is clearly a function of the resin, in this case the epoxy. The

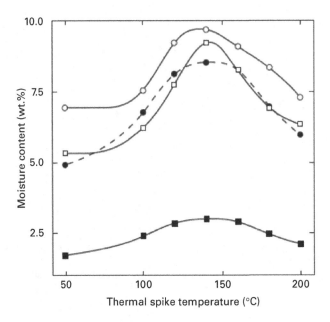

Figure 8.11 Enhanced moisture concentrations in Fibredux 924 matrix resins and carbon-fibre composites after 22 thermal spikes at 96% RH at 50 °C: ○ base epoxy (924E), ☐ thermoplastic toughened epoxy (924T), ● calculated values for the matrix from the composite, ■ carbon-fibre composite (924C) [7,24,25].

thermoplastic is less polar and also reduces M_∞. Other modifiers shown in Table 8.2 also reduce the maximum moisture content during thermal spiking [8,25,27].

8.4   Differentiating Plasticization from Hydrolytic Degradation

Plasticization is the phenomenon in which the diluent interacts with the glassy polymer, lowering T_g and E′, as exhibited by epoxy resins on absorption of moisture. However, the shear modulus of a rubber is inversely proportional to the average network chain length, according to the Gaussian theory of rubber elasticity [28]. Therefore, measurement of G_m provides a means of establishing whether hydrolysis has occurred. If we neglect the presence of incomplete crosslinks at chain ends in the network, then the shear modulus is given by

$$G_R = \frac{RT\rho}{M_c},\qquad(8.19)$$

where R is the gas constant, T is the temperature, ρ is the density, and M_c is the number average molecular weight of a network chain.

If hydrolysis occurs, leading to network chain scission and a decrease in the degree of crosslinking, then the value of M_c will increase and that of G_R will decrease. DMTA can be used to measure an estimate of G_R above T_g.

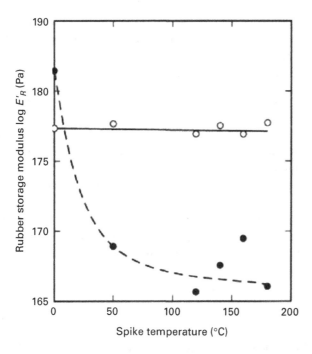

Figure 8.12 The rubber storage moduli, E'_R, of two cyanate ester (AroCy L10) resins as a function of moisture absorption at 50 °C and 96% RH, and under thermal spiking to temperatures of 100–180 °C, illustrating the hydrolytic degradation of the thermally cured resin (●) compared to the epoxy hardened resin (o) [8,29].

An example of this approach is given in Figure 8.12, where it is shown how E'_R, the storage modulus in the rubbery region (i.e. above T_g), is significantly reduced when hydrolysis occurs while remaining constant in the absence of hydrolysis.

8.5 Anomalous Effects

8.5.1 Role of Impurities and Unreacted Resin Components

Cured resins behave as semi-impermeable membranes, allowing water to diffuse but acting as a barrier to larger molecules so that osmosis can occur. Therefore, if the resin contains water-soluble impurities, thermodynamics will drive the water into the resin. Osmotic pressure (π) is a colligative property and is directly proportional to the molal concentration of the impurity:

$$\pi = RTc, \tag{8.20}$$

where c is the molal (mol kg^{-1}) concentration of solubilized inclusions, T is the temperature, and R is the gas constant.

Since moisture also plasticizes and hence softens the matrix resin, localized pressure at inclusion sites causes blisters to form. This is referred to as 'boat pox' in glass-fibre reinforced unsaturated polyesters (GRP). The difference in concentration across the matrix membrane is higher in freshwater than in saline environments, so this is more prevalent in the former.

Typical impurities leading to osmosis are:

- *Unsaturated polyester resins*: residuals from the synthesis such as acids, anhydrides, or glycols;
- *GRP*: residuals in the glass-fibre sizing. Choice of fibre 'finish' is critical, therefore employ reinforcements with a size destined for use in aqueous (e.g. chemical plant or marine) environments, such as powder-bound matt.
- *Epoxy resins* are less susceptible, but residual DICY curing agent can result in osmotic blistering. The solution is to avoid the formation of encapsulated DICY particles during the curing process. This is achieved by fine dispersion of the DICY curing agent throughout prepreg material by *micronization*. The hydrolysis products from DICY are alkaline, so care is required since glass fibres are prone to degradation [25,30].

8.5.2   Residual Stress

In hand lay-up, GRP residual stresses can be induced at the interface between the gel coat and the structural resin. According to Chen and Birley [31,32] this provides an additional thermodynamic mechanism for the absorption of water and blister formation, which is akin to osmosis. Differential shrinkage of these resins either thermally or during curing of the structural resin can lead to residual stress formation. Thus, the thermomechanical properties of the gel coat or surfacing resin should be matched to those of the structural resin.

8.6   Effect of Moisture on Composite Performance

8.6.1   Thermal Stresses

Laminates from plies of continuous fibres at different orientations are used to deal with the anisotropy of the individual plies. We saw in Chapter 6 that residual thermal strains develop in the plies as a result of constrained shrinkage during cooling from the curing temperature. Considering the strain in the longitudinal direction of the transverse ply of a $0°/90°/0°$ ε_{tl}^{th} at temperature T_2, we have

$$\varepsilon_{tl}^{th} = \frac{E_l b (\alpha_t - \alpha_l)(T_1 - T_2)}{(E_l b + E_t d)}, \tag{8.21}$$

where α_l is the expansion coefficient of the ply parallel to the fibres, α_t is the value transverse to the fibres, E_l and E_t are Young's moduli of longitudinal and transverse plies of thickness, b and $2d$, respectively, and T_1 is the strain-free temperature.

The induced thermal strain reduces the first ply cracking strain (ε_{tlu}) according to eq. (8.22), where ε_{tu} is the failure strain of an isolated transverse ply:

$$\varepsilon_{tlu} = \varepsilon_{tu} - \varepsilon_{tl}^{th}. \tag{8.22}$$

Factors that affect the value of α_m (which determines α_t) and T_1 modify the micromechanical behaviour.

Application at elevated temperatures could promote further crosslinking and an increase in T_1 and reduce α_m so that on cooling $(T_1 - T_2)$ will tend to increase while α_t decreases. The cooling cycle during thermal cycling can lead to thermal cracking, and the number of transverse cracks multiplies with duration and number of thermal steps. Generally, thermal cycling of dry composite materials to temperatures lower than the cure temperature does not lead to thermal fatigue.

8.6.2 Effect of Moisture Absorption

The resin matrix in a composite absorbs water in service. As discussed above, this process is kinetically slow, so the concentration of water in the resin will vary with time and location. Thermal cycling will tend to move the water into the structure.

Absorption of water causes the polymer to expand, but in a composite the swelling is constrained by the higher stiffness of the plies in the fibre direction, similarly to that described for thermal contraction. Thus, conditioning in moist environments leads to a reduction in magnitude of the thermal strain present in the laminate. The degree of swelling can be described quantitatively using analogous equations. For example, the swelling strain in the transverse ply in the longitudinal direction (ε_{tl}^{s}) is given by

$$\varepsilon_{tl}^{s} = \frac{E_l b(\beta_t - \beta_l)(M_1 - M_2)}{E_l b + E_t d}, \tag{8.23}$$

where M_1 and M_2 are the initial and final moisture concentrations, respectively, and β_t and β_l are the moisture swelling coefficients in the transverse and longitudinal directions of a unidirectional ply.

8.6.2.1 Thermal Cycling of Wet Laminates

We see above that moisture absorption causes the swelling of the matrix resin. A result of this is that the density of the matrix, ρ_m, decreases so that the expansion coefficient, σ_m, of the resin can be expected to increase. α_t can therefore be expected to increase after moisture absorption. Therefore, on cooling from a thermal cycle we could expect ε_{tl}^{th} to be higher, but that will depend on the effect of water absorption on the value of T_1. The actual value of $(\alpha_t - \alpha_l)(T_1 - T_2)$ or $\Delta\alpha\Delta T$ for the wet laminate will determine whether ε_{tl}^{th} increases or decreases and whether or not transverse cracks are induced. α_m tends to follow the temperature profile of $\tan\delta$. Figure 8.8 shows typical DMTA plots of an epoxy resin containing absorbed water. Figure 8.13 shows

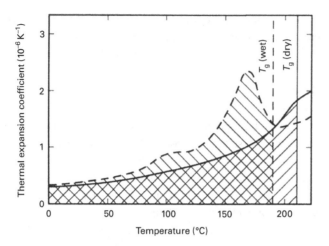

Figure 8.13 Temperature dependence of the transverse thermal expansion coefficients of a carbon-fibre composite (Narmco Rigidite 5245C): dry (continuous curve) and wet (dashed curve). The hatching illustrates the differing values of $\Delta\alpha\Delta T$ and hence thermal strain, for wet and dry laminates [7,33].

how water absorption influences the α_t/T profile of a carbon-fibre composite. Also shown is T_g for the wet and dry material. The hatched areas represent $\Delta\alpha\Delta T$ for the wet and dry composites, assuming T_1 equates to T_g. This shows that, despite the reduction in the strain-free temperature, T_1, $(\Delta\alpha\Delta T)_{wet} > (\Delta\alpha\Delta T)_{dry}$.

Thus, the result of moisture absorption and thermal cycling will have a complex outcome. Moisture absorption can reduce thermal strains in the plies of a laminate and hence reduce the propensity for transverse cracking. However, cooling from the temperature achieved in a thermal event will tend to increase the magnitude of the thermal strain at room temperature or below. This is illustrated in Figure 8.14, where we see the changes in thermal strain as a result of moisture absorption at 50 °C and 96% RH. The thermal strain was monitored from the curvature of an unbalanced 0°/90° beam. Isothermal moisture leads to a reduction in thermal strain while the enhanced moisture absorption arising from a thermal excursion at 140 °C has much less effect on thermal strain. Higher-temperature thermal spikes can lead to larger thermal strain and thermal cracking on cooling [7].

However, we should remember that the rate of water absorption is low, so that initially the outer plies will have a reduced thermal strain while the inner plies will be largely unaffected. Thus, under load differential transfer cracking can be expected.

Under thermal cycling the outer plies will have an enhanced potential for transverse cracking on cooling. Therefore, thermally induced transverse cracking without external load becomes possible with time. Further, as the stiffness of the individual plies changes, the degree of constraint alters and hence the mechanism of transverse cracking. These effects will appear over time in the form of a thermal fatigue phenomenon.

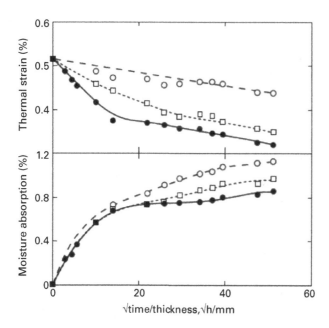

Figure 8.14 The effect of moisture absorption at 50°C and 96% RH on the thermal strain in a balanced 0°/90°/0° Fibredux 927 carbon-fibre laminate estimated from the curvature of an unbalanced 0°/90° beam. Control (●), 120 °C spike (□), 140 °C spike (○) [7].

8.7   Compressive Properties

In compression the matrix has an important role providing the fibres with support so that fibre buckling is inhibited. Fibre buckling occurs in regions of high compressive stress, where the matrix is unable to support the load and 'kink' bands form. The modulus of the matrix is a critical factor preventing kinking (of the fibres) in a shear band. Soutis [34] discussed the modelling of the mechanism of failure using eqs (8.24) and (8.25):

$$\sigma_c = \frac{\tau_y \left[1 + \left(\frac{\alpha_{ty}}{\tau_y}\right)^2 \tan^2 \lambda\right]^{1/2}}{(\Phi_o + \Phi)}, \tag{8.24}$$

where τ_y and σ_{ty} are the in-plane shear and transverse yield stress of the composite, respectively, Φ_o is the angle of alignment in the kink band, Φ is the additional fibre rotation in the kink band under a remote compressive stress, σ_c, and λ is the kink band orientation angle.

$$\tau(\gamma) = \tau_u \left[1 - \exp\left(-\frac{G_{12}\gamma}{\tau_u}\right)\right], \tag{8.25}$$

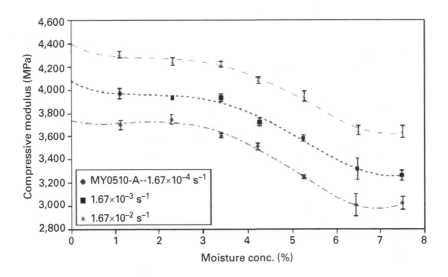

Figure 8.15 The compressive modulus of a 4,4′-DDS cured TGAP (MY0510) epoxy resin after moisture absorption, measured at 22 °C with three different strain rates [7,35].

where G_{12} is the shear modulus of the composite (normally measured at a shear strain, γ, of 0.5%), γ is the shear strain, and τ_u is the shear strength.

Figure 8.15 shows how the compressive modulus of a TGAP/4,4′-DDS epoxy resin is reduced by moisture absorption. At the highest strain rate, the compressive modulus fell from 4.4 to 3.6 GPa on absorption of 7% water. It is also noticed that initial moisture absorption is not so deleterious. This is because the unoccupied volume will be filled first, before plasticization occurs.

The reduction in compressive modulus will have an effect on the compressive yield strength, as shown in Figure 8.16. The trend is different in that it is a continuous decrease. This is most likely a result of the load state near the yield point when the water molecules become mobilized.

The compressive properties of unidirectional carbon-fibre composites have been studied by Soutis [34], who found that the reduction in shear yield strength and compressive modulus causes fibre kinking to occur at lower stresses.

Table 8.8 gives the measured compressive strengths of 0° carbon-fibre (T800) Fiberdux 924 epoxy composites in the presence of absorbed moisture in comparison to predictions from eqs (8.24) and (8.25). This confirms that absorbed moisture results in reduced compressive properties.

The compressive properties of a composite under all loading conditions are strongly affected by moisture absorption because of the reduction in the shear properties of the matrix polymer. The design of artefacts with PMCs needs to reflect the limitations of these materials in compression, especially in service where environmental conditions are likely.

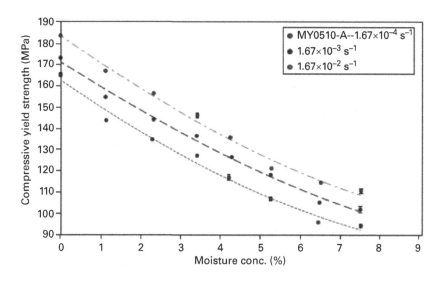

Figure 8.16 Compressive yield strength of 4,4′-DDS cured TGAP (MY0510) epoxy resin after moisture absorption, measured at 22 °C with three different strain rates [7,35].

8.8   Non-aqueous Environments

Composite materials come into contact with organic solvents, which are used in paints, paint strippers, and degreasing agents. Crosslinked thermoset resins or crystalline thermoplastics do not dissolve in common organic solvents. Swelling is therefore the critical issue. The degree of swelling will be maximized in contact with fluids of similar solubility parameters to the polymer matrix. The solubility parameters of the matrix and the solvent can be estimated using simple additive principles [36]. The concept of solubility parameters is described simply by eq. (8.26):

$$\delta_{solvent} = \delta_{polymer} \pm 2\text{MPa}^{1/2}, \tag{8.26}$$

where $\delta_{solvent}$ and $\delta_{polymer}$ are the solubility parameters.

Maximum plasticization of the matrix will occur when the values of $\delta_{solvent}$ and $\delta_{polymer}$ are equal. However, full compatibility is considered to occur over the range given in eq. (8.26). Table 8.9 gives typical values for resins used in composite manufacture. Table 8.10 provides the values for fluids that may come into contact with artefacts. Most epoxy-based systems, according to eq. (8.26), should be resistant. However, Skydrol hydraulic fluids are potentially of concern, so care should be taken to avoid contact. Absorption of solvents will lead to a high potential to creep under off-axis loads. Schulte and coworkers [37,38] have demonstrated how different organic solvents, such as hydraulic fluids encountered in aerospace structures, lead to a reduction in the secant modulus of ±45° glass-fibre laminates, and the number of cycles to failure under flexural fatigue.

Table 8.8. Compressive strengths of T800/924C unidirectional (0°) laminates over a range of temperatures at 95% RH [7,34]

Test temp. (°C)	Condition	Compressive strength (MPa)		Tensile modulus[a] (GPa)	Shear strength (MPa)		Shear yield stress (MPa)		Shear modulus[b] (GPa)	
		Exp.	Pred.		Exp.	Pred.	Exp.	Pred.	Exp.	Pred.
20	Dry	1,415	1,411	160	110		40		6.0	
20	Wet	1,060	1,040			89		29.5		5.4
50	Dry	1,230	1,235	155	105		35		5.8	
50	Wet	930	917			78		26		5.4
80	Dry	1,137	1,129	149	98		32		5.4	
80	Wet	828	829			69		23		4.9
100	Dry	973	953	136	90		28		4.9	
100	Wet	654	653			54		18.5		4.5

Moisture content, $M_\infty = 1.42\%$; assumed initial fibre misalignment, $\phi_o = 1.75°$; kink band inclination angle, $\lambda = 15°$; the value of ϕ_o is not affected by the environmental test conditions. Predictions from eqs (8.22) and (8.23) and experimental unidirectional compressive or shear strengths. [a] Secant axial modulus measured at 0.25% axial strain. [b] Secant shear modulus measured at 0.5% shear strain.

Table 8.9. Typical solubility parameters of epoxy and polyester resins

Monomeric unit	Definition	$\delta_{polymer}$ (MPa$^{1/2}$)
Epoxy:		
DGFBF	Diglycidyl ether of bisphenol F	22.36
DGEBA	Diglycidyl ether of bisphenol A	21.26
TGAP	Triglycidyl p-amino phenol	22.92
TGDDM	Tetraglycidyl diamino diphenyl methane	21.84
DDS	p-diamino diphenyl sulphone	27.18
TGAP/DDS (2:1)		24.06
TGDDM/DDS (5:1)		22.30
TGDDM/TGAP/DDS (19:13:10)		22.89
DGEBA/NMA (1:2)	NMA= nadic methylene anhydride	24.00
UPE: Styrene segment		18.7
UPE: polyester segment		21.8

Table 8.10. Typical solubility parameters (δ_l) of potential fluids in contact with composites

Liquid	Function	δ_l (MPa$^{1/2}$)
Hydraulic fluid		≈14
Paraffins	Fuel: kerosene	15–16
Oil (lubricating)	Lubrication	≈16
Trichloroethylene	Degreasing	18.7
Methylene dichloride		20.2
Organophosphate esters (Skydrol)	Aerospace (fire-resistant) hydraulic fluid	≈22
Ethylene glycol	Antifreeze	34.9

The reduction in matrix modulus is concurrent with a reduction in matrix shear strength, so that the mechanism of failure in flexure will change from matrix fracture to delamination. In compression the matrix will be less able to support the fibres and buckling failure will occur at lower stress [37,38].

Weatherhead [39] provides a survey of the durability of resins against a range of environments.

8.9   Design of Durable Composites

8.9.1   Choice of Fibres

E-glass fibres are susceptible to stress corrosion in acidic environments, and more recent formulations of glass have improved resistance. As discussed in Chapter 2, these fibres are also subjected to corrosive degradation in both acid and alkaline environments. Therefore, the role of the resin matrix in protecting the fibres is critical in the design of a composite for durability in aqueous environments.

Other fibres also have to be protected from aqueous environmental degradation by careful choice of resin. The amide groups in **aramid fibres** are potentially hydrolysable in aqueous environments, both alkaline and acidic. Exposure to ultraviolet (UV) light promotes oxidation, but the degradation products tend to be protective since they absorb UV light [40]. When used in ropes these fibres are normally coated with a protective polymer.

Carbon fibres are generally resistant to aqueous environments, but strong oxidizing acids are degrading.

For durability, the interface needs to be optimized to ensure maintenance of interfacial integrity during service.

8.9.2   Durable Glass-Fibre Composites

Figure 8.17 illustrates the environmental stress corrosion cracking (ESCC) of glass-fibre composites in aqueous acidic environments. There are three fracture mechanisms at differing levels of stress:

1. **stress-dominated stage** at high stresses, in which crack propagation is more rapid than the corrosion of the glass at the crack tip;
2. **stress corrosion cracking**, in which the rate of corrosion of the glass is similar to the rate of crack propagation – thus the crack remains sharp and propagates into the weakened glass, leading to a synergistic effect;

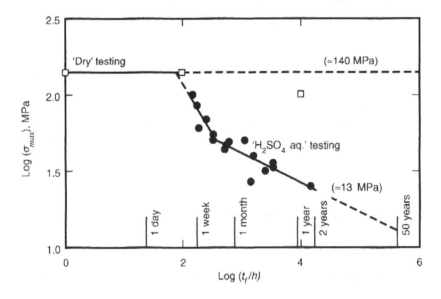

Figure 8.17 Environmental stress corrosion cracking failure times (t_f) of circumferentially cut coupons of GRP, taken from sewer linings, in 0.5 M H_2SO_4. The data in brackets represent the maximum stress for a 50-year lifetime [11,41].

3. **stress-assisted corrosion**, in which the rate of corrosion of the fibres is larger than the rate of crack propagation, so that the crack tip is blunt.

In this case eq. (8.27) applies:

$$\sigma_{max} = 2\sigma_a \left(\frac{x}{\rho}\right)^{1/2},$$
(8.27)

where σ_{max} is the stress at the crack tip, σ_a is the applied stress, and x and ρ are the flaw depth and radius, respectively. As seen, σ_{max} decreases as the radius of curvature of the flaw, ρ, increases during corrosion.

Figure 8.18 shows the matrix variables that need to be controlled for durable GRP pipes in acidic environments:

1. select fibres to maximize strength under non-corrosive conditions for resistance to stage I;
2. maximize matrix and barrier resin fracture toughness to optimize ESCC in stage II; and
3. optimize interfacial stability and barrier resin diffusion characteristics for resistance to stage III.

There is a compromise between fracture toughness and diffusion constant because highly crosslinked matrix resins have low rates of diffusion of aqueous environments, but also low resistance to fracture (i.e. low fracture toughness), and vice versa.

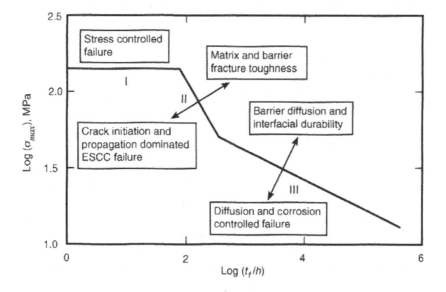

Figure 8.18 Schematic showing how to control the various material variables on the three mechanisms of ESCC at differing levels of stress [11].

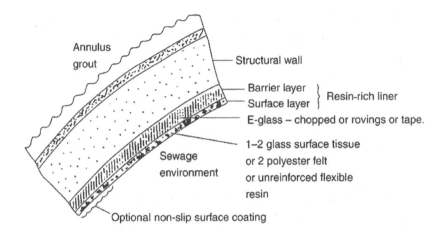

Figure 8.19 Typical design of a durable GRP pipe for corrosive environments [42,43].

Design of a durable pipe or container needs to address these requirements, which is achieved through a number of technological means, depending on the manufacturing process. A ductile resin can be employed for the barrier resin to provide resistance to crack initiation and growth. Alternatively, reinforcement veils or tissues can be included. The design also recognizes the stress state of the structure. For example, pipes are usually loaded in flexure so that the mid-pipe material can employ cheaper fillers. A typical construction design of a durable GRP pipe for corrosive environments is shown in Figure 8.19 [41,42].

As described in Chapter 4, different manufacturing routes can be employed to provide glass-fibre composites with durability in ESCC conditions. The role of the barrier layer is to ensure that a brittle crack does not form. This can be achieved by employing a resin-rich surface. Fracture toughness is introduced using ductile resins and/or glass surface tissue or polyester felt. The barrier layer can use short glass fibres that ensure that a propagating crack will be deflected from a straight path and spread the load across several randomly arranged fibres.

8.9.3   Thermal Environments

Subjecting a composite to elevated temperatures can be detrimental, not least because of the 'rule of thumb' arising from the Arrhenius equation. Thus, the rate of a chemical reaction is doubled with each 10 °C rise in temperature. We can expect a reduced life at elevated temperatures. Organic polymers are thermally stable up to temperatures of ≈ 300 °C, but more critically E-glass fibres tend to have a lower strength at $T > 250$ °C, while carbon fibres are susceptible to thermal oxidation at $T > 300$ °C. Aramid, polybenzimidazole, and polybenzoxazole fibres are thermally stable at temperatures in excess of 300 °C. Oxidized acrylic fibres, such as Panox, have sufficient thermal resistance for use as fire-resistant textile fibres, but since they are precursors to carbon fibres they are not truly thermally stable in excess of 220 °C.

8.10   Conclusions

The environments that composite materials are likely to come into contact with are discussed. The role of these agents is also considered. It is shown that extreme conditions can lead to degradation of mechanical performance. However, understanding of these mechanisms provides the technology for designing durable structures.

8.11   Discussion Points

1. Consider the nature of diffusion of aqueous environments into composite materials. How does the orientation of the fibres influence the rate of diffusion?
2. Describe a configuration of fibre composite coupons for determining diffusion constants.
3. How does crosslink density in a matrix resin influence its diffusion constant?
4. Discuss the factors that influence water absorption by a matrix resin.
5. Which property of a matrix resin is affected most by absorbed water?
6. How can you differentiate plasticization from hydrolytic degradation of a thermoset resin?
7. What do you understand by *thermal spiking*?
8. Which reinforcing fibres suffer from stress corrosion cracking?
9. Describe the factors that need to be addressed in the design of a pipe or container using glass-fibre composites.

References

1. J. Crank, *The mathematics of diffusion*, 2nd ed. (Oxford: Clarendon Press, 1975).
2. G. S. Springer, ed. *Environmental effects in composite materials* vols. 1–3 (Westport, CT: Technomic, 1981–1988).
3. J. Crank and G. S. Park, eds, *Diffusion in polymers* (New York: Academic Press, 1968).
4. A. H. Windle, Case II sorption. In *Polymer permeability*, ed. J. Comyn (London: Chapman and Hall, 1985), pp. 75–118.
5. N. L. Thomas and A. H. Windle, A theory of case II diffusion. *Polymer* **23** (1982), 529–542.
6. T. A. Collings, Moisture absorption: Fickian diffusion kinetics and moisture profiles. In *Handbook of polymer-fibre composites*, ed. F. R. Jones (Harlow: Longman, 1994), pp. 366–371.
7. C. H. Shen and G. S. Springer, Moisture absorption and desorption of composite materials. *J. Compos. Mater.* **10** (1976), 2–20.
8. F. R. Jones and J. P. Foreman, The response of aerospace composites to temperature and humidity. In *Polymer composites in the aerospace industry*, ed. P. E. Irving and C. Soutis (Cambridge: Woodhead Publishing, 2015), pp. 335–369.
9. T. A. Collings and S. M. Copley, On the accelerated ageing of CFRP. *Composites* **14** (1983), 180–188.

10. B. Ellis and M. S. Found, The effects of water absorption on a polyester/chopped strand mat laminate. *Composites* **14** (1983), 237–243.

11. F. R. Jones, The effects of aggressive environments on long-term behaviour. In *Fatigue in composites*, ed. B. Harris (Cambridge: Woodhead Publishing, 2003), pp. 117–146

12. S. K. Karad, F. R. Jones, and D. Attwood, Mechanisms of moisture absorption by cyanate ester modified epoxy resin matrices. Part I: Effect of Spiking Parameters. *Polymer* **43** (2002), 5209–5218.

13. L. G. E. Struik, *Physical ageing in amorphous polymers and other materials* (Amsterdam: Elsevier, 1978).

14. F. N. Kelley and F. Bueche, Viscosity and glass temperature relations for polymer-diluent systems. *J. Polym. Sci.* **50** (1961), 549–556.

15. T. G. Fox, Influence of diluent and of copolymer composition on the glass temperature of a polymer system. *Bull. Am. Phys. Soc.* **1** (1956), 123.

16. D. Porter, *Group interaction modelling of polymer properties* (New York: Marcel Dekker, 1995).

17. F. R. Jones, Molecular modelling of composite matrix properties. In *Multi-scale modeling of composite material systems*, ed. C. Soutis and P. W. R. Beaumont (Cambridge: Woodhead Publishing, 2005), pp. 1–32.

18. V. R. Gumen, F. R. Jones, and D. Attwood, Prediction of the glass transition temperatures for epoxy resins and blends using group interaction modeling. *Polymer* **42** (2001), 5717–5725.

19. J. P. Foreman, D. Porter, S. Behzadi, K. P. Travis, and F. R. Jones, Thermodynamic and mechanical properties of amine-cured epoxy resins using group interaction modelling. *J. Mater. Sci.* **41** (2006), 6631–6638.

20. J. P. Foreman, S. Behzadi, D. Porter, P. T. Curtis, and F. R. Jones, Hierarchical modelling of a polymer matrix composite. *J. Mater. Sci.* **43** (2008), 6642–6650.

21. J. P. Foreman, S. Behzadi, D. Porter, and F. R. Jones, Multi-scale modelling of the effect of a viscoelastic matrix on the strength of a carbon fibre composite. *Philos. Mag.* **90** (2010), 4227–4244.

22. W. W. Wright, The effect of diffusion of water into epoxy resins and their carbon-fibre reinforced composites. *Composites* **12** (1981), 201–205.

23. V. R. Gumen and F. R. Jones, Modelling of the effect of moisture absorption on the thermomechanical relaxation spectra for thermoset/thermoplastic resin blends. In *Thermoplastics-based blends and composites*, ed. I. Chodak and I. Lacik (Weinheim: Wiley-VCH, 2001), pp. 139–148.

24. J. A. Hough, Enhanced moisture absorption in advanced composites, PhD thesis, University of Sheffield, UK (1997).

25. F. R. Jones, Durability of reinforced plastics in liquid environments. In *Reinforced plastics durability*, ed. G. Pritchard (Cambridge: Woodhead Publishing, 1999), pp. 70–110.

26. S. K. Karad and F. R. Jones, Mechanisms of moisture absorption by cyanate ester modified epoxy resin matrices: the clustering of water molecules. *Polymer* **46** (2005), 2732–2738.

27. S. K. Karad, J. Hough, and F. R. Jones, The effect of thermal spiking on moisture absorption kinetics, mechanical and viscoelastic properties of carbon fibre reinforced epoxy laminates. *Compos. Sci. Technol.* **65** (2005), 1299–1305.

28. L. R. G. Treloar, *The physics of rubber elasticity*, 3rd ed. (Oxford: Clarendon Press, 1975).

29. S. K. Karad, F. R. Jones, and D. Attwood, Mechanisms of moisture absorption by cyanate ester modified epoxy resin matrices. Part III: Effect of Blend Composition. *Composites A* **33** (2002), 1665–1675.

30. F. R. Jones, The effects of aggressive environments on long-term behaviour. In *Fatigue in composites*, ed. B. Harris (Cambridge: Woodhead Publishing, 2003), pp. 117–146.
31. F. Chen and A. W. Birley, *Plast. Rub. Compos. Process. Appl.* **15** (1991), 161–168.
32. F. Chen and A. W. Birley, *Plast. Rub. Compos. Process. Appl.* **15** (1991), 169.
33. P. M. Jacobs and F. R. Jones, The mechanism of enhanced moisture absorption and damage accumulation in composites exposed to thermal spikes. In *Composite design, manufacture and application (ICCM VIII)*, ed. S. W. Tsai and G. Springer (Corina, CA: SAMPE, 1991), paper G.
34. C. Soutis, *Plast. Rub. Compos. Process. Appl.* **38** (2009), 55–60.
35. S. Behzadi, Role of matrix yield in composite performance, PhD thesis, University of Sheffield, UK (2006).
36. D. W. van Krevelen and K. Te Nijenhuis, *Properties of polymers* (Amsterdam: Elsevier, 2009).
37. K. L. Schulte, Cyclic mechanical loading. In *Reinforced plastics durability*, ed. G. Pritchard (Cambridge: Woodhead Publishing, 1999), pp. 151–185.
38. K. L. Schulte, A. Mulkers, H. D. Berg, and H. Scholke, Environmental influence on the fatigue behaviour of amorphous glass/thermoplastic matrix composites. In *Proceedings of the International Conference on Fatigue of Composites*, ed. S. Degallaix, C. Bathias, and R. Fongeres (1997), pp. 339–346.
39. R. G. Weatherhead, *FRP technology: fibre reinforced resin systems* (London: Applied Science, 1980).
40. J. L. J. van Dingenen, Gel spun high performance polyethylene fibres. In *High performance fibres*. ed. J. W. S. Hearle (Cambridge: Woodhead Publishing, 2001), pp. 62–92.
41. S.-W. Tsui and F. R. Jones, Evaluation of the performance of the barrier layers of sand-filled GRP sewer linings. *Plast. Rub. Process. Applicat.* **11** (1989), 141–146.
42. F. R. Jones, Glass fibres. In *High performance fibres*, ed. J. W. S. Hearle (Cambridge: Woodhead Publishing, 2001), pp. 191–238.
43. F. R. Jones, Designing composites for durability in aqueous and corrosive environments. In *Designing cost effective composites* (London: Institution of Mechanical Engineers, 1998), pp. 65–82.

9 Joining, Repair, Self-Healing, and Recycling of Composites

Protocols for repair and recycling of composites are described. Future developments that embrace self-healing systems are also considered. End-of-life options such as fibre and matrix recovery are also discussed. The economics of differing approaches are briefly considered using a whole-life cost model.

9.1 Joining

One of the main advantages of composite materials is the reduced number of parts required to assemble into an artefact or structure. In many cases the components can be moulded in a single piece. One example is the monocoque chassis and body of a sports car. However, in many applications this is not always feasible and the joining of components is necessary. There any several techniques employed which have differing advantages. They can be divided into (a) mechanical fastenings and (b) bonded joints, as shown in Figure 9.1.

Table 9.1 provides the advantages and disadvantages of these techniques. There are a number of joint designs that can be employed, and Figure 9.2 provides a description of available joint configurations.

9.1.1 Mechanically Fastened Joints

Apart from the properties of the composite parts to be assembled using mechanical fasteners, there are a number of other parameters to consider, including type, size, clamping force, washer diameter, and hole size.

The clamping force is clearly critical. This is defined as the through-thickness force exerted in closing the fastener. Thus, three-dimensional stresses need to be considered, but this is analytically complex and difficult to model. We will only consider the fundamental net section stresses in order to understand the joint failure mechanisms. Readers are referred to the texts by Matthews and Rawlings [1] and Camanho and Tong [2] for more detailed discussions.

Figure 9.3 illustrates the distribution of force (P) within double and single shear loading. The applied force is transmitted through the fastener from one side of the joint to the other.

Table 9.1. Advantages and disadvantages of mechanically and bonded joints

	Mechanically fastened joints	Bonded joints
Fastener type	1. Screws 2. Rivets 3. Bolts	Adhesive
Advantages	1. No surface preparation 2. Disassembly possible 3. Straightforward inspection	1. Stress concentrations minimized 2. Absence of drilled holes and consequential damage 3. Small weight penalty
Disadvantages	1. Holes provide stress concentrations 2. High weight penalty	1. Surface preparation 2. Disassembly impossible 3. Environmental effects (e.g. moisture absorption) on adhesives 4. Inspection of joint integrity is difficult

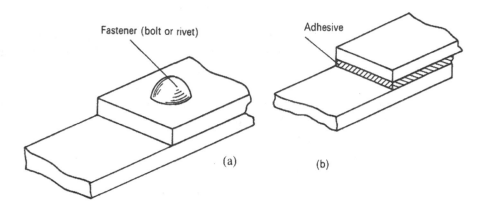

Figure 9.1 Schematics of (a) mechanically fastened and (b) adhesive-bonded joints [1].

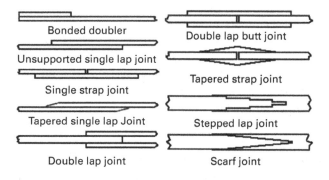

Figure 9.2 Definitions of joint configurations [1].

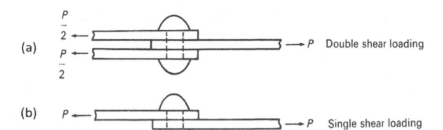

Figure 9.3 Loading configurations for (a) single and (b) double lap shear joints [1].

The failure modes of mechanically fastened joints are described in Table 9.2. We see that many joint failures occur as a result of complex stresses, which lead to mixed-mode failures.

Tensile failure can be related to the net section stress, as shown in eq. (9.1):

$$P_u^{ten} = \sigma_u^{ten}(w - d)t, \tag{9.1}$$

where P_u^{ten} is the failure load, $(w - d)t$ is the net area of the laminate (i.e. member width) under load and is related to the diameter of the fastener hole in a single-hole joint or pitch in an array of fasteners, t is the plate thickness, w is the plate width, d is the diameter of the hole, and σ_u^{ten} is the net tensile strength of the joint. Thus, narrow laminates will carry a higher stress and joints are most likely to fail by tensile fracture.

Shear-out failure is related to the shear stresses generated at the edge of the hole parallel to the load over the end distance, e_d, which is the length from the end of the laminate to the midpoint of the hole, which is typically approximately $4d$. Equation (9.2) provides a measure of shear-out strength, σ_u^{shear} of the joint. Low values of e_d lead to a high probability of shear-out failure:

$$P_u^{shear} = \sigma_u^{shear} 2e_d t. \tag{9.2}$$

Bearing failure is related to the projected area of the hole, which is a function of its diameter, as shown in eq. (9.3):

$$P_b = \sigma_u^{bearing} dt, \tag{9.3}$$

where P_b is the failure load under a bearing stress, $\sigma_u^{bearing}$ is the bearing strength, and the bearing stress is given by $\sigma^{bearing} = P/dt$.

Cleavage failure is fundamentally shown in Table 9.2 as a transverse tension, which results from bending loads; however, in reality *bursting* occurs in a mixed mode involving tensile and bending loads and cannot be expressed simply.

Irrespective of the failure mode, joint strength is usually expressed as bearing strength.

Fastener failure can involve pull-through, which is a function of the through-thickness strength of the composite laminate. Bolt failure results from shear stresses in the fastener. In general, these failures can be avoided by ensuring that extreme values

Table 9.2. Typical failure modes of mechanically fastened joints [3]

Failure mode		Comment
Shear out		Caused by shear stresses and occurs along shear-out planes on hole edge. Typical failure mode when end distance is short.
Tension (net section)		Caused by tangential tensile or compressive stresses at the edge of the hole. For uniaxial loading conditions, failure occurs when bypass/ bearing stress ratio is high (or d/w is high).
Bearing		Occurs in the area adjacent to the contact area due to compressive stresses, likely when bypass/bearing stress ratio is low (or d/w is low). Strongly affected by through-thickness clamping force.
Cleavage		Seen in composite joints with unidirectional plies. Otherwise mainly mixed mode.
Bearing/shear out		
Bearing/tension/shear out		Mixed-mode failure
Tension/shear out		
Bolt pull-through		Due to low through-thickness strength of composite material.
Bolt shear failure		Caused by high shear stresses on the bolt.

of d/t are not employed. Low values of d/t (i.e. large plate thickness) lead to bending of the fastener, while large values (i.e. low plate thickness) result in high shear stresses in the bolt. Values of d/t in the range $1 < d/t < 3$ are generally satisfactory.

With composite laminates the fibre orientation is clearly a major factor in the formation of a successful bolted joint. Unidirectional composites in which the load is parallel to the fibres will have a low shear-out strength, and the segment indicated in

Table 9.3. Specific bearing strengths for single-hole joints from composites and metals [1]

Material	Bearing strength, $\sigma_u^{bearing}$ (MPa)	Specific bearing strength (MPa/ Mg m^{-3})	Density (Mg m^{-3})
CFRP ($V_f = 0.6$)	1070	695	1.54
GFRP ($V_f = 0.6$)	900	428	2.10
GRP-Woven Rovings ($V_f = 0.6$)	682	324	2.10
GRP-Woven Rovings ($V_f = 0.44$)	490	258	1.9
CSM ($V_f = 0.19$)	390	240	1.62
	(260)*	(160)*	
Aluminium alloy L72	432	160	2.70
Steel S96	973	124	7.85

* Tensile failure.

Table 9.2 will pull out at low loads. For some fibre–matrix systems it may not be possible to achieve a tensile failure before shear failure occurs even for very narrow specimens with $w < 2d$ and high end distances with $w > 10d$. Laminates with fibres at $\pm 45°$ have high shear strengths and reduced stress concentrations at the hole.

Bearing strength is controlled by the compressive strength of the material and through-thickness constraint introduced by the fastener. Therefore, support for the load-bearing 0° fibres is essential, and inclusion of $\pm 45°$ and 90° fibres is needed for good bearing performance.

As described in Chapter 6, ply dispersion is beneficial to joint performance. Similar geometries are also beneficial to the performance of a composite joint. For example, in the case of glass-fibre composites, chopped strand mat (CSM) material in which the fibres have randomly distributed in-plane angles performs better than glass-reinforced polymers (GRP) made from woven rovings. The former material has more isotropic properties. However, it is essential to employ the highest V_f possible. Table 9.3 gives typical bearing strengths of composite materials and shows the benefits of composite bolted joints over aluminium and steel. The CSM material with low V_f is more likely to fail in tension, and the bearing stress is not reached.

9.1.2   Mechanical Fasteners

The common mechanical fasteners used with composites are self-tapping screws, rivets, and bolts. Self-tapping screws are not recommended because of the low through-thickness strengths of laminates and conventional GRP. For demountable joints, an insert would need to be employed. Rivets can be used on 3-mm thick material, but care must be taken with the drilling of the holes so that minimal damage is introduced. Countersunk rivets place a limitation on the laminate thickness, and the angle should be at least 120° to avoid pull-through. Bolts provide the most efficient method of joining, and carefully tightened washers should be used. The laminate needs a significant fraction (~55%) of $\pm 45°$ fibres for carbon-fibre-reinforced polymers (CFRP), and higher for GRP.

9.1.3   Adhesive-Bonded Joints

Adhesive joints can be accomplished in four ways:

1. Using an adhesive to bond two pre-cured laminates (*secondary bonding*).
2. *Co-curing* two prepreg assemblies into a joint. Often an unreinforced resin film is used as the adhesive, which is cured at the same time as the laminate. In this case, the adhesive is the same compounded matrix resin.
3. *Co-bonding* of a pre-cured laminate with an uncured laminate using an adhesive layer.
4. *Multi-material bonding* of a composite material with a metal using an adhesive.

As shown in Figure 9.4, the failure of an adhesive-bonded composite joint occurs in a number of modes:

1. debonding of the adhesive film at the adhesive–composite–adherend interface (not shown);
2. tensile fracture of the laminate adherend;
3. interlaminar delamination of the composite adherend under shear;
4. cohesive failure of the adhesive; and
5. transverse fracture of the adherend under loads.

The mechanism of failure is often a function of the configuration of the bonded joint resulting from the choice of assembly and the presence of resin-rich layers, arising, for example, from the prepreg. Mechanism 1 should not normally be possible in a correctly assembled joint and can be avoided by careful surface preparation. Several methods are available for preparing adhesive-bonded joints, discussed in the following.

9.1.3.1   Peel-Ply Method

In this technique a ply of fabric, such as one from polyester fibres, is included in the lay-up surface, which can be stripped to expose an uncontaminated composite surface

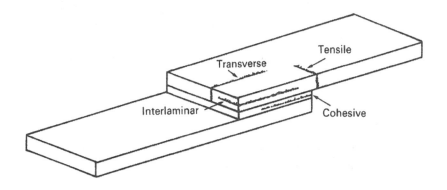

Figure 9.4. Bonded joint with illustrative failure modes [1]

immediately before bonding. However, it is reported that it is extremely difficult to ensure that release agents used in the manufacture of the 'textile' peel-ply are absent.

Flinn and coworkers [4,5] showed that of three fabrics studied – polyamide (nylon), siloxane-coated polyester, and polyester – only the polyester provided a contaminant-free surface. Residuals from the polyamide and siloxane coating were detected by XPS (X-ray photoelectron spectroscopy). As-moulded composites are often contaminated with 'release' agents, which prevent the formation of a strong adhesive bond.

9.1.3.2   Abrasion and Solvent Cleaning

After removal of the peel-ply the surface is abraded with a light grit blast or by careful hand abrasion, followed by a solvent wipe to remove abrasion products. The choice of solvent is also critical, since most 'solvating' solvents may well contaminate the surface and be absorbed into the resin.

9.1.3.3   Plasma or Related Treatments

Recently, atmospheric plasma treatment torches have been developed and marketed which provide rapid surface cleaning and surface oxidation [6,7]. The potential of these techniques is very high, but careful use is essential since overtreatment is easily possible. In principle this technique is ideal, since treatment times are a few seconds and the relevant jig and protocol will need to be employed.

9.1.4   Adhesives

Since most matrix materials are based on epoxy resins, adhesives will also have a similar chemistry (see Chapter 5 on interfacial response). Joints need to withstand shear stresses, and most adhesives are designed to exhibit elastic–plastic properties. In this way, deformation of the adhesive will arise before debonding occurs. Thus, the shear stress–strain properties of the adhesive are critical, and the constant shear stress model of stress transfer will be applicable.

Figure 9.5 shows that epoxy resins can exhibit elastic and plastic components in the stress–strain response. In this case, at room temperature (22 °C) the initial compressive modulus at this strain rate is 4.05 GPa [8]. Typical adhesives have a tensile modulus of $\approx$3 GPa, but generally the adhesive deforms in shear and so there is a shear modulus of 1 GPa with a shear strength of 50 MPa.

Structural adhesives have been discussed by Black in two relevant articles in *Composites World* [9,10].

9.1.5   Joints

It is observed that the maximum adhesive stresses in a joint can be reduced and the joint strength increased by the following design principles:

1. use identical adherends where possible;
2. use the highest possible in-plane adherend stiffness;

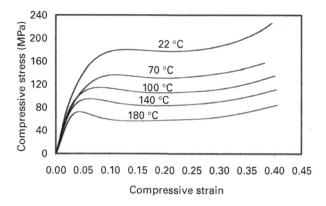

Figure 9.5 True compressive stress–strain curves of an epoxy resin (MY0510/DDS) over a range of temperatures at a strain rate of $1.67 \times 10^{-3} \text{ s}^{-1}$ [8].

3. use homogeneous ply lay-up with a large number of plies – the lay-up should ensure that the fibres in the outer ply, which are in contact with the adhesive, are unidirectionally aligned to the joint length;
4. use a large overlap; and
5. use adhesive with lowest possible tensile and shear elastic moduli.

Consider the *double lap joint* shown in Figure 9.6, where the shear stress–strain response of the adhesive is schematically represented in Figure 9.6a, which leads to shear stress distribution in a double lap shear joint as described in Figure 9.6b. It is common to assume that the adhesive behaves in an elastic–plastic manner. Therefore, the shear strain in the adhesive reaches a maximum at the joint ends with a value of γ_e, the yield strain as defined in Figure 9.6a. With a sufficiently long joint, an 'elastic trough' of length s will exist. The plastic zone is independent of joint lap length, l, and other material parameters. The joint strength is based on the failure criterion of total adhesive shear strain, which occurs when l is just long enough for the elastic trough to form when $s = 0$. A consequence of this failure criterion is that no significant increase in joint strength will accrue from a longer overlap. Typically, for a CFRP composite joint, the critical overlap length is given by $30t_a$, where t_a is the adherend thickness. Any increase above this value will not result in a significant increase in joint strength. As t_a increases, the peel stresses (the through-thickness tensile component) are enhanced, thereby limiting joint strength; an adhesive with a low through-thickness tensile strength will fail as a result.

In contrast, in single lap joints the shear adhesive properties have little effect on performance, and strength is limited by adherend properties, peel stresses, and overlap length.

For the single lap joint in Figure 9.7 the average shear stress is given according to eq. (9. 4):

$$\tau = \frac{P}{bl}, \tag{9.4}$$

where P is the in-plane tensile load for a joint of dimensions lap length l and width b.

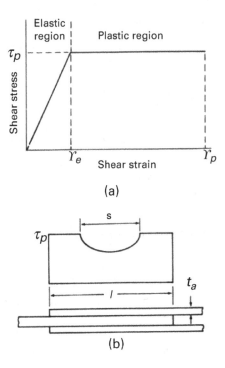

(a)

(b)

Figure 9.6 (a) Idealized adhesive shear stress–strain curve; (b) shear stress distribution for a double lap joint. τ_p is the plastic yield shear stress, Y_e, Y_p are the limiting elastic and plastic shear strains, and l is the total length of the lap joint. The plastic zone is $\frac{1}{2}(l - s)$ and t_a is the adherend thickness [1,11].

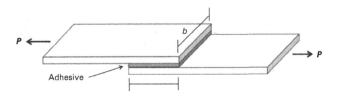

Figure 9.7 Single lap joint under a tensile force, P. b is the width of the specimen and l is the overlap length of the joint.

This is an approximation of the joint stress; a better solution, which provides for shear-lag distribution and takes into account the differential shear stress in the adhesive, which is induced as shown in Figure 9.8, was given by Volkersen [11–13]. Transverse stresses will also develop in the joint, for which Goland and Reissner [14] provided the first analysis. For single lap joints, peel stresses are an order of magnitude greater than in double lap joints. These *peel stresses* will often be responsible for failure of the joint, since they form at the joint ends, as shown in Figure 9.9. These stresses should be minimized in joint design since additional stresses from stress

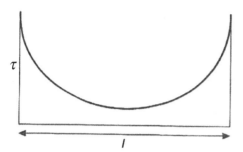

Figure 9.8 Volkersen prediction of shear stress, τ, distribution within the joint given in Figure 9.7.

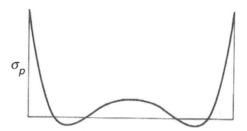

Figure 9.9 Goland and Reissner prediction of the distribution of the transverse or peel stress (σ_p) in the single lap shear joint shown in Figure 9.7 [13].

concentration effects associated with parameters such as corners and the elastic properties of the adhesive are likely to be critical.

The induction of thermal stresses through curing of the adhesive or service will also play an important role. Similarly, hygrothermal (moisture absorption, resin swelling, etc.) stresses need to be fully understood (see Chapter 5).

Scarf joints, as described in Figure 9.2, provide a mechanism of ensuring that the shear stress in the adhesive will be uniform. As a result, the strength of these joints increases with lap length and a small scarf angle is beneficial. Stronger and more brittle adhesives can provide a stronger joint.

Stepped joints are a variant of the scarf joint, and the elastic limit is reached over a small number of steps so that, as with double lap joints, once this is reached a further increase in step length is not beneficial. Joint strength can only be improved by using more steps of smaller thickness. Thus, the geometry approaches that of the scarf joint.

For a detailed discussion of the design and analysis of adhesive-bonded joints, the reader is referred to Zhu and Kedward's report to the Office of Aviation Research in the USA [15]. A recent paper by Budhea et al. [16] provides a review of all aspects of the use of adhesive-bonded joints.

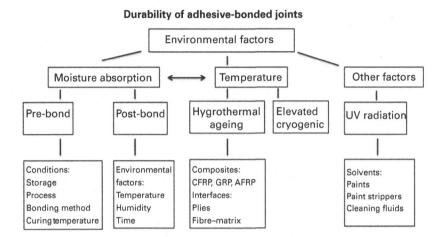

Figure 9.10 Factors determining the environmental durability of a bonded joint.

9.1.6   Environmental Factors

As discussed in detail above, polymeric adhesives are also affected by the presence of absorbed moisture or water or other liquids, such as solvents in paints, paint strippers, and cleaning fluids. Figure 9.10 lists these factors [16]. It is clearly shown that careful control of the adhesive-bonding materials and processes is critical to the durability of an adhesive-bonded joint. Similar considerations apply for repair, which we discuss in Section 9.2.

9.2   Repair [17–19]

Damage in composite structures occurs through a number of mechanisms:

- manufacturing defects;
- accumulation of damage arising in service or from occasional excursions to abnormal service loads or environmental conditions;
- impact or localized mechanical overload from misuse, such as impact from a dropped tool.

In these cases repair may be possible, but inspection is needed to identify the extent of the damage. A common technique in the aerospace industry is ultrasonic inspection using a C-scan arrangement. For individual components, X-radiography with penetrant enhancement may be useful. The inspection technique needs to be capable of establishing the dimensions and location of any defective material.

Repair should restore the integrity and structural performance of the artefact. Clearly, the most effective repairs will involve a 'bonded joint technology' that should return the structure to its original form. Mechanically bolted patches may be possible, but this discussion will be limited to bonded repair approaches:

1. *Minor damage*, which has not affected the structural performance – cosmetic repair. This type of damage can be repaired using a 'filler' or a filled resin paste, which can be sculpted to repair the material and restore its surface finish. An alternative technique is to inject a resin into the minor debonded regions and delaminations. This can be achieved by providing drilled holes into which the heated filler resin can be injected until the resin emerges from the adjacent vent-hole. The 'filler' needs to have a low viscosity to facilitate flow into the damage. This is normally achieved by raising the temperature of the 'filler'. Pressure should be applied during resin curing to ensure integrity is achieved.

2. *Major damage* requires a more rigorous procedure involving bonded flush scarf patches or external patches. Figure 9.11 illustrates these techniques for a laminate and a sandwich panel with an expanded core. 'Doubler' refers to the need to adhere a patch to either side of the repair. In many cases, access to the *hidden* side may not be possible and single patches are used. For aerodynamic reasons, a scarf patch would be preferred.

Figure 9.12 illustrates repair using a flush scarf patch with a step or contoured adhesive zone. This is the preferred technique to achieve the strongest joint with

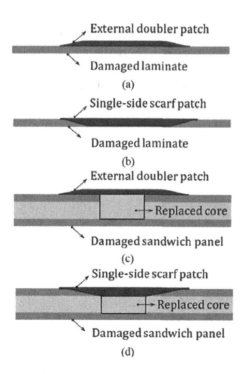

Figure 9.11 Illustration of external and scarf patch repair of composite laminates and sandwich panels: (a) external patch; (b) scarf patch; (c) external patch for sandwich panel; (d) scarf patch for a sandwich panel with a replaced core. Doubler refers to the need for patch repair on both sides of the structure [19].

surface smoothness. It is important to use the following stages to achieve a good result:

1. Careful preparation of the damaged area with the correct scarf angle. Square angles should be avoided to prevent stress concentrations in the repair.
2. Drying of the laminate (slowly over, say, 24–48 h at temperatures of 80–90 °C) to remove absorbed moisture. Moisture can affect subsequent resin cure, but also it is also necessary to avoid complications with thermal expansion and contraction of wet resin.
3. The fibre orientation in the patch has to be matched to that of the parent laminate (i.e. the removed material). However, it is best to avoid unidirectional fibres in the surface layer, which should preferably have a ±45° or woven fibre configuration to provide protection against future scratches and abrasions.
4. Co-curing of the patch is preferred and consideration of the initial cure cycle can provide the best reintegration of the patch. For example, all epoxy resins will have a slight under-cure, which can be used in combination with cure of the patch.

External patches can be simply bonded to the exterior surface(s) of the structure when there is no need to retain the original smooth contoured surface, as shown in Figure 9.11. These repairs require less preparation than scarf repairs. Limited access to the hidden side or substructure interference tends to favour the use of external patch repair. The complex load path around the patch means it should be designed to minimize the bending strains. Therefore, the patch repair must withstand peel forces at the edges within the 'joint' and a step down in the patch diameter is essential. It may also be preferable to include small fasteners to resist peel. Since composite laminates

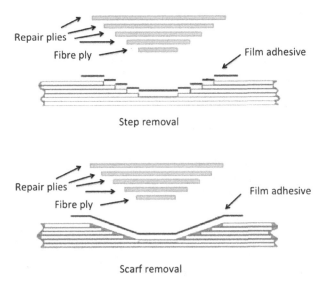

Figure 9.12 Representation of flush patch repair techniques using a stepped or contoured approach.

have low through-thickness strength and peel forces dominate, this technique is limited to the repair of 'thin-section' laminates.

9.3   Self-Healing Repair Strategies

It is clearly valuable to introduce self-healing capability into composite structures. Since micromechanical damage generally occurs within the matrix (transverse cracking), at the fibre–matrix interface (debonding), and/or the interply interface (delamination), it is possible to envisage that molecular interventions can be used to provide a self-healing capability to the matrix within the composite. This can be achieved through techniques employing different mechanisms:

1. *Mendable matrix resin*, which exhibits reversible depolymerization and polymerization with a ceiling temperature well above the service temperature of the composite. Healing can be thermally stimulated at the ceiling temperature of the resin. The term *mendable polymers* has been used [20].
2. *Diffusion healing* is where a soluble linear polymer molecule that can diffuse by reptation through the free volume of the thermoset matrix to bridge a hairline crack in the matrix is included. It is often referred to as *thermal self-healing* [21].
3. In *autonomic healing*, an isolated liquid resin and catalyst are incorporated, which can be activated when a crack propagates. This is commonly achieved by dispersing microencapsulated resin or monomer throughout the resin matrix. The monomer or resin is released when a crack impinges the microcapsule. Activation of the healing chemistry is induced by a dispersed catalyst [22].
4. *Vascular systems* are a variant of autonomic healing in which hollow fibres (usually glass fibres) are incorporated in the composite with the reinforcing fibres [23]. Some contain the resin or monomer healing agent, while others contain the liquid hardener (or catalyst). Fracture of the hollow fibres releases the liquid resin and hardener, which can mix and cure within the cracks. An alternative to this approach is to create a true vascular system by curing the matrix around a number of long, randomly contoured 'fibrous' formers, which can be removed to leave a vascular system of 'veins' through which the liquid healing agent can be circulated [24].
5. Other healing methodologies have been reported in the literature, which combine different chemistries, ionomers [25] and molten thermoplastic infusion [26], to repair damaged areas.

The advantages and disadvantages of these systems are described in Table 9.4. Many systems involve a resin or composite that employs innovative chemistry, and as such have limited the introduction of self-healing technologies into composite structures. Aerospace structures, for example, require extensive testing before application, so new matrix resins and systems are only introduced when additional extensive research is unnecessary because the modification is small and does not affect the mechanical properties. Only thermal self-healing [2] employs essentially the same chemistry as the unmodified matrix.

Table 9.4. Self-healing strategies

Self-healing system	Mechanism	Examples	Comment	Advantages	Disadvantages
1. Re-mendable polymeric matrix	1. Reversible polymerization 2. Reversible Diels–Alder chemistry of backbone and crosslinks 3. Self-assembly	1. PMMA thermoplastic 2. Furan–maleimide polymers 3. Ionomers	PMMA not normally employed as a matrix	1. Thermal activation 2. Multiple heal cycles	1. Intrinsic degradation chemistry at high temperatures 2. Special polymer not normally used for composites
2. Thermal self-healing; single-phase system	Reptational diffusion with entanglement	Linear end-capped polyepoxide dissolved in crosslinked epoxy resin	Ionomer or other soluble self-assembling linear oligomers can be used	1. Self-healing capability of conventional structural epoxy matrix resins without loss of matrix properties 2. Solid-state 3. Multiple healing cycles 4. Thermal activation	1. Lack of space-filling capability 2. High viscosity of epoxy resin with dissolved healing agent limits application to prepreg processing; ionomers provide a self-assembling healing agent and hence lower resin viscosity 3. Restricted to hairline cracks
3. Autonomic	Dispersed microcapsules and/or 'isolated' catalyst	1. Metathesis polymerization.	Dicyclopentadiene/ Grubbs catalyst (ruthenium IV compound)	Space-filling capability of healing agent	1. High concentrations of microcapsules required for healing 2. Separate protected dispersion of catalyst 3. Expensive ruthenium catalyst 4. Elevated temperatures are still beneficial 5. One cycle
4. Vascular systems	1. Hollow fibres 2. Venus system	Step-growth polymerization of epoxy resin/hardeners		1. Delivery system for matrix resin and healing agent 2. Matched resin repairs	1. Large diameter of hollow fibres 2. Thermal curing 3. Circulating pump for vascular system
5. Thermoplastic additives; phase-separated system	Infusion of molten thermoplastic	Dispersion of particulate thermoplastics	Mechanism for pressure infusion involves in situ water vapour	1. Highly applicable technique 2. Solid state	Thermoplastic dispersion and repair limits mechanical performance before and after healing

9.3.1   Re-mendable Matrix Polymers

To understand the background to this methodology, we briefly discuss the relevant concepts of polymerization and depolymerization.

9.3.1.1   Chain Addition Polymerization–Depolymerization

Fundamentally, many linear polymers have the potential for thermal healing because chain addition polymerization has a thermodynamic equilibrium nature. Providing 'active' chains still exist, polymerization/depolymerization is reversible, as shown in eq. (9.5):

$$P_n^* + M \rightleftarrows P_{(n+1)}^* + M \rightleftarrows P_{(n+2)}^* + (m-2)M \rightleftarrows P_{(n+m)}^*, \tag{9.5}$$

where P_n^*, P_{n+m}^* are active polymers of degree of polymerization n and $n+m$, and M is the monomer. The propagating centres can be free radical, anionic, cationic, or other covalent but reactive end-groups.

An equilibrium constant, K_p, can be defined:

$$K_p = \frac{[P_n^*]_e}{[P_n^*]_e [M]_e} = \frac{1}{[M]_e}, \tag{9.6}$$

where $[M]_e$ is the equilibrium monomer concentration, and $[P_n^*]_e$ and $[P_{n+m}^*]_e$ are the equilibrium concentrations of polymer and are identical. The subscripts n and m are retained so that the chain propagation mechanism is clear.

The potential for a chemical reaction is provided by the Gibbs free energy, G, which is described as:

$$\Delta G_p = \Delta H_p - T\Delta S_p, \tag{9.7}$$

where ΔG_p, ΔH_p, ΔS_p are the free energy, enthalpy, and entropy of polymerization, respectively, and T is the temperature. For polymerization, ΔG_p has a negative value; when the value is positive the polymer chain will depolymerize.

Since ΔS_p will have a negative value because the degrees of freedom will be lost as the chain propagates, ΔH_p will need to have a negative value for ΔG_p to also be negative and for a polymer to form. Inspection of eq. (9.7) shows that $T\Delta S_p$ becomes large and negative as the temperature increases, and $-T\Delta S_p$ becomes increasingly positive. Thus, as the temperature is increased monomer is favoured over polymer. The temperature at which only monomer exists is called the *ceiling temperature*, T_c.

The standard state Gibbs free energy for polymerization, ΔG_p°, is related to the equilibrium constant, K_p:

$$\Delta G_p^\circ = -RT\ln K_p \tag{9.8}$$

and

$$T_c = \frac{\Delta H_p^\circ}{\Delta S_p^\circ + R\ln[M]_\circ}, \tag{9.9}$$

where ΔH_p° and ΔS_p° are the standard state values of $\Delta H_p, \Delta S_p$, respectively, $[M]_\circ$ is the concentration of monomer at T_c where the polymer is unstable, and $[M]_e = [M]_\circ$.

Polymers such as the acetals, which polymerize/depolymerize through a cationic mechanism to exhibit a ceiling temperature at 127 °C, can be expected to show some healing capability. Indeed, process temperatures exceed this temperature. Polymers based on polyformaldehyde need to be thermally stabilized by either end-capping or copolymerization for processing. Therefore, scission and reformation of skeletal bonds can be expected during thermally activated healing. In the case of polymethylmethacrylate (PMMA), cast sheet contains occluded free radicals, which provide a mechanism for healing capability. The major technical problem with these polymers is the speed of chain depolymerization above T_c, so that a means of locating the released monomer in the damaged region is required. In principle, copolymers and/or branches or crosslinks provide an option for limiting the lengths of chains, which can unzip.

9.3.1.2   Step-Growth Polymerization

This mechanism of polymer synthesis involves the statistical reaction of multi-functional monomers, which contain two or more reactive functional groups. With bi-functional monomers, linear polymers are formed. Considering linear polymer synthesis, monomers can be either bi-functional with the same groups (A———A) and (B———B) or different reactive groups (A———B). Thus, either monomers A———A and B———B or A———B can be used for linear polymer synthesis:

$$
\begin{aligned}
&A---A+B---B \rightarrow A---AB---B \\
&A---AB---B+A---A \rightarrow A---AB---BA---A, \text{ etc.} \\
&A---AB---B+A---AB---B \rightarrow A---AB---BA---AB---B, \text{ etc.}
\end{aligned}
$$

(9.10)

or

$$
\begin{aligned}
&A---B+A---B \rightarrow A---BA---B+A---B \rightarrow \\
&A---BA---BA---B.
\end{aligned}
$$

(9.11)

These reactions can occur with or without the loss of a small molecule. The Carothers equation provides a prediction of the average degree of polymerization, $\overline{P}_n$:

$$
\overline{P}_n = \frac{1}{1-\rho},
$$

(9.12)

where ρ is the extent of reaction of the functional groups. For a high molecular weight polymer, a large $\overline{P}_n$ can only be achieved when an extremely high extent of reaction of the functional groups occurs. This can only be achieved using the bi-functional monomer A———B, where the concentrations of A and B are equal. Alternatively using A———A and B———B monomers requires techniques to ensure that the concentrations of A and B are equal.

With functionality $f > 3$, branched polymers form, as shown in Figure 9.13 and as the extent of reaction increases crosslink form.

Some healing capability can be available in this polymer because the extent of reaction cannot reach 100% and annealing of the microcracked thermoset leads to further reaction (i.e. 'post-curing'). Certain functional groups can be involved in

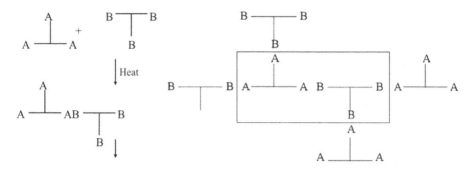

Figure 9.13 Schematic of step-growth polymerization of three-functional monomers, illustrating branch formation.

reversible skeletal bond formation and a healing mechanism. For example, in polyester systems ester interchange provides the potential for healing:

$$COO---R' + HO---R \rightleftarrows ---COO---R + HO---R', \quad (9.13)$$

where AB is represented by –COO– and B by HO–.

The problem with this mechanism is the limited mobility of the molecular segments, which is required to bring the groups into a reactive volume.

Chen and coworkers [20] used retro Diels–Alder chemistry to provide reversible bond formation. In this case, the A and B groups are respectively a diene and dienophile. The dimerization of 1,3-cyclopentadiene at room temperature is a classic example of this chemistry, as shown in eq. (9.14):

$$\text{(9.14)}$$

diene   dienophile   dicyclopentadiene

The reaction can be thermally reversed at an elevated temperature. The so-called retro Diels–Alder reaction is:

$$\text{(9.15)}$$

In order to achieve re-mendable properties in a polymer, Chen et al. [20] synthesized a number of polymers based on differing diene–dienophile combinations. For example, furan-maleimide chemistry provided a demonstrable re-mendable polymer. An example is given in Figure 9.14.

The polymer in Figure 9.14 exhibited 57% recovery in strength, while an improved system based on bismaleimide isopropane (2MEP) and tetra-furan (4F) exhibited 83% recovery in strength.

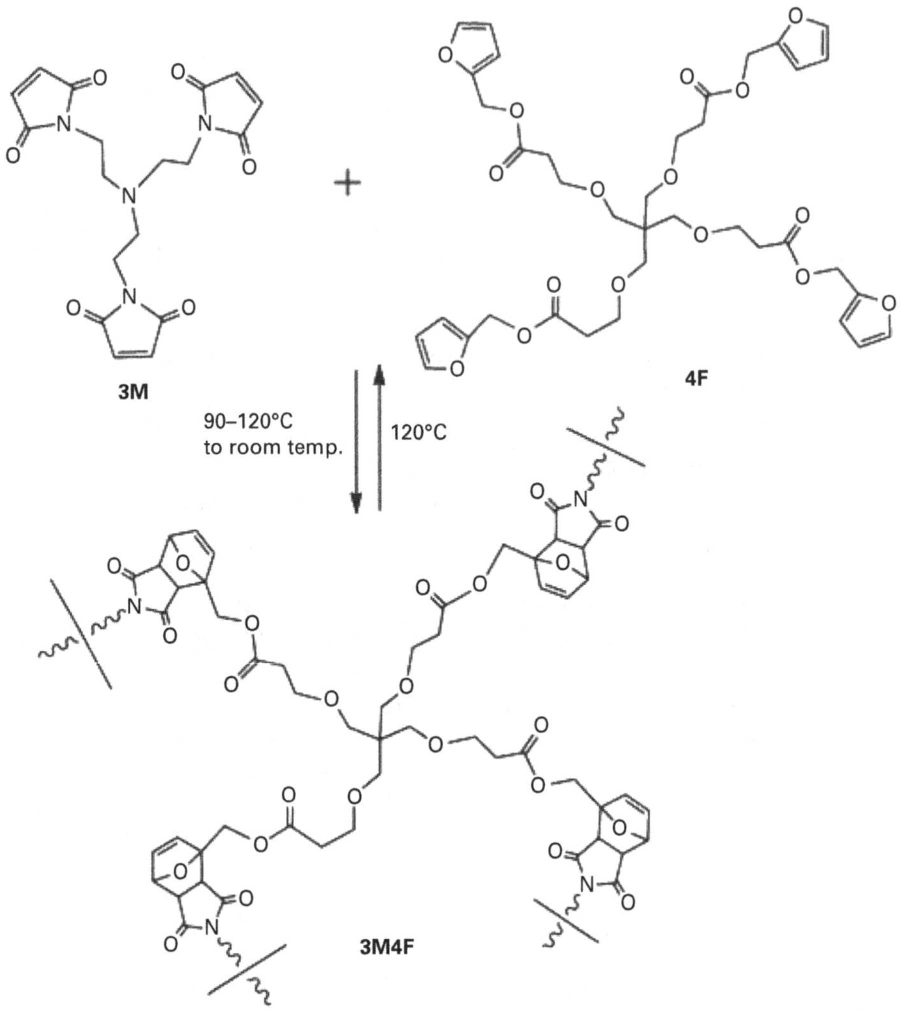

Figure 9.14 Example of a re-mendable polymer that exploits a retro Diels–Alder reaction between a tris-maleimide (3M) and a tetra-furan (4F) in a step-growth system.

9.3.1.3 Ionomers

Ionomers are polymers that have reversible crosslinking, enabling elastomers to be moulded with conventional thermoplastic techniques. The ionomer bonds are reversible, as shown in Figure 9.15, so that the 'rubbery' properties associated with crosslinking can be achieved without the permanent crosslinks associated with vulcanization.

Varley [25] showed that commercial ionomers exhibit recovery of mechanical performance after ballistic perforation. The reformation of ionomeric clusters (Figure 9.16), together with the flow of the non-crosslinked polymer, provides a mechanism for healing.

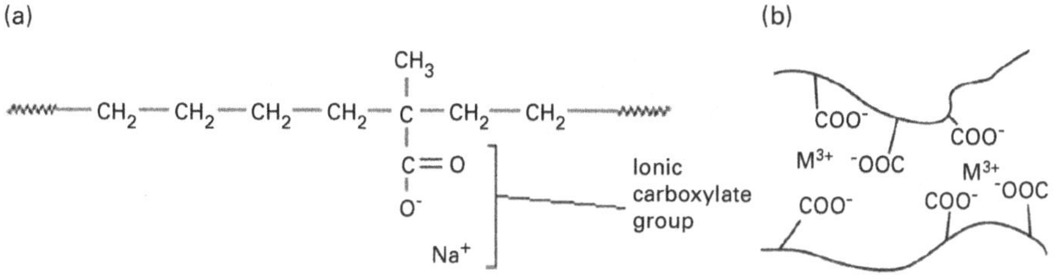

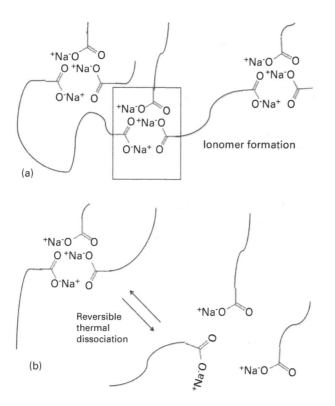

Figure 9.15 (a) Typical structure of an ethylene-co-methacrylic acid copolymer neutralized with Na⁺; (b) clustering of ionic groups illustrating ionomer crosslinking in a thermoplastic elastomer.

Figure 9.16 Schematic of (a) ionomer formation via association of Na carboxylate polymer end-groups; (b) reversible thermal dissociation.

9.3.2   Thermal Self-Healing Single-Phase Systems

The autonomic healing strategies involve embedding micro-containers into the composite, which release a polymerizable monomer into a matrix containing a dispersed catalyst when a crack forms. Variants involve hollow fibres containing both monomer or resin and catalyst or hardener separately. The disadvantage is that

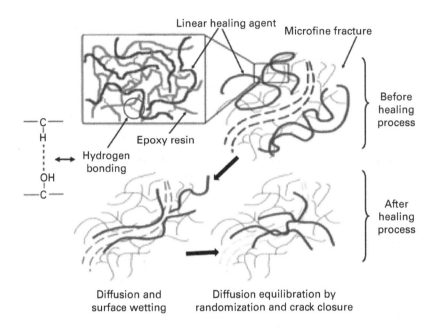

Figure 9.17 Schematic of diffusional solid-state healing showing the role of a linear polymeric healing agent [26,28].

the microcapsules or hollow fibres are of larger dimension than the reinforcing fibres, and hence introduce a stress concentration into the structure. Ideally, the conventional epoxy resins used as matrices should be easily modified to ensure multiple healing capability without a reduction in mechanical and thermomechanical performance.

Jones and coworkers [21] recognized that polymers contain a so-called unoccupied or free volume. Thus, the density of a polymer is commonly less than that of a 100% crystalline morphology. With crosslinked thermoset epoxies used as matrices for many composite structures, the unoccupied volume is related to the cure schedule and cooling protocol [27]. Healing capability was introduced by dissolving a linear polymer molecule, which can diffuse through the unoccupied or free volume to bridge crack surfaces. The concept is described in Figure 9.17.

The requirements of a healing agent for diffusional healing can be defined as follows [21,29]:

1. should be *soluble* in the continuous matrix phase;
2. should be *reversibly* bonded to the crosslinked epoxy resin through intermolecular or hydrogen bonding;
3. should become mobile above a minimum temperature for healing by diffusional crack bridging;
4. the dissolution of a linear chain molecule healing agent at the required concentration should not reduce the thermomechanical and mechanical properties significantly.

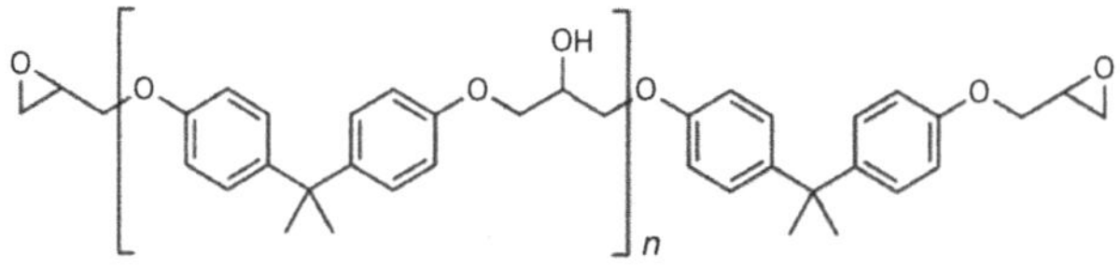

Figure 9.18 Structure of a typical bisphenol A epoxy resin. The end-groups are involved in the curing reactions. n has values of 0–34.

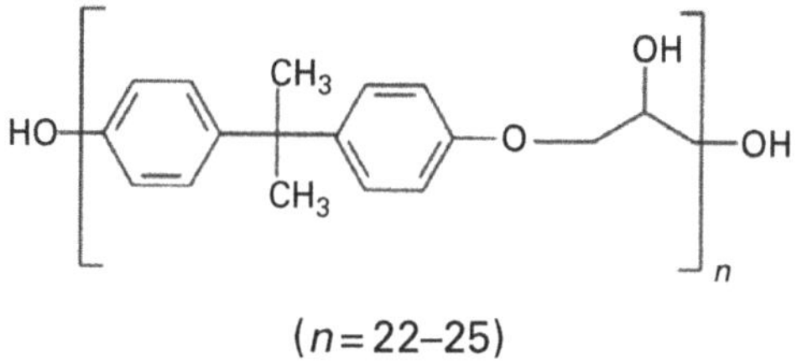

$(n = 22–25)$

Figure 9.19 Structure of a phenoxy polymer which is an hydroxyl end-capped bisphenol A epoxy. n can range from 22 to 25. Longer-chain polymers with $n = 150$ are also available. $M_1 = 294$ g mol^{-1}.

9.3.2.1 Design of Healing Agent

Solubility can be identified by comparing the solubility parameters of the matrix resin with the healing agent. The rule of thumb is that molecules with the same solubility parameter (δ) should be equal or within ± 2 MPa$^{1/2}$:

$$\delta_{HA} = \delta_m \pm 2 \text{MPa}. \tag{9.16}$$

This means that molecules with similar molecular structures will be completely miscible. A full discussion of solubility parameters is given elsewhere [30]. From the point view of providing healing capability to cured epoxy resins, Hayes et al. [31] dissolved a linear bisphenol A epoxy without epoxy end-groups. The end-groups can be capped with non-functional groups to ensure that the healing agent is *not* co-reacted into the cured network and remains mobile. Figures 9.18 and 9.19 illustrate how this can be achieved.

Jones [21,26] has calculated the effect of the structure of a hardener on the solubility parameters of a range of cured epoxy resins. The solubility parameter of DGEBA is 21.36 MPa$^{1/2}$, while a TGAP resin cured with DDS has a value 24.06 MPa$^{1/2}$ and TGDDM cured with DDS has a value of 22.30 MPa$^{1/2}$. While these values satisfy eq. (9.17), we could expect a limit to the concentration of healing agent that can be included. The other factor determining the solubility of the healing agent is its molecular weight. It is observed [26,28] that the reduced healing efficiency correlates with the phase separation of the healing agent with molecular weight of 44,000 at concentrations between 7.5 and 10%.

The *molecular weight* of the healing agent is an important parameter that controls both its effectiveness and healing efficiency. We can explore this by briefly examining the physics of polymer diffusion. The diffusion of a polymer molecule within a

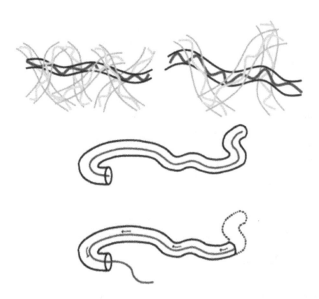

Figure 9.20 The 'polymer chain in a tube' model illustrating diffusion of a polymer molecule by reptation. (a) Segments of randomly coiled polymer molecules representing the unoccupied volume; (b) reptation mechanism [26,38].

polymer is considered to occur by a reptation mechanism, as shown in Figure 9.20 by the 'chain in a tube' model [32–37].

Polymer molecules can diffuse through the bulk material because of the occluded unoccupied and free volume. Free volume is defined as the minimum space required for the chain segments within the randomly coiled polymer molecules to crossover. Crosslinked polymers have additional unoccupied volume associated with glass formation on curing and/or cooling of the network. According to Edwards [32,36], an individual molecule is confined to a virtual tube along which it can diffuse in a translational manner, as shown in Figure 9.20. De Gennes [37] was the first to describe diffusion of a free chain by reptation along the virtual tube. The time to diffuse out of the temporary tube was given by:

$$\tau_{rep} \approx \frac{L^2}{D_p} \approx N^3, \qquad (9.17)$$

where τ_{rep} is the time required to leave the virtual tube, L is the contour length of the tube, N is the number of links in the chain, and D_p is the self-diffusion constant for the polymer. D_p is related to the structure of the polymer as given in eq. (9.18):

$$D_p \approx \frac{r^2}{\tau_{rep}} \approx N^{-2}, \qquad (9.18)$$

where r is the average end-to-end distance of the random coil.

Experimentally, τ_{rep} and D_p are found to be related to the molecular weight of the polymer (M) according to $\tau_{rep} \approx M^{3.4}$ and $D_p \approx M^{-2.3}$. Thus, efficient healing requires a healing agent of low molecular weight, but we need to consider the minimum entanglement length of the healing agent chains to ensure a strong repair is achieved. Wool [39] has provided an estimate of the critical molecular weight for entanglement of a chain, M_c:

$$M_c \approx 30 C_\infty M_\circ, \tag{9.19}$$

where $M_\circ$ is the molecular weight of the monomeric unit, and C_∞ is the characteristic ratio of the polymer chain and is a measure of the extent of coiling of the chain, as given in eq. (9.20):

$$C_\infty = \frac{<r^2>_\circ}{nl^2}, \tag{9.20}$$

where n is the number of bonds of bond length l, and $<r^2>_\circ$ is the square of the unperturbed average end-to-end distance of the chain. The characteristic ratio is a function of n, so C_∞ represents the value for the chain as n approaches ∞.

'Rigid' chains have larger values than 'flexible' chains: a flexible freely rotating tetrahedrally bonded chain, C_∞, has a value of 2.0, whereas the more rigid polystyrene chain, with a pendant phenyl ring, has a value of 10.2.

Wool [39] has calculated the value of M_c for a series of linear polymers and for polycarbonate, which has a structure based on bisphenol A, with an M_c of 4,300–4,800 g mol^{-1}. Therefore, the healing agent should have a minimum molecular weight of 5,000 g mol^{-1} to achieve effective healing. This has been shown to be a good representation of the requirements of healing agents for diffusional healing, as indicated by the variation of healing efficiencies of epoxy resin with molecular weight [26].

9.3.2.2 Self-Assembly

One of the disadvantages of the diffusion method of introducing healing functionality is the reduced control over processing technologies. This arises from the fact that dissolution of a polymer increases the viscosity of the resin. Thus, liquid moulding techniques such as resin transfer moulding (RTM) require higher pressures for flow. Control over the viscosity of toughened epoxies is achieved, for example, by dispersing the thermoplastic toughening agent as a fibre within the reinforcing fibres [40,41]. On curing of the impregnated composite, the thermoplastic dissolves in the matrix resin and then phase separates into a co-continuous phase required for high fracture toughness. A similar technology could be applied to the healing agent. An alternative approach is to employ a low molecular weight healing agent that self-assembles after processing into the molecular weight required for efficient healing. Jones and Varley [42,43] synthesized an ionomer from a bisphenol A glycidyl ether which self-assembles in the cured resin to provide high strength recovery in a healing cycle. Figure 9.21 gives an example of how this is achieved. Figure 9.16 shows how the ionomer assembles to extend the chain length of the polymeric healing agent.

Healing efficiency is determined from the load to fracture of a healed fracture toughness specimen compared to the control:

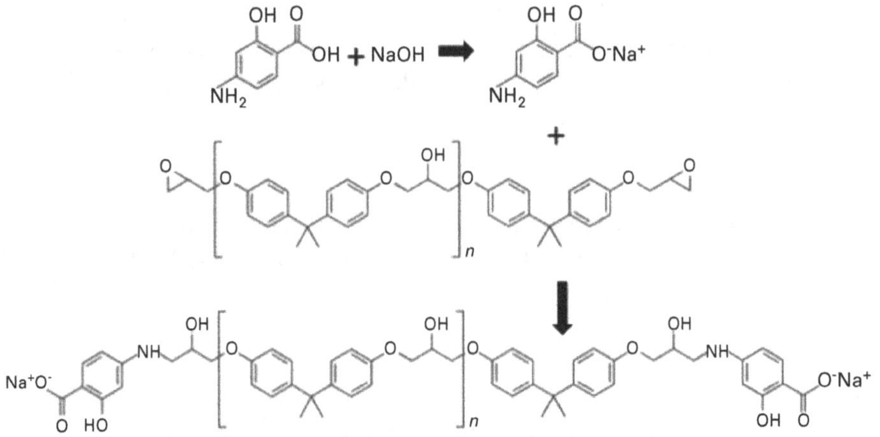

Figure 9.21 Synthesis of an ionomeric healing agent: bisphenol A epoxy end-capped with sodium salicylic acid [42].

$$\text{Healing efficiency} = \frac{K_{1c}\,(\text{healed})}{K_{1c}\,(\text{control})} \times 100 \approx \frac{P_u(\text{healed})}{P_u(\text{control})} \times 100 \qquad (9.21)$$

In Figure 9.22 the fracture loads (P_u) of single edge notch (SEN) coupons were compared. The values given are an average of seven individual tests. The figure shows the healing efficiency of fractured SEN epoxy resin specimens. The ionomer-based system (Na ionomer) has a healing efficiency that compares favourably to the 'long-chain' healing agents (DGEBA 1 and 2, molecular weight 44,000). The commercial Phenoxy (PKFE, molecular weight 16,000) achieved a similar efficiency. Phenoxy PKHP (molecular weight 13,000) achieved a higher efficiency. However, it has been demonstrated that the weight concentration of 7.5% is close to the maximum solubility before phase separation [26]. The differences may arise from differing degrees of phase separation of the healing agent, which reduces the healing ability available via this mechanism. The low molecular weight healing agents have similarly low efficiencies to the controls. Thus, it is concluded that ionomer formation is an effective technique for extending the molecular weight of healing agents and hence thermal healing. Healing can be accomplished over more than seven cycles of fracture and heal. The experiments were conducted to establish the repeatability of the healing effect. The control sample (without healing agent) shows some healing capability over two cycles resulting from post-curing of the resin. The healing agents with low molecular weights exhibited only a minimal healing improvement over that associated with additional curing. Since the non-reactive analogue of the ionomer is not efficient, it is concluded that self-assembly of the ionomeric healing agent is a good mechanism for extending the effective molecular weight of a low molecular weight DGEBA for healing.

9.3.3   Autonomic Systems

The term *autonomic* refers to systems in which healing is initiated immediately on fracture. This can be achieved by incorporating containers for the healing agent(s) that

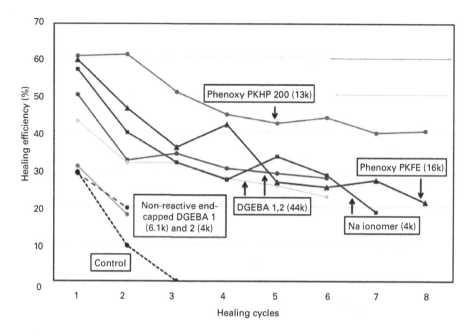

Figure 9.22 Thermal healing of an un-toughened simulated aerospace epoxy resin based on 4,4-DDS cured DGEBF and TGAP using SEN specimens containing 7.5% DGEBA-based healing agents in the epoxy component. (a) Control without healing agent; (b) non-reactive end-capped DGEBA (molecular weight 6,100; end-cap 1); (c) non-reactive end-capped DGEBA (molecular weight 4,000; end-cap 2); (d) DGEBA (molecular weight 44,000; DGEBA 1 and 2); (e) Phenoxy PKHP 200 (molecular weight 13,000) and PKFE (molecular weight 16,000); (f) Na ionomer of DGEBA (molecular weight 4,000).

fracture on impingement. Much of the skill is in the chemistry of synthesizing microcapsules containing the monomer. This is often done using microemulsion techniques and employing urea–formaldehyde to create hollow 'microspheres' containing monomers (Figure 9.23). In the original work the monomer was dicyclopentadiene, which can be polymerized by ring opening metathesis polymerization (ROMP) at room temperature using a Grubbs catalyst, as shown in Figure 9.23.

Grubbs catalysts are based on ruthenium compounds (Figure 9.24). In this seminal work the healing agent, dicyclopentadiene, was employed with different Grubbs catalysts. First-generation Grubbs catalysts have been superseded by second- and third-generation (Hoveyda–Grubbs) catalysts, which provide more efficient healing and stable products.

To achieve healing of composite materials, a number of practical aspects need to be addressed:

1. effective containment of the healing agent and catalyst:
   – microencapsulation of the healing agent
   – dispersion of the catalyst in the resin;
2. a compromise between the dimensions of the micro-containers and volume fraction is needed:

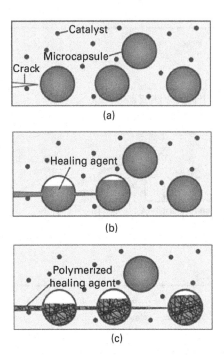

Figure 9.23 (a) Microencapsulated healing agent embedded in a matrix containing a dispersed catalyst capable of polymerizing the healing agent. (b) Cracks that form in the matrix propagate and rupture the microcapsules, releasing the healing agent into the crack plane through capillary action. (c) The healing agent contacts the catalyst, initiating polymerization, and bonds the crack faces [22].

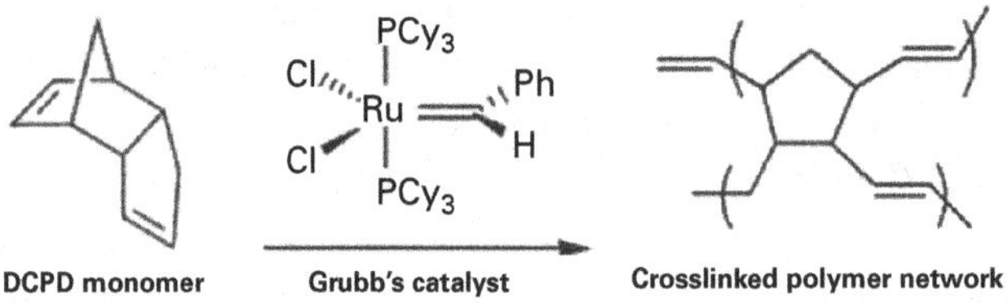

Figure 9.24 Ring opening polymerization of dicyclopentadiene with Grubbs catalyst [22].

- Dimensions should be similar to the reinforcing fibres (7 µm for carbon fibres and 15–20 µm for glass fibres), otherwise stress concentration associated with the fractured microcapsules can reduce the fracture toughness of the matrix resin
- sufficient volume of healing agent needs to be released when the microcrack impinges the microcapsule(s);
3. monomeric healing agents are preferred so that capillary transport occurs;
4. cost of ruthenium catalysts;

5. the introduction of components of the healing system should not reduce the performance of the composite material and structure. For application to aerospace structures the new material needs to be fully validated.

Many studies have identified the best conditions to achieve efficient healing using microcapsules. It was observed that employing wax dispersion, which improved the rate of dissolution in the monomeric healing agent, could reduce the optimum concentration of dispersed catalyst.

Other variants include the use of 'solvent' healing agents and mixed microcapsules of resin monomers and hardeners. This technique has been reviewed in detail elsewhere [43,44].

9.3.4   Vascular Healing

9.3.4.1   Hollow Fibres

There are two major disadvantages to the use of microencapsulated healing systems:

1. the limited volume of healing agent available for healing of cracks; and
2. the components of matrix resin cannot be readily used as healing agents – for epoxy resins an epoxy monomer and hardener in the correct proportions need to be released.

These disadvantages can be partially addressed by using *hollow fibres* as the containers [23]. Figure 9.25 is a micrograph of the hollow glass fibres (HGFs).

Hollow glass fibres can be filled with healing agents under vacuum with either

1. pre-mixed epoxy components; or
2. individual hollow fibres containing either epoxy resin or catalyst or hardener:
   – stoichiometric quantities of hardener and resin can be achieved by matching the volume fractions of the epoxy-filled HGF with hardener or catalyst-filled HGFs;
   – higher volumes of healing agent can be potentially released on fracture.

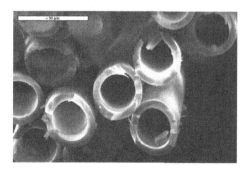

Figure 9.25. Micrograph of HGFs with an external diameter of 60 ± 3 μm and internal diameter of ≈ 40 μm [23].

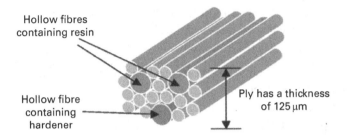

Figure 9.26 Schematic showing alignment and packing of HGFs containing resin and hardener within a 90° E-glass ply of a 16-ply balanced angle-ply laminate [23].

As with microcapsules, the dimensions of the hollow fibres (35–60 μm with internal diameter of 25–40 μm) are significantly larger than the reinforcing fibres. However, they can be aligned with the reinforcement fibres, ensuring that minimum disturbance of the microstructure occurs. Figure 9.26 shows how this is achieved.

Using this technique, Bond et al. [23] have reported 87% retention of flexural strength after damage induced by indentation and healing at 200 °C for 2 h.

9.3.4.2   Vascular Systems

In analogy with a human venous system, a micro-vascular network needs to be introduced into the composite to enable fluid transport to the damaged region. To achieve this, a pumping arrangement is required. The design of these networks is discussed elsewhere [24,45,46]. The limitation of this approach is currently defined by manufacturing technology for achieving vascular arrangements. There are a number of approaches in development:

1. *Continuous sacrificial fibres* embedded in a random or 2D or 3D network arrangement within the matrix resin, which can be removed after curing to provide a venous system. Traditional technologies would employ a low-melting temperature fibre (wax, polymer, or metal) that could be melted for removal by flow.
2. *Continuous hollow fibres* embedded in a random arrangement.
3. Machining of *capillaries* into the matrix using, for example, laser ablation.
4. Additive manufacture (or dot matrix printing) provides the best current technology for designing a composite with a vascular system. Whether this can be combined with an arrangement for reinforcing fibres for high performance needs further research.

9.3.5   Thermal Self-Healing Phase-Separated System: Thermoplastic Infusion

An alternative to the single-phase solid-state diffusional model described above, in which diffusion of a soluble linear molecule provides effective healing, is a two-phase solid-state system. In this case, thermoplastic polymers, mainly as particles or inter-mingled fibres, are dispersed in the thermosetting polymer matrix. Healing occurs by infusion of the molten thermoplastic into the microcracks and cracks. Three aspects of the thermoplastic are critical;

1. particle size and degree of dispersion of the healing agent: control ensures that sufficient quantities of healing agent can be delivered to the damaged region;
2. viscosity of the molten healing agent: a minimum viscosity facilitates sufficient flow into the damage for healing, especially for larger cracks;
3. chemical interaction between the healing agent and matrix: optimum interaction ensures restoration of mechanical properties of the healed material. The so-called, *pressure delivery mechanism* is also a function of the interaction between the matrix and the thermoplastic.

Figure 9.27 shows how the healing mechanism involves bleeding of the molten healing agent into the damage. The original work of Zako and Takano [47] employed epoxy-terminated thermoplastic particles with a melting point of 200 °C. These solid particles had a size <50 μm and could be included at a volume fraction of 40%, and were found not to compromise the stiffness of the composite. Luo et al. [48] used poly (ε-caprolactone) that phase-separated during curing of the epoxy to provide a so-called 'bricks and mortar' methodology. Healing occurred at 190 °C, which is much above the melting point of poly(ε-caprolactone) (59 °C), so bleeding into the damage occurred. Alternative strategies use shape-memory polymers to close the crack prior to thermoplastic infusion [49,50]. Poly(ethylene-co-methacrylic acid) (EMAA) particles were used with a room-temperature-cured epoxy amine resin by Meure et al. [51], which was effectively healed at 150 °C for 30 minutes. Further research [52,53] identified a pressure delivery mechanism. Microscopy revealed the presence of a single large bubble within each EMAA particle, which expanded upon heating at 150 °C to push and expand the molten EMAA (T_m ~85 °C) into cracks between the fracture surfaces. This was attributed to the chemistry given in Figure 9.28, whereby water molecules are generated on the surface of the EMMA particles, which coalesce

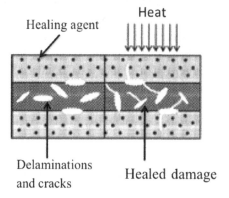

Figure 9.27 Schematic of two-phase solid-state healing illustrating how the dispersed healing agent flows into delamination and cracks [26].

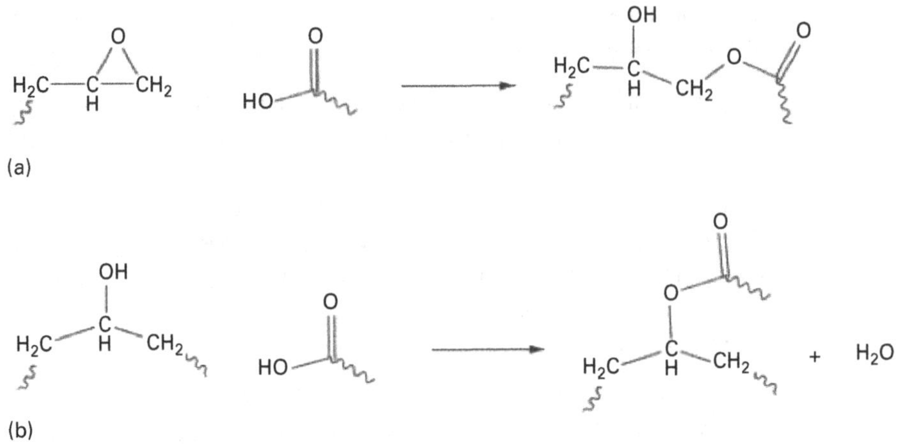

(a)

(b)

Figure 9.28 Potential amine-catalysed reactions occurring between surface COOH groups on EMAA particles. (a) Residual epoxy groups provide an adhesion mechanism for the thermoplastic within the resin cracks; (b) OH groups formed during cure for introduction of the local pressure delivery mechanism.

into a large bubble during curing to provide a dormant pressure mechanism that can be activated during healing.

Wang et al. [54] used thermoplastic films such as poly(ethylene-co-methyl acrylate) (EMA) and EMAA as interlayer toughening agents in an epoxy amine carbon composite. It is therefore attractive to employ the healing agent in a form that can also function as a toughening agent. The company Cytec [41] used intermingled PES fibres within the textile reinforcement, which dissolve in the epoxy resin and phase-separate into an optimum morphology during curing for maximum fracture toughness. This technology enabled the increase in viscosity associated with dissolving the toughening agent used for prepreg manufacture to be avoided. In this way, RTM moulding techniques could be employed. Meure et al. [55] reported the use of comingled EMAA fibres in carbon-fibre epoxy laminates. Thermoplastic fibres with diameters of 50–75 μm and 100–150 μm at two different concentrations were compared with EMAA particle modified composites. It was shown that the pressure delivery healing mechanism was unaffected. Figure 9.29 illustrates the healing mechanism for the healing agent stitched textile composite.

Subsequently, a greater number of thermoplastics have been examined for solid-state healing. The use of this technology in structural composites requires the demonstration that mechanical performance is not compromised by the incorporation of the thermoplastic healing agent. Readers are referred to a recent review of solid-state healing [26].

9.4   Smart Composites

These are materials that have multi-functional properties combining damage detection and healing capability. Thus, regular scanning of the structure would identify damage,

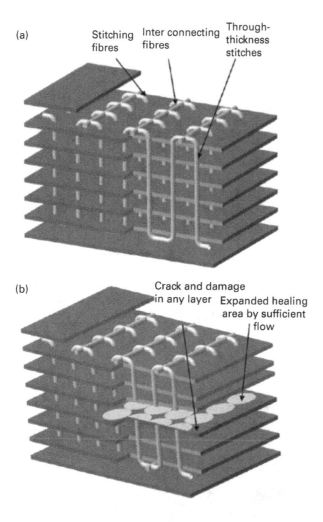

(a)

Stitching fibres

Inter connecting fibres

Through-thickness stitches

(b)

Crack and damage in any layer

Expanded healing area by sufficient flow

Figure 9.29 Representation of 3D stitched composite showing the intermingled fibrous healing agent (a) before and (b) after healing [26].

which can be healed by initiating the appropriate mechanism. It should be noted that *autonomic* systems should not need the detection technique since, by definition, healing is triggered when the damage occurs. However, only one healing cycle is possible. But with multiple healing capability, a damage-detection technique and the means of inducing healing are required.

9.4.1 Carbon-Fibre Composites

Jones and coworkers [21,29] provided the fundamental technology for carbon-fibre composites. It was observed that the electrical conductivity of carbon fibre could be used to detect barely visible impact damage (BVID), while the thermal conductivity

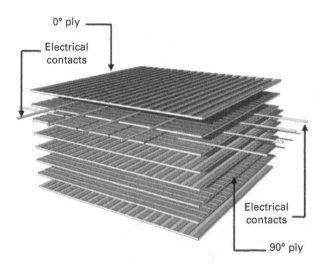

Figure 9.30 Schematic of a self-sensing arrangement for a cross-ply carbon-fibre composite. Adjacent 90° and 0° plies have electrical contacts implanted so that the electrical conductivity of the carbon fibres can be monitored and used to create a map.

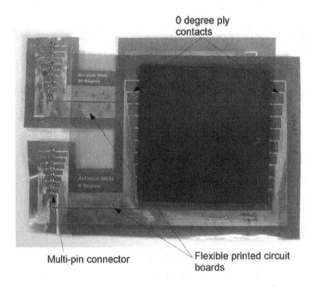

Figure 9.31 A CFRP panel containing flexible printed circuit boards as interleaves for connection with two plies of the laminate [57].

and electrical resistance could be employed to provide the stimulus for thermal healing. Figure 9.30 illustrates the concept for detecting impact damage using carbon fibres. From a practical point of view, the plies need to be electrically isolated, usually with an interlayer, although some reinforcement systems may result in an *in situ* resin-rich interlayer or specifically include a toughening interlayer. Figure 9.31 shows a 'flexible circuit board' technique for simply introducing the contacts and interlayer.

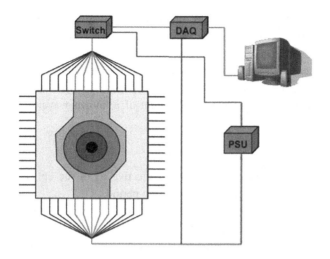

Figure 9.32 The concept of sense-and-heal in a carbon-fibre composite with fibre self-sensing of BVID and intrinsic thermal self-healing using the fibres to provide localized heating to the damaged area. Data acquisition (DAQ) maps the damage then switches to the power supply unit (PSU) for healing.

Hou and Hayes [56] demonstrated that BVID could be detected from measurements of electrical conductivity of the carbon fibres. Surprisingly, the absolute resistance of the embedded fibres decreased as a result of an impact event, which introduced BVID, while higher energy events, which caused fibre fracture, showed an increase in absolute resistance. Thus, matrix damage could be detected because of changes in stress-related piezo-resistive behaviour of the carbon fibres. They also demonstrated that damage could be identified over a 10 m distance. The concept of sense-and-heal was discussed by Hayes and coauthors [29,57] and is described schematically in Figure 9.32 [57].

9.4.2 Glass-Fibre Composites

Optical self-sensing can be achieved by ensuring that the reinforcing fibres act as light guides. Clearly, light transmission is reduced when fibres break so that light transmission of a bundle can indicate progressive damage. The major requirement for sensing is that the refractive index (RI) of the matrix is less than that of the fibres. Conventional E-glass has an RI of 1.54 at 500 nm, but high-performance resins have similar values (1.57–1.6). Thus, cladding the fibres with a low-RI coating or the modification of the matrix polymer is necessary for sensing. Alternatively, the composition of the fibres could be modified to increase the RI. Chalconide and rare earth (e.g. niobium) element doped glasses, referred to as H-glasses, have been examined [58]. These solutions are not so easily introduced commercially. However, these fibres with micrometre dimensions could be included as sensors within the E-glass reinforcement, which is a significant improvement over traditional optical fibre sensors. *Traditional optical fibres* are polymer-clad, doped-silica fibres of diameter in excess of 125 μm. These are reviewed in detail elsewhere [59]. In order to use optical fibres as sensors, sections of the cladding are removed and etched with a Bragg grating. A Bragg grating consists of a series of parallel lines of designed spacing, which act as selective mirrors that reflects light of a known wavelength when the wavelength satisfies the Bragg equation:

$$\lambda_B = 2n_e\Lambda, \tag{9.22}$$

where λ_B is the Bragg wavelength, n_e is the effective RI of the optical fibre in the region of the Bragg grating, and Λ is the grating period.

The pitch of the grating is modulated by strain, so the wavelength of reflected light changes correspondingly and therefore the extension and thus strain in the fibre and the composite can be recorded. Furthermore, under impact loads the strain waves can be monitored via the modulation of the strain in the presence of an acoustic wave. More than one Bragg grating can be written onto an optical fibre, with differing pitch providing individual monitoring of different areas. The disadvantages including the following:

1. The sensor measures strain but not damage; it is not a direct measure of damage, and so requires understanding for interrogation.
2. The typical diameter of an optical fibre is 125 μm, so disruption of the packing of the reinforcement fibres of diameters of 5–7 μm (carbon) or 15–20 μm (glass) can be anticipated. Mis-orientation of the reinforcement can have a major impact on the strength and failure mechanism of a composite.
3. Optical fibres will have a higher load to failure than the individual reinforcement fibres because of their much larger cross-section, so they cannot detect damage directly through loss of light transmission on fracture.

9.4.2.1   Optical Self-Sensing

Ideally we could utilize the transmission characteristics of the glass fibres. To make E-glass act as a light guide, the matrix needs to have a lower RI. This can be achieved by introducing a low-RI component into the matrix so that total internal reflection occurs within the fibre. Under such conditions, a change in light intensity can be used to identify fibre fracture directly. Research in this area is in its infancy. Rauf et al. [60] employed propylene carbonate to modify the RI of a bisphenol A epoxy resin matrix in order to use E-glass fibres to detect fibre fracture.

9.4.3   Other Damage Sensors

A description of damage-detection techniques is given in Chapter 6, but here we are concerned principally with self-sensing techniques that provide direct damage detection as a precursor for self-repair. Most other detection techniques require a test protocol followed by external assessment, and may utilize coupling agents (e.g. C-Scan), which may impact the damaged material and hence the repair method.

9.5   Recycling of Composites

As the number of applications of composite materials increases, the potential volume of waste material also increases. Fortunately, many applications utilize the long-term durability of the material, but its disposal at end-of-life presents a major problem in need of a solution. In addition, we also recognize that production waste (virgin material) also occurs, despite one of the major advantages of composites being the ability to mould large components without excessive machining and assembly of multiple parts. From an environmental perspective, *recycling* or *reuse* is now a major issue. Thus, the following guidelines need to be employed, in order of priority:

1. design manufacturing routes to prevent and reduce waste;
2. reuse products;
3. recycle materials;
4. incinerate waste and utilize the calorific value of the polymer (and reinforcement?);
5. incinerate waste with recovery of material and energy;
6. incinerate waste with recovery of energy;
7. incinerate waste without recovery of energy;
8. landfill.

9.5.1   Thermoplastic Composites

Thermoplastic composites have the potential for *reuse*, and this is an advantage that has promoted the development of thermoplastic processing techniques. A cascade of processing techniques exist in which the fibre length in the composite decreases. This is illustrated in Figure 9.33.

During reuse of thermoplastic prepreg waste, the mechanical performance of the moulded composite artefacts will be reduced accordingly, not least from the loss in fibre length and control of fibre orientation.

Waste from short-fibre-reinforced thermoplastics is generally reused in combination with virgin material to minimize its effect on mechanical performance.

Recycling of used thermoplastic waste requires inclusion of additional sorting and washing procedures, which adds cost and makes its use more expensive than virgin material. As a result, recovering the energy by incineration is often more economically feasible.

9.5.2   Thermosetting Composites

Thermoset polymers cannot be re-moulded, so a number of approaches to recycling have been considered. The techniques for recycling of thermoset composite scrap materials are described in Figure 9.34.

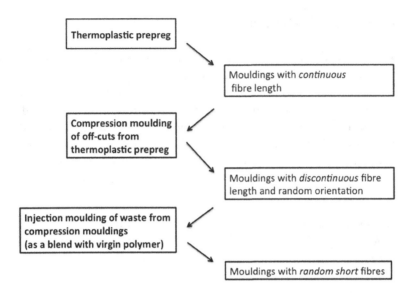

Figure 9.33 Cascade in the reprocessing of thermoplastic prepreg waste. As the fibre length is reduced the mechanical performance decreases.

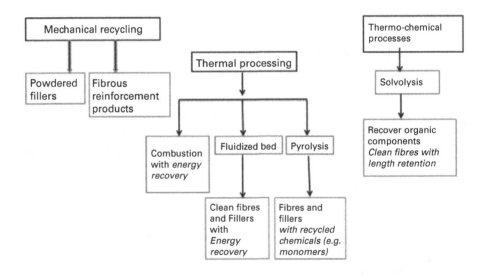

Figure 9.34 Recycling processes for thermoset composite materials [59,61].

The principal techniques are:

1. *Mechanical recovery* of reinforcements as either highly comminuted powdered *filler* or as a *longer fibrous* reinforcement, which can be used as a filler or reinforcement and usually included within virgin moulding material.

2. *Incineration* of waste polymeric material and the calorific value of the material recovered. To gain value from the incombustible components, such as fillers and fire retardants, burning scrap in a cement kiln is effective because the glass reinforcement and mineral fillers can be incorporated in cement.

3. *Pyrolysis:* recovery of the fibres and useful organic molecules by thermal processing. In pyrolysis, combustible material is heated in the absence of oxygen, where it degrades the polymeric material into lower molecular weight organic molecules in liquid or gaseous form, which is a potential feedstock for chemical processing and a solid carbonaceous char. Ideally, the organics would be monomers, which can be reused in the production of new resins and polymers. The evolved gases can provide fuel for the process.

4. *Fluidized bed thermal process:* the techniques for recovering the fibrous reinforcements from reinforced plastics involve either solvolysis or thermolysis of the matrix polymer. The latter is best carried out in a fluidized bed of silica sand created with a stream of hot air at temperatures of 450–550 °C. Figure 9.35 provides a description of the fluidized bed technique. In the fluidized bed, the polymer volatilizes, releasing the fibres and fillers, suspended as individual particles in the gas stream. The fibres and fillers can then be separated from the gas stream. However, the strength of glass fibres is strongly temperature-dependent and the temperatures employed for the pyrolysis of the matrix polymer will severely compromise the strength of the recovered fibres. As a result, recovered glass fibres are not currently employed in applications in which strength is critical, severely restricting the application of recovered glass fibres to non-structural parts. However, Thomason et al. [62] have patented a surface treatment process for recovering the strength of these recycled glass fibres. The inclusion of silane coupling agents in the process is expected to also provide good interfacial properties.

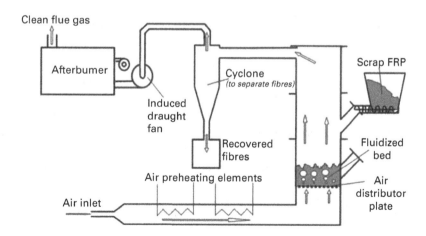

Figure 9.35 A fluidized bed thermal process for recovering glass and/or carbon fibres from composite scrap [59,61].

The strength of recovered carbon fibres is less affected by thermal processing but it was found necessary to use a 'paper' manufacturing process to prepare carbon-fibre reinforcement in the form of a mat for use in composites. An additional surface treatment was found to be necessary for good interface formation [63].

5. *Solvolysis:* recovery of the reinforcements using reactive solvents is achieved by hydrolysis and/or thermal degradation of the crosslinked polymer matrix. A typical solvent is supercritical water at temperatures >374 °C and pressure >221 bar [64]. The rate of hydrolysis of the polyester resin within GRP composites by subcritical water determines the efficiency of the recovery process. The development of supercritical processing needs further understanding of the conditions required to optimize the process [65]. Since water is damaging to glass fibres, this technique might be more appropriate for carbon-fibre composites. However, carbon fibres are usually combined with more hydrolytically stable matrix resins so that thermal processing is probably preferred.

9.6   Cost–Life Analysis

From the foregoing, we see that repair, self-healing, and recycling approaches are being developed to provide more environmentally friendly materials and processes for sustainable composite structures. To achieve this, the costs need to be considered throughout the life of a structure. These ideas can inform the design and choice of materials for an application.

Whole life cycle cost is an economic assessment that considers all relevant projected costs and revenues associated with a particular asset, structure, or project over its lifetime [66]. With regard to high-performance materials, various methods have been proposed for applying life cycle costs, such as those of Ehlen [67], Richard et al. [67], and Hastak and Haplin [69], to enable comparison of composite materials with more conventional construction materials.

The relatively high specific strength and stiffness and corrosion resistance of composite materials make them favourable alternatives to more conventional materials such as steel and reinforced concrete in civil engineering [70] and aluminium in aerospace applications. The development of composites for repair and maintenance of existing bridge structures has been identified as a priority [68]. As we have discussed elsewhere, composite materials are capable of outperforming more conventional materials in terms of structural performance, maintenance, and cost. However, high initial capital construction and manufacturing costs means that life cycle cost analysis is essential for the promotion of composite usage [71].

When assessing the feasibility of a project, there is often a significant focus on its initial costs, which is often a relatively small proportion of the total cost of a project over its duration [68]. Three key stages of the life of a structure should be considered in the costing process [71]:

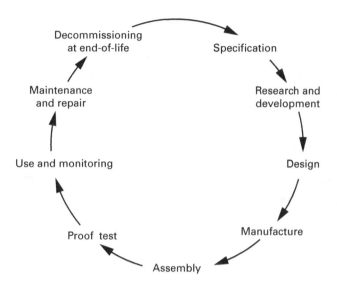

Figure 9.36 Stages in the use of a composite artefact included in whole-life costing.

1. manufacturing (including research and development);
2. installation, assembly, or construction and use (including maintenance and operation); and
3. end-of-life (decommissioning or replacement).

For composites, there is limited knowledge and case studies on maintenance and costs, which has limited life cycle cost analysis to date. This is an area of current research and development as material sustainability is embraced. Figure 9.36 describes the stages of production and use that need to be included in the whole-life costing of an application of a composite material.

9.7 Conclusions

The joining and repair of damaged composite materials is discussed. The latter is extended to include potential methods for repairing damage or providing mechanisms of self-healing during service. We examine the potential techniques of damage sensing, which can provide composites with smart performance such that a healing mechanism is automatically triggered (autonomic methods involving encapsulated healing agents) or where external healing methods are initiated.

Recycling techniques, whereby the reinforcing fibres are recovered for further use, are discussed. To stimulate recycling and/or recovery, the use of whole-life costing needs to be employed. The stages of whole-life costing of an artefact from design through manufacture to service life and recovery are discussed. Energy recovery through controlled incineration should be the final stage after multiple reuse or recovery cycles.

9.8   Discussion Points

1. What are the principal methods of joining composite materials during manufacture of an artefact? Give advantages and disadvantages of mechanical and adhesive bonding methods.
2. Describe the available repair techniques for damaged composites.
3. Review the methods used to provide a self-healing mechanism. Discuss the differences between autonomic and thermally stimulated healing techniques.
4. What is understood by a smart healing composite?
5. Which costing method should be used in the design and use of a composite artefact to encourage recycling?

References

1. F. L. Matthews and R. D. Rawlings, *Composite materials: engineering and science* (London: Chapman and Hall, 1994).
2. P. Camanho, L. Tong, ed., *Composite joints and connections: principles, modelling and testing* (Oxford: Woodhead Publishing, 2011)
3. W. R. Broughton, L. E. Crocker, and M. R. L. Gower, *Design requirements for bonded and bolted composite structures* (Teddington: HMSO, 2002).
4. B. Flinn and M. Phariss, *The effect of peel-ply surface preparation variables on bond quality*, DOT/FAA/AR-06/28 (Federal Aviation Administration, 2006).
5. B. D. Flinn, B. K. Clark, J. Satterwhite and P. J. Van Voast, Influence of peel ply type on adhesive bonding of composites. In *Proceedings of SAMPE 2007* (Covina, CA, 2008).
6. Y. Kusano, T. L. Andersen, H. L. Toftegaard, et al., Plasma treatment of carbon fibres and glass-fibre-reinforced polyesters at atmospheric pressure for adhesion improvement. *Int. J. Mater. Eng. Innov.* **5** (2014), 122–137.
7. Y. Kusano, Atmospheric pressure plasma processing for polymer adhesion: a review. *J. Adhesion* **90** (2014), 755–777.
8. S. Behzadi and F. R. Jones, Yielding behaviour of model epoxy matrices for fibre reinforced composites: effect of strain rate and temperature. *J. Macromol. Sci. B Phys.* **44** (2005), 993–1005.
9. S. Black, Structural adhesives, part i: industrial. *Composites World*, 4 November 2016.
10. S. Black, Structural adhesives, part iI: aerospace. *Composites World*, 7 November 2016.
11. L. J. Hart-Smith, Design of adhesively bonded joints. In *Joining fibre reinforced plastics*, ed. F. L. Matthews and R.D. Rawlings (Oxford, Elsevier Applied Science, 1987).
12. Volkersen, O., Die Niektraftverteilung in Zugbeanspruchten mit Konstanten Laschenquerschritten. *Luftfahrtforschung* **15** (1938), 41–47.
13. R. D. Adams and W. C. Wake, *Structural adhesive joints in engineering* (Oxford: Elsevier Applied Science, 1984).
14. M. Goland and E. Reissner, The stresses in cemented joints. *J. Applied Mech.* **11** (1944), A17–A27.
15. Y. Zhu and Keith Kedward, *Methods of analysis and failure predictions for adhesively bonded joints of uniform and variable bondline thickness, Final Report, DOT/FAA/AR-05/12* (Washington, DC: Office of Aviation Research, 2005).

16. S. Budhea, M. D. Baneaa, S. de Barrosa, and L. F. M. da Silva, An updated review of adhesively bonded joints in composite materials. *Int. J. Adhesion Adhesives* **72** (2017), 30–42.

17. Composites World. Composites repair (2014). www.compositesworld.com/articles/composites-repair.

18. K. B. Katnam, L. F. M. da Silva, and T. M. Young, Bonded repair of composite aircraft structures: a review of scientific challenges and opportunities. *Progr. Aerospace Sci.* **61** (2013), 26–42.

19. Wikipedia. Composite repair (n.d.). https://en.wikipedia.org/wiki/Composite_repairs#/media/File: Figure-4-typical-laminate-and-sandwich-repairs.jpg.

20. X. Chen, F. Wudl, A. Mal, et al., New thermally remendable highly crosslinked polymeric materials. *Macromolecules* **36** (2003), 1803.

21. F. R. Jones, W. Zhang, and S. A. Hayes, Thermally induced self healing of thermosetting resins and matrices in smart composites. In *Self healing materials: an alternative approach to 20 centuries of materials science*, ed. S. van der Zwaag (Dordrecht: Springer, 2007), pp. 69–93.

22. S. White, N Sottos, P. H. Geubell, et al., Autonomic healing of polymer composites. *Nature* **409** (2001), 794–797.

23. I. P. Bond, R. S. Trask, H. R. Williams, and G. J. Williams, Self healing fibre- reinforced polymer composites: an overview. In *Self healing materials: an alternative approach to 20 centuries of materials science*, (Dordrecht: Springer, 2007), pp. 115–138.

24. R. S. Trask and I. P. Bond, Bioinspired engineering study of Plantae vascules for self-healing composite structures *J. R. Soc. Interface* **7** (2010), 921–931.

25. R. Varley, Ionomers as self healing polymers. In *Self healing materials: an alternative approach to 20 centuries of materials science*, ed. S. van der Zwaag (Dordrecht: Springer, 2007), pp. 95–114.

26. F. R. Jones and R. J. Varley, Solid-state healing of resins and composites. In *Recent advances in smart self-healing polymers and composites*, ed. G. Li and H. Meng (Cambridge: Woodhead Publishing, 2015), pp. 53–99.

27. M. T. Aronhime, X. Peng, J. K. Gillham and R. D. Small, Effect of time–temperature path of cure on the water absorption of high T_g epoxy resins. *J. Appl. Polym. Sci.* **32** (1986), 3589–3626.

28. M. S. Md. Jamil, F. R. Jones, N. N. Muhamad, and S. M. Makenan, Solid-state self-healing systems: the diffusion of healing agent for healing recovery. *Sains Malaysiana* **44** (2015), 843–852.

29. S. A. Hayes, W. Zhang, M. Branthwaite, and F. R. Jones, Self-healing of damage in fibre-reinforced polymer-matrix composites. *J.R. Soc. Interface* **4** (2007), 381–387.

30. D. W. van Krevelen and K. te Nijenhuis, *Properties of polymers,* 4th ed. (Oxford: Elsevier, 2009).

31. S. A. Hayes, F. R. Jones, K. Marshiya, and W. Zhang, A self healing thermosetting composite material. *Composites A* **38** (2007), 1116–1120.

32. M. Doi and S. F. Edwards, Dynamics of concentrated polymer systems. Part 1, Brownian motion in the equilibrium state. *J. Chem. Soc. Faraday Trans 2* **74** (1978), 1789.

33. M. Doi and S. F. Edwards, Dynamics of concentrated polymer systems. Part 2, molecular motion under flow, *J. Chem. Soc. Faraday 2* **74** (1979), 1802.

34. M. Doi and S. F. Edwards, Dynamics of concentrated polymer systems. Part 3, the constitutive equation. *J. Chem. Soc. Faraday Trans 2* **74** (1979), 1818.

35. M. Doi and S. F. Edwards, Dynamics of concentrated polymer systems. Part 4, rheological properties, *J. Chem. Soc. Faraday Trans 2* **75** (1979), 1838.

36. M. Doi and S. F. Edwards, *The theory of polymer dynamic* (New York: Oxford University Press, 1986).

37. P. G. De Gennes, Reptation of a polymer chain in the presence of fixed obstacles. *J. Chem. Phys.* **55** (1971), 572–579.

38. http://mr.chem.unc.edu/research/dynamics

39. R. P. Wool, Polymer entanglements. *Macromolecules* **26** (1993), 1564–1569.

40. J. Lowe, Aerospace applications. In *Design and manufacture of textile composites*, ed. A. C. Long (Cambridge: Woodhead Publishing, 2005), pp. 405–423.

41. www.cytec.com/sites/default/files/datasheets/CYCOM_977–20_w_PRIFORM_031912 .pdf.

42. R. J. Varley, B. Dao, C. Pillsbury, S. J. Kalista, and F. R. Jones, Low-molecular-weight thermoplastic modifiers as effective healing agents in mendable epoxy networks. *J. Int. Mat. Syst. Struct.* **25** (2014), 107–117.

43. F. R. Jones, R. Varley, B. Dao, C. Pillsbury, and S. J. Kalista, Self assembling healing agents for mendable epoxy networks. In *Proceedings of the 16th European Conference on Composite Materials, ECCM 16* (Seville: ECCM, 2014).

44. S. D. Mookhoek, S. C. Mayo, A. E. Hughes, et al., Applying SEM-based X-ray micro-tomography to observe self-healing in solvent encapsulated thermoplastic materials. *Adv. Eng. Mater.* **12** (2010), 228–234.

45. K. S. Toohey, N. R. Sottos, J. A. Lewis, J. S. Moore, and S. R. White, Self-healing materials with microvascular networks. *Nat. Mater.* **6** (2007), 581–585.

46. C. J. Norris, J. A. P. White, G. McCombe, et al. Autonomous stimulus triggered self-healing in smart structural composites. *Smart Mater. Struct.* **21** (2012), 1–10.

47. M. Zako and N. Takano, Intelligent material systems using epoxy particles to repair microcracks and delamination damage in GFRP. *J. Int. Mat. Syst. Struct.* **10** (1999), 836–841.

48. X. F. Luo, R. Q. Ou, D. E. Eberly, et al. A thermoplastic/thermoset blend exhibiting thermal mending and reversible adhesion. *ACS Appl. Mater. Interfaces* **1** (2009), 612–620.

49. J. Nji and G. Li, A biomimic shape memory polymer based self-healing particulate composite. *Polymer* **51** (2010), 6021–6029.

50. J. Nji and G. Q. Li, Damage healing ability of a shape-memory-polymer-based particulate composite with small thermoplastic contents. *Smart Mater. Struct.* **21** (2012), 1.

51. S. Meure, D. Y. Wu, and S. Furman, Polyethylene-co-methacrylic acid healing agents for mendable epoxy resins. *Acta Mater.* **57** (2009), 4312–4320.

52. S. Meure, R. J. Varley, D. Y. Wu, et al., Confirmation of the healing mechanism in a mendable EMAA-epoxy resin. *Eur. Polym. J.* **48** (2012), 524–531.

53. S. Meure, D. Y. Wu, and S. A. Furman, FTIR study of bonding between a thermoplastic healing agent and a mendable epoxy resin. *Vib. Spectrosc.* **52** (2010), 10–15.

54. C. H. Wang, K. Sidhu, T. Yang, J. Zhang, and R. Shanks, Interlayer self-healing and toughening of carbon fibre/epoxy composites using copolymer films. *Compos. A Appl. Sci. Manuf.* **43** (2012), 512–518.

55. S. Meure, S. Furman, and S. Khor, Poly[ethylene-co-(methacrylic acid)] healing agents for mendable carbon fiber laminates. *Macromol. Mater. Eng.* **295** (2010), 420–424.

56. L. Hou and S. A. Hayes, A resistance-based damage location sensor for carbon fibre composites. *Smart Mater. Struct.* **11** (2002), 966–969.

57. T . J. Swait, F. R. Jones, and S. A. Hayes, A practical structural health monitoring system for carbon fibre reinforced composite based on electrical resistance. *Compos. Sci Technol* **72** (2012), 1515–1523.

58. S. A. Hayes, T. J. Swait, and A. D. Lafferty, Self-sensing and self-healing in composites. In *Recent advances in smart self-healing polymers and composites*, ed. G. Li and H. Meng (Cambridge: Woodhead Publishing, 2015), pp. 243–261.

59. F. R. Jones and N. T. Huff, The structure and properties of glass fibres. In *Handbook of properties of textile and technical fibres*, ed. A. R. Bunsell (Cambridge: Woodhead Publishing, 2018), pp. 757–803.

60. A. Rauf, R. J. Hand, and S. A. Hayes, Optical self-sensing of impact damage in composites using E-glass cloth. *Smart Mater. Struct.* **21** (2012), 045021.

61. S. J. Pickering, Recycling technologies for thermoset composite materials: current status. *Compos. A. Appl. Sci. Manuf.* **37** (2006), 1206–1215.

62. J. L. Thomason, E. Sáez-Rodríguez, and L. Yang, Glass fibre recovery. European patent, EP 3024793 A1, 2013.

63. J. Howarth and F. R. Jones, Interface optimisation of recycled carbon fibre composites. In *Proceedings of the 15th European conference on composite materials, ECCM 15* (Venice: ECCM, 2012), p. 5.

64. G. Oliveux, J.-L. Bailleul, and E. Le Gal La Salle, Chemical recycling of glass fibre reinforced composites using subcritical water. *Composites A* **43** (2012), 1809–1818.

65. G. Oliveux, L. O. Dandy, and G. A. Leeke, Current status of recycling of fibre reinforced polymers: review of technologies, reuse and resulting properties. *Prog. Mater. Sci.* **72** (2015), 61–99.

66. Constructing Excellence, Whole life costing factsheet, 2004.

67. M. A. Ehlen, Life cycle costs of fibre reinforced polymer bridge deck. *J. Mater. Civil Eng.* **11**(1999), 224–230.

68. D. Richard, T. H. M. Hong, A. Mirmiran, and O. Salem, Life cycle performance model for composites in construction. *Composites B* **38** (2007), 236–246.

69. M. Hastak and D. W. Haplin, Assessment of life-cycle benefit-cost of composites in construction. *J. Compos. Construct.* **4** (2000), 103–111.

70. L. Hollaway, A review of the present and future utilisation of FRP composites in the civil infrastructure with reference to their important in-service properties. *Construct. Build. Mater.* **24** (2010), 2419–2445.

71. P. Ilg, C. Hoehne, and E. Guenther, High performance materials in infrastructure: a review of applied life cycle costings and its drivers – the case of fibre reinforced composites. *J Cleaner Prod.* **112** (2016), 926–945.

10 Case Histories

Since the discovery of carbon fibres in the 1960s, the applications have grown. Because of the high specific strength and stiffness, aerospace applications have dominated, especially initially in military aircraft. The intent here is to demonstrate how the choice of material has been identified. Most critical demonstrators have come from the field of aerospace because of the benefits of carbon fibres and the development of confidence in their use in safety-critical designs. The latter has involved much testing and durability studies. Middleton has provided several case histories detailing the development of composite applications in aircraft structures [1]. The use of composite components has increased with improved confidence in the durability and reliability of these materials and structures. The Airbus A380 was introduced in 2006 using a carbon-fibre-reinforced polymer (CFRP) centre wing box, while the fuselage employed an aluminium–glass fibre composite laminate (GLARE). The centre wing box is a critical carbon-fibre composite structure that joins the wings to the fuselage. Together with several other composite components, such as the horizontal and vertical stabilizers, keel beam, and pressure bulkhead, the total composite usage is 22% w/w. In 2011 the Boeing 787 Dreamliner employed carbon-fibre materials for the fuselage and wings. In total, the latter used 80–90% by volume or 50% by weight of composite materials. The Airbus A350, introduced in 2015, also uses CFRP for the fuselage and wings, and in total composite usage is 53% w/w.

To understand the benefits associated with the use of composite materials, we examine selected case histories of composite applications starting with the classic introduction of helicopter rotor blades. Here, the excellent fatigue resistance of a composite material meant that there was no need to regularly replace the metal-equivalent rotor blades after a fixed lifetime.

Future developments are set to be in civil and automotive engineering and energy generation.

10.1 Rotor Blades for Helicopters

Rotor blades need to operate in extremely harsh environmental conditions arising from rotational tip velocities of about 200 m s^{-1} (~480 mph), and bending moments during flight. Extremes in both humidity and temperature are encountered. The temperatures can range from –40 °C to +90 °C. Efficient and effective rotor blades require particular material properties in order to manufacture durable components. Steel and aluminium blades suffered from various design and structural problems. Most critical for these metals were poor fatigue resistance and low specific strength. Composite materials provide solutions to these disadvantages and enabled an improved design of rotor blades. Manufacturing techniques employed for construction using composites provide the following benefits:

1. weight saving over aluminium alloys;
2. high strength and stiffness (3–6 times higher than Al–Zn–Mg alloy);
3. excellent fatigue resistance;
4. tailored directional mechanical properties;
5. complex shapes and contours are easily achieved;
6. a reduced number of parts compared to the metallic equivalent;
7. reduced machining; and
8. corrosion resistance in aggressive environments.

Initially, metal tooling was employed for laying up the composite prepreg. Tape-laying with ultrasonic cutting was developed to accurately arrange and profile the prepreg into the correct configuration. The removal of the tooling spar proved to be complex and expensive, requiring double the workshop space for extraction. Subsequently this was replaced with a polymethacrylamide rigid foam mandril, which could be left in place. An example is given in Figure 10.1, which illustrates the use of the materials in the rotor blade [2]. The relevant aspect is the use of a range of glass fibre and carbon composite prepreg to achieve the required mechanical performance. In Chapter 5 we saw that the failure strain of carbon fibre is less than that for glass fibre, so the probability of fibre-break accumulation is high in their composites. To optimize fatigue resistance, a hybrid arrangement of prepreg proved beneficial.

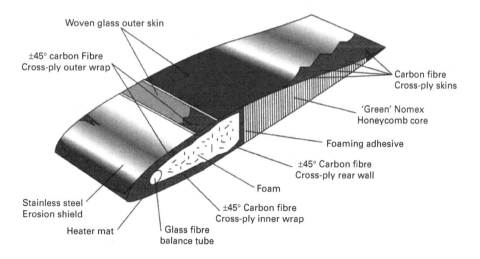

Figure 10.1 Typical cross-section of a helicopter blade, illustrating the use of composite materials [2].

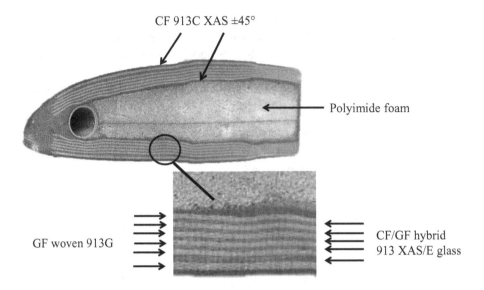

CF 913C XAS ±45°

Polyimide foam

GF woven 913G

CF/GF hybrid
913 XAS/E glass

Figure 10.2 Cross-section of a helicopter composite spar showing a typical composite lay-up. (further information is given by Middleton [1]).

The Fibredux 913 resin system is a PEES toughened epoxy resin cured with DICY, which can be cured at 125 °C using compression moulding at about 100 psi, where the tools ensure that the correct shape is created. The core foam is often over-sized precisely to allow for crushing during the process. The lay-up of the prepreg was given by Middleton [1] and is shown in Figure 10. 2. The trailing edge of the blade is constructed similarly and consolidated onto a Nomex aramid honeycomb core. The skins are usually pre-cured to ensure that the barrier performance to moisture absorbance is optimum.

10.2   Propellers

Extensive development work at Dowty Rotol [3] since 1978 has led to the use of fibre composites for manufacturing propellers destined for a number applications, including aircraft and marine vehicles.

Hub and blades are subjected to a complex stress state due to steady and oscillatory forces. Both centrifugal and bending forces arise in achieving thrust. Any change in the pitch of the blade leads to aerodynamic twisting moments and adds to the flexural loads. The frequency of rotation affects the harmonic and torsional components of flexure. Carbon-fibre composites have been chosen for the spars because of a modulus that provides resistance to the main service loads. Early blades used a lay-up of dry-assembled carbon fibre and/or glass fibre in a textile form with a binder to stabilize the preform. A polyurethane foam core is introduced together with a lightning conductor braid and leading-edge reinforcement. The complete assembly is placed in a blade

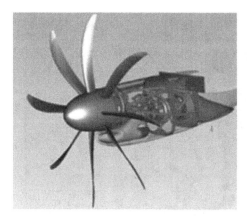

Figure 10.3 Typical Dowty propeller [3].

mould for resin transfer moulding (RTM) of the liquid resin. The viscosity of the resin is reduced using heated tooling. Careful control of the resin transfer ensures that impregnation is efficient. A polyurethane coating is applied to protect the blade surface from erosion by water and natural particulates in use. The thickness of the coating is tailored to the intended service environment. At the leading edge, the coating might be ineffective so a nickel sheath is bonded to the polyurethane to enhance erosion resistance. Compared to the original aluminium blades, there is 50% weight reduction.

Adding more blades in an assembled propeller require a heavier hub, which introduces a much larger centrifugal twisting moment, which means the rotating blade tends to elongate to a finer pitch (Figure 10.3). A reduction in blade weight, which can be achieved with composite materials, is highly beneficial.

One other major advantage of composite materials is the shape, which can be formed using the variety of available process methods. The shape of a propeller lends itself to using a 'fibre placement' technique onto a foam core. A braiding approach is ideally suited to the fabrication of propeller blades, as shown in Figure 10.4. The braiding is achieved by winding the fibres from a number of spools, as illustrated. Carbon fibre and/or other fibrous preforms can be consolidated using RTM.

10.3 Aerospace Components: Rocket Module from Thermoplastic CFRP

The advent of the use of composite materials in airliners, especially for the fuselages of the Boeing 787 Dreamliner and the Airbus A350, has identified the need for 'huge' autoclaves. As a result, non-autoclave processing has become an important requirement. Thermoplastic matrices present a potential way forward since there is no chemistry involved in the process. Control of the cooling is essential to ensure that the required crystalline microstructure is consistently achieved. Thermoplastics with sufficiently high service temperatures are polyether ether ketone (PEEK), polyether

Figure 10.4 Dowty braiding machine for composite blades [4].

ketone (PEK), and polyether ketone ketone (PEKK). PEEK has a crystalline melting point, T_m, of 340 °C and glass transition temperature, T_g, of 143 °C, while for PEK $T_m = 372$ °C and $T_g = 153$ °C and for PEKK $T_m = 400$ °C and $T_g = 190$ °C. The problem is that while T_m provides the maximum service temperature at T_g, there is a small reduction in the modulus of the polymer. Despite the typical V_f of 60%, a fall in the modulus of the composite provides a limitation to structural performance above T_g. The other disadvantage is the temperatures involved in the impregnation of the fibres to produce prepreg or tape. The viscosity of the molten high molecular weight polymers is significantly higher than non-polymeric thermosetting resins used for composites. Thus, processing costs are higher than for thermosetting resins such as epoxies.

A recent report from the Technical University of Munich [5] illustrates the benefits of this technology. Using thermoplastics-based composites, a 40% weight reduction was achieved in the design and manufacture of a research rocket scientific payload module. This section of the REXUS research rocket flies at a vertical speed of 1,200 m s^{-1}, at a maximum acceleration of $\approx$20 G, reaching a maximum height of 80–100 km. The baseline structure uses aluminium with an outside diameter of 356 mm and length of 300 mm, which is replaced by a similarly sized composite shell together with thermoplastic load input rings, which provide bolt connections to adjacent modules. The connecting rings (male and female) were compression moulded from 'long' (2–3 mm) carbon-fibre PEEK granules at 390 °C using an increasing force of 50–200 kN, and de-moulded at 100 °C.

The module shell was manufactured from unidirectional Tenax carbon-fibre PEEK prepreg tape using automatic fibre placement (TP-AFP) as shown in

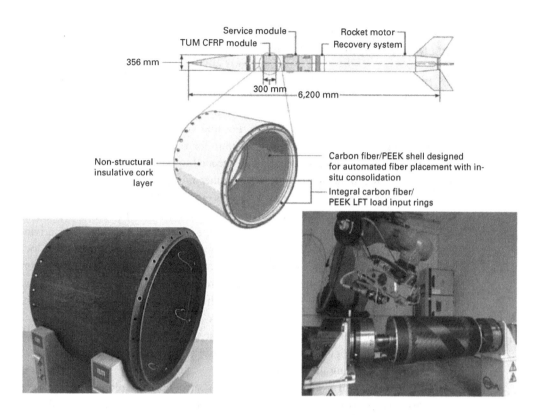

Figure 10.5 An example of a rocket module demonstrator manufactured from PEEK-based CFRP. Automated fibre placement for the shell consolidated onto pre-manufactured connecting rings by compression moulding from long-fibre PEEK moulding compound [5]. (with permission from Technical University of Munich and Karl Reque for the drawing (top)).

Figure 10.5 The dimensions required are set by the need to replace the aluminium version of the artefact of similar stiffness. The fibre laminate lay-up was optimized in a finite element structural analysis and consisted of symmetrical lay-up of 34 plies of 0°/±15°/±45°/90° fibre orientation. Embedded fibre-optic sensors were included in the lay-up at different positions and depths, and subsequently connected to a measurement system inside the module. The TP-AFP enables *in-situ* consolidation of the thermoplastic tape at room temperature onto the CFRP load input rings. Autoclave consolidation was not required and direct consolidation onto the previously manufactured rings ensures that additional mechanical fasteners or adhesives could be avoided. Full-scale testing to meet qualification loads was undertaken successfully. In a subsequent version, the rings were manufactured by centrifugal casting from the same materials.

Thermal measurements in service will be used to identify whether a thermoplastic polymer with a T_g of 143 °C is essential. Thus, a cheaper option might be possible. This application is a good example of how composites design and manufacture is progressing with the development of thermoplastics composites.

10.4   Automotive Applications: Crush Tube Members for Energy Absorption in Vehicular Crashes

The uses of composite materials in automotive applications have developed over many years. Early attempts employed hand lay-up of glass fibre using unsaturated polyester resins for body components, especially in the Reliant models. Clearly this is only possible in low-volume production. For mid-volume production, reaction injection moulding of polyurethane short glass fibre for vertical body panels for Corvette cars was utilized, but the horizontal panels needed to be made from sheet moulding compound (SMC; higher glass fibre V_f and other mineral fillers) which have higher stiffness to avoid sagging. These developments enabled SMC panels to be commonly employed for lorry cabs. Aerospace investment in composite structures has provided the confidence to develop structural artefacts for automotive vehicles. The driving force for these developments is a reduction in weight for lower fuel usage. Friedrich and Almajid [6] have reviewed applications in automotive structures. The recent development of composite structures for the absorption of impact energy is of special interest. For a more general discussion of crashworthiness of automotive composite structures, the reader is referred to the article by Lukaszewicz [7].

One development is the use of composite energy-absorbing structures to improve safety in a collision. Figure 10.6 illustrates the use of composites in automobile front-end structures, including the crashbox [6].

Crash resistance of composite structures is also being explored in helicopter structures. Figure 10.7 shows the location and structure of a protective sub-floor in a helicopter fuselage [8].

The energy-absorbing structure employs an array of composite tubes. Hull [9] provided the fundamentals of crushing mechanisms, which have spurred the development of these structures. In the early work, glass-fibre composites were used to examine the crushing mechanisms. Subsequently, carbon-fibre structures were shown to absorb higher amounts of impact energy. Hu et al. [10] have reviewed the performance of CFRP tubes with differing fibre configurations.

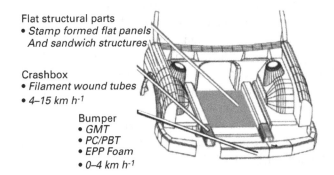

Flat structural parts
• *Stamp formed flat panels
And sandwich structures*

Crashbox
• *Filament wound tubes*
• *4–15 km h⁻¹*

Bumper
• *GMT*
• *PC/PBT*
• *EPP Foam*
• *0–4 km h⁻¹*

Figure 10.6 Automotive front structure showing crash-absorber elements. GMT, glass mat thermoplastic; PC/PBT, is polycarbonate/polybutylene terephthalate; EPP-foam, expanded polypropylene foam [6].

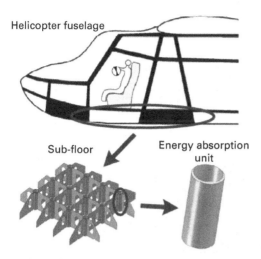

Helicopter fuselage

Sub-floor

Energy absorption
unit

Figure 10.7 Schematic diagram of the energy absorption structure of the sub-floor in a helicopter fuselage [8].

Tubes made from metallic and thermoplastic materials that are homogeneous, isotropic, and ductile collapse by progressive plastic folding. The geometry of the folds and the loads for collapse depend on the shape and dimensions of the tubes. Axial crushing of tubes from composites materials is more difficult to model, but the energy absorbed is higher. Most structural composites artefacts are made from brittle fibres, such as glass and carbon, in a brittle or ductile resin, where collapse by plastic folding is not possible. Since the properties of composite materials are a strong function of fibre arrangement, V_f, and the quality of the fibre–matrix interface, progressive collapse also depends on many of these variables.

Two types of progressive collapse of composite tubes are observed:

1. progressive folding in:
   - thin-walled tubes with continuous carbon and glass fibres;
   - a range of tubes of varying wall thickness with *tough polymer fibres* such as aramid; and
   - glass-fibre composites from unsaturated or epoxy resins tested at *high temperatures*;
2. progressive crushing, which involves the formation of a zone of microfracture at one end of the tube, which propagates along the tube at the same rate as the crushing platen. Interacting variables include:
   - microfracture processes at the crush zone;
   - forces acting at the crush zone;
   - microstructure of the composite;
   - shape and dimensions of the tube;
   - crush initiation and trigger mechanisms; and
   - test variables such as crush speed and temperature.

Figure 10.8 The nose cone of a Formula 1 racing car before and after a simulated crash impact showing crushing over a specified distance for impact-energy absorption (from Advanced Composite Group Ltd. Heanor, UK).

10.4.1   Progressive Crushing

Traditionally, buckling of automotive body parts was used for absorption of impact energy in a crash situation. However, the observation that progressive crushing of a composite absorbed more energy than progressive buckling of a metal component has led to significant development of 'crush' members for incorporation in cars and other applications for energy absorption. Figure 10.8 illustrates the application in a Formula 1 racing car, which shows the crush zone. By carefully designing the shape and fibre arrangement in the nose cone, a controlled crush zone can be built in, which absorbs the impact energy over a specified length, thereby providing the driver with crash protection.

10.4.1.1   Design of Tubes

Square-ended tubes made from brittle materials probably will fail by a catastrophic brittle fracture. The brittle fracture strength, σ_{cu}, under compression provides an upper limit to the strength of the tube. A typical load–displacement curve for a square-ended tube exhibits fracture in which interpenetration of the fractured halves provides some residual load-bearing capacity, but this is insufficient for structures, which need to collapse controllably to absorb large quantities of energy.

Progressive crushing can be induced in tubes from brittle materials by initiating or triggering fracture at one end at stresses below σ_{cu}. A stable microfracture zone is able to propagate down the tube. The simplest trigger is achieved by *chamfering one end of the tube*, often at 45°, where crushing is initiated at the high-stress region at the tip of

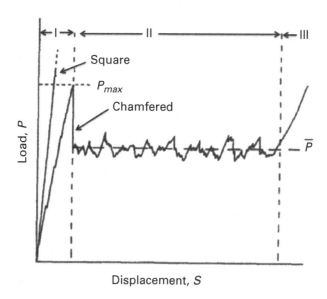

Figure 10.9 Typical load–displacement (P–S) curves of tubes with chamfered and square ends. The former exhibits progressive crushing. I, formation of crush zone; II, progressive crushing; III, compaction of debris [9].

the chamfer and develops into a stable crush zone. Figure 10.9 shows typical load–displacement curves for square-ended and chamfered tubes. The initial slope of the latter is lower than that of the former because crushing occurs at the chamfer in addition to the elastic deformation. Local fracture occurs at P_{max}, causing a relaxation in the crush front, which is followed by the formation of the crush zone at a displacement, S_i. The magnitude of the drop in load is a function of the chamfer angle becoming approximately zero at some angles. Continued crushing occurs at approximately constant load, $\bar{P}$. The rise in crush load in stage III represents the compaction of the debris inside the tube.

The specific progressive crushing stress, Σ_{crush}, is given by

$$\Sigma_{crush} = \frac{\bar{\sigma}}{\rho} = \frac{\bar{P}}{A\rho}, \tag{10.1}$$

where $\bar{\sigma}$ is the mean crush stress, ρ is the density of the material, and A is the cross-sectional area of the wall of the tube.

Since the depth of the crush zone is independent of crush distance, the specific energy absorption, U_{crush}, is

$$U_{crush} = \frac{\bar{\sigma}}{\rho} = \Sigma_{crush}. \tag{10.2}$$

The formation of a crush zone involves two mechanisms, splaying mode and fragmentation mode. A stable crush zone depends on the arrangement of the fibres in the wall, as well as the choice of resins and fibres.

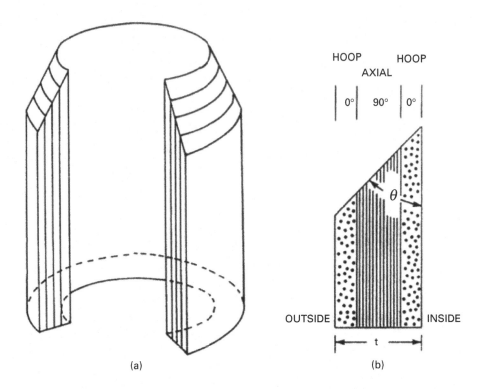

Figure 10.10 Arrangement of fibres in 0°/90°/90°/0° tubes: (a) section through a tube showing the chamfer which means the maximum stress is on the inside wall; (b) chamfer geometry [9].

Figure 10.10 illustrates a possible arrangement of fibres in a tube where $V_f \approx 0.5$. In the work of Hull [9], the wall consisted of axial fibres, and fibres at 90° had a thickness of 3.7 mm. The tube was prepared by hoop winding of the fibres so the direction of the fibres in the wall are defined by the hatching in Figure 10.10.

Figure 10.11 illustrates the microfracture mechanisms observed in the crushing of a 0°/90°/90°/0° glass fibre–unsaturated polyester resin. The following stages were observed in the *splaying mode*:

1. crushing of the inner hoop-wound layer (Figure 10.11b), where microfracture involves shear failure in compression at 90° to the fibres;
2. when the platen meets the axial layers, compressive stresses cause the fibres to buckle and kink (Figure 10.11b);
3. a wedge of crushed material (w in Figure 10.11c) is formed; and
4. the wedge of crushed material forces the axial layers to the inside and outside of the tube (Figure 10.11d), causing progressive compressive microfracture of the inner hoop layers and tensile fracture of outer hoop layers as splaying develops.

Fragmentation mode is observed in tubes prepared from glass-fibre textile reinforcements. The orientation of the fibres follows the weave of the selected woven cloth. Progressive crushing involves fragment formation in the crush zone. The

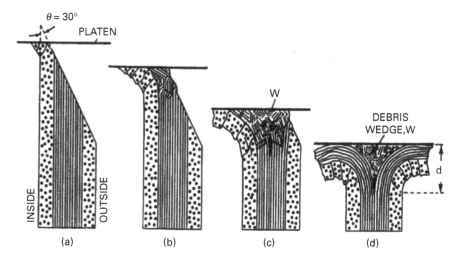

Figure 10.11 Schematic of formation of a splaying-mode crush zone from microscopic observation [9].

arrangement of the fibres in the wall of the tube is controlled by the nature of the weave. The fragmentation mode is critically dependent on shear failure parallel to the layers of cloth in resin-rich regions. Further details are discussed elsewhere [9].

A competition between splaying and fragmentation microfracture mechanisms determines the eventual crush mode. Thus, the crush load will depend on laminate configuration, elastic properties, and strengths of individual laminae.

10.4.1.2   Design of Tubes: CFRP

Many commercial applications involve carbon-fibre laminates (from prepreg), so the configuration of the lay-up is a critical variable. Figure 10.12 shows that the specific crushing stress increases with the increasing fraction of axial fibres. These are only slightly dependent on the rate of impact. The anomalous result at 4 mm s^{-1} can be attributed to a change in crushing mechanism.

Table 10.1 summarizes the effect of ply configuration on the specific crushing stress of CFRP tubes prepared from unidirectional prepreg. The fraction of axial fibres was shown by Hull to determine the absorption of energy [9]. Higher volume fractions of axial fibres provide the highest values of crushing stress. Hu [8,10] has reviewed the performance of a range of CFRP tubes, and provides further design detail. In the manufacture of tubes using continuous fibres, hoop winding is often employed to provide surface stability. Prepreg prepared tubes will also require over-winding with prepreg or continuous fibres. Thus, the effect of configuration is critical to design.

Lukaszewicz [7] has reviewed the selection of tube shape, fibre arrangement, and polymer matrix in an article of crashworthiness of automotive composite structures. Table 10.2 compares the specific energy absorption of a range of carbon-fibre composites that show potential for application in automotive structures.

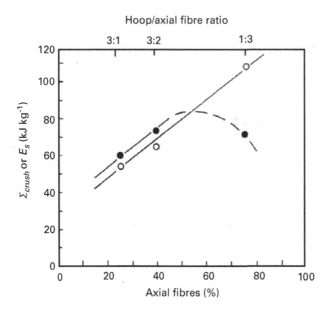

Figure 10.12 Effect of hoop-to-axial ratio on specific crushing stress (Σ_{crush}) and specific energy absorption (E_s) of carbon epoxy resin tubes tested at ($\bullet$) 4 mm s^{-1} and ($\circ$) 2 m s^{-1} [9].

10.4.1.3   Design of Tubes: Glass-Fibre Composites

Many applications are cost-dependent, so glass fibres provide a design option that benefits automotive and industrial sectors. Table 10.3 provides the typical values of specific energy absorption for glass-fibre composites in a range of thermosetting resin matrices with differing fibre arrangements. Polyamide (PA), which is often referred to as nylon, is included as a control.

10.4.1.4   Design of Tubes: Thermoplastic Matrices

Manufacturing is moving towards thermoplastic matrices because of the drive for recycling. Many developments in manufacturing technologies are exploring the benefits of thermoplastics over thermosets, where the chemistry of cure of the latter can limit processing speed. Friedrich has reviewed the crash potential of a range of glass-fibre thermoplastics in comparison to a carbon-fibre thermoplastic and the thermoset, SMC (Table 10.4).

10.5   Civil Engineering Infrastructure

10.5.1   Composites in Bridge Construction

An application of composites of growing commercial significance that has benefited from the extensive developments in aerospace for safety-critical structures is bridge

Table 10.1. Effect of fibre distribution and crushing speed on the specific crushing stress of carbon-fibre epoxy tubes made from unidirectional prepreg [9]

Hoop–axial fibre ratio	Layer configuration		Specific crushing stress, Σ_{crush} $(kN\ m^{-2}/kg\ m^{-3})$ Crush rate	
			4 mm s^{-1}	2 m s^{-1}
1:3	Symmetrical		71	109
1:3	All hoop outside		88	64
1:3	All hoop outside		6	16
3:1	Symmetrical		60	54
3:2	Symmetrical		74	65
3:2	Interleaved		43	37

construction. Since the demonstration bridge in Aberfeldy, Scotland in 1992, confidence in designing bridges from composite materials has intensified. The following advantages of the use of glass and carbon fibre for modern bridge-building are reported:

Table 10.2. Effect of tube shape and fibre arrangement on the specific energy absorption during axial loading of carbon-fibre composites [7]

Shape	Fibre arrangement	Matrix	Specific energy absorbed (kJ kg^{-1})
Round	Unidirectional	Epoxy/PEEK	57–127
Round	Knitted	Epoxy	26–60
Round	Unidirectional	PEEK	171
Round	Unknown	PA-PEEK	160
Round	Unidirectional cloth	Epoxy	73–82
Round	Braid	Epoxy	51
Conical	Fabric	Epoxy	70
Conical	Woven	Epoxy	65
Square	Unidirectional	Epoxy	37
Square	Braid	Epoxy	30
Square	Fabric	Epoxy	15
Square	Braid	Vinyl ester	20

Table 10.3. Effect of tube shape and fibre arrangement on the specific energy absorption during axial loading of glass-fibre composites [7]

Shape	Fibre arrangement	Matrix	Specific energy absorbed (kJ kg^{-1})
Round	Mat	Unsaturated polyester	32–51
Round	Random mat and NCF	Unsaturated polyester/vinyl ester	35–77
Round	Chopped strand mat	Vinyl ester	62
Round	Unidirectional	Epoxy	58–76
Conical	Chopped strand mat	Epoxy	68
Square	Cloth	Unsaturated polyester	32
Square/hexagonal	Woven cloth	Polyamide	40–49
Square/I-beam	Unidirectional	Unsaturated polyester	36–44

Table 10.4. Specific absorbed energy of thermoplastic composites compared to SMC thermoset [6]

Fibre	Matrix	Specific energy absorbed (kJ kg^{-1})
Carbon-fibre fabric	Polyamide 12	90
Glass-fibre, unidirectional 0°	Polyamide 12	67
Glass-fibre unidirectional 90°	Polyamide 12	52
Aramid fabric	Polyamide 6	50
Glass-fibre knitted fabric	Polyethylene terephthalate	40
Glass mat reinforced thermoplastics 30%	Polypropylene	31
Glass-fibre SMC	Unsaturated polyester	24

1. freedom from maintenance;
2. benefits of lightweight – lighter constructions with high strength, smaller foundations, and faster erection;
3. renovation of existing bridges damaged by corrosion;
4. reduced transportation and erection costs;
5. factory construction is possible, with lightweight structures being transported for installation;
6. rapid machining and assembly is possible, leading to high-speed installation;
7. the correct selection of matrix resin provides resistance to corrosion and weathering; and
8. good thermal and electrical insulation.

10.5.2   Aberfeldy Footbridge

The Aberfeldy footbridge (Figure 10.13), which was a major advance in the large-scale application of composites in bridge engineering, has a main span of 63 m. It has a cable-stayed design using pultruded cellular glass-fibre construction for the bridge deck.

The deck employed an advanced composite construction system (ACCS) (Figure 10.14), while the Parafil cables use embedded aramid fibres (Kevlar) composite. Resistance to wheel loading of the surface deck is provided by two layers of cellular fibre-reinforced plastic arranged in orthogonal directions. The cellular voids are filled with structural foam.

The main span is 63 m long and 2.2 m wide, and is supported by 40 cable stays from two 17.5-m high towers. The cable stay structure is 113 m long [12]. A unique method of erecting the towers, cables, and deck was employed, which negated the need for heavy lifting gear. Because of low tolerances for the A-frames, the legs and cross beams were delivered to site preassembled. The lightweight construction meant that only a winch and telescopic forklift truck were required to lift the frames into the

Figure 10.13 The Aberfeldy, Scotland foot bridge constructed from glass-fibre composites [11].

Figure 10.14 The advanced composite component system using interlocking GRP pultruded sections [11].

final vertical position. The crossbeams were bonded together, then the cables and guys were attached to the ends of the crossbeams, which were positioned every 6 m across the river. A temporary 65 m ramp, from standard scaffolding, was used below the deck for bonding. It was essential to keep the components dry during curing, so a protective cover was placed over the ramp [11,13].

The alignment of the sections during the bonding needed to be accurate as tolerances of 1 mm m^{-1} in length could result in a significant horizontal misalignment and the deck missing its bearings. Suspended construction with incremental launching using a winch was employed to provide the construction accuracy.

10.5.2.1 Monitoring

The bridge has been monitored for more than 25 years, and the following points observed:

1. Mould and moss growth was a considerable problem. The composite materials are capable of absorbing 1.5% w/w. Poor detailing was responsible for the growth of mould, lichen, moss, and algae on the primary structure and parapets. Addition of mould inhibitors to the resin is a potential solution.
2. Weathering of the bridge is one of the major components of a bridge's durability. The erosion of the resin has exposed the glass-fibre reinforcement of the parapet and handrails.
3. The standard ACCS components performed well, showing little wear [14,15].
4. The bridge design did not account for loading of a golf cart or a small tractor, and the bridge was overloaded on several occasions. In 1997, strengthening was achieved by bonding GRP plates to the top side of the deck and GRP sheets to the deck-edge beams, near the stay connections.
5. Impact damage to hand railings after collision with a golf cart and protective kickboards may have improved the design.

10.5.3 Future Designs

Lessons learned over 25 years of examining the Aberfeldy bridge were applied to the Wilcott suspension bridge, which combined a GRP deck with steel cables and stainless steel parapets [11]. Alternative fibre-reinforced plastics (FRP) bridge decking systems, such as DuraSpan, have been developed [16].

The Storck Bridge in Winterthur benefited from the observations of Aberfeldy [17]. It was erected in 1996 over 18 rail tracks at the station, and has a central A-frame tower supporting two approximately equal spans of 63 and 61 m. The CFRP cables (Figure 10.15) consist of 241 wires, each with a diameter of 5 mm. They were subjected to a load three times larger than permissible for >10 million load cycles, which is much greater than expected during the bridge's life [18,19]. The cables are built from parallel bundles of carbon-fibre wire. These CFRP wires are corrosion-resistant, so corrosion-inhibiting grout is not required. To minimize the loss of strength, the wires are protected against wind erosion and ultraviolet radiation using a polyethylene (PE filled with carbon black) pipe. Two of the 24 cables are equipped with conventional sensors and also with state-of-the-art glass fibre-optic sensors to provide permanent monitoring of stress and strain. Fibre-optic Bragg gratings (FBGs) and electrical resistance strain gauges were surface-adhered to loaded wires and to

Figure 10.15 The Storck Bridge and CFRP stay cable consisting of 241 'wires' of 5 mm diameter. Cable load capacity: 12 MN [18].

unloaded dummy wires for temperature compensation. The average strain of a single CFRP wire is only 0.12%.

Two years later, for the Kleine Emme bridge near Lucerne, all the Bragg grating sensors were embedded in the CFRP wires during the pultrusion process. Some FBGs were pre-strained on dummy wires (not loaded) to a level of 0.25% to monitor creep resulting from delamination of the fibre coating or the epoxy adhesive.

10.5.3.1   The Anchorage of CFRP Cables

An important issue limiting the application of CFRP cables is the anchoring system. The outstanding mechanical performance of CFRP is in the longitudinal direction. The transverse properties, for example interlaminar shear, are relatively low, making it difficult to anchor CFRP wire bundles for full static and fatigue strength. Meier [18] reports the development of a casting material to fill the space between the steel cone and the CFRP wires. The casting material, which is referred to as a load transfer medium (LTM), has the following requirements:

1. maintenance of long-term static and fatigue strength of the CFRP wires;
2. avoid galvanic corrosion between the carbon fibre and the steel cone;
3. the LTM must be an electrical insulator; and
4. the cone should provide radial pressure to support the interlaminar shear strength of the CFRP wires.

To avoid high shear stress concentration on the surface of the CFRP wire at the entrance to the anchoring system, which could result in pull-out of the wire, a graded filled epoxy resin has been developed.

The LTM has a low modulus but continuously increases to a maximum at the foot of the anchor. The LTM is composed of alumina (Al_2O_3) granules of average diameter 2 mm. To achieve a low modulus, the granules are coated with a thick layer of pre-cured epoxy resin. For a medium modulus, the granules are coated with a thin layer of pre-cured resin. For a high modulus, uncoated granules are used. The graded fillers are placed in the socket infused with epoxy resin by vacuum-assisted RTM (VARTM). In this way, the graded modulus of the LTM can be tailor-made.

Fatigue tests performed on cables anchored by this system, at EMPA in Switzerland, confirmed the superior performance of CFRP under cyclic loading [19].

10.5.4   Future Developments

The CFRP cables are five times lighter than those made from steel, with higher strength. A high strength to low weight ratio will enable bridges to be built in future with considerably longer spans than are currently possible.

Long-span footbridges using CFRP have been designed with spans of 200–300 m. Designs omitting intermediate support structures become practical when the high-modulus carbon-fibre reinforcement is fully used. By avoiding masts and cables and use of durable fibre-reinforced structures, the through-life costs are significantly reduced. The initial costs associated with the foundations and installation of conventional bridge structures can be reduced when employing lightweight designs [12]. However, the supply and cost of carbon fibres still limits their incorporation in bridge design. Future supply of cheaper carbon fibres manufactured from alternative precursors is being researched. There are designs that explore the use of recycled composite materials or recovered fibres. Until that time, these modern designs will be limited by the supply and availability of carbon fibres. As with other applications, there is much benefit from the use of thermoplastic matrices and the development of processing of thermoplastics for bridge structures is predicted.

10.5.5 Advantages and Disadvantages of Carbon-Fibre Composites in Bridge Structures

Advantages

1. high specific strength and stiffness;
2. resistance to corrosion and stress corrosion;
3. outstanding performance under fatigue loading;
4. inhibited relaxation;
5. lightweight: stiffness to weight ratio enables longer stays; enables extremely long-span bridges; reduced cable sagging; and
6. reduced need for regular replacement and lower maintenance than steel hanger cables in suspension bridges.

Disadvantages

1. high initial costs limits their introduction;
2. reduced carbon-fibre price would not necessarily help since CFRP cables can only compete when the whole-life costs are considered. Use of CFRP strip and sheet bonding techniques for rehabilitation of structures has been implemented using cost-effectiveness criteria.

10.5.6 Bridge Across the Straits of Gibraltar [20]

There has been a proposal for a cable-stayed CFRP bridge across the Straits of Gibraltar [20]. To avoid compression forces in the deck, a cable-net concept was introduced, using large CFRP main cables integrated into the deck and anchored to the ground at the abutments. This can only be achieved with advanced composites, especially with CFRP for the cables, comprising $\approx 70\%$ w/w of the superstructure, where their high performance can be used to full advantage.

10.6 Bridge Repair

Following the earthquakes in California in 1994, when 57 people were killed and 8,700 injured, and which caused $13–40 billion worth of damage, interest in using composite rehabilitation systems intensified.

Carbon fibres are best suited for bridge repair because of their resistance to the alkaline environment, associated with cement and concrete, and, in contrast to glass fibres, stress corrosion. Adhesive-bonded external CFRP strips or sheets can replace steel plate reinforcement.

The advantages are:

1. overall cost savings arising from this simple strengthening method –lower costs and cost efficiency, both initial and over the life cycle.
2. corrosion free;
3. on-site handleability: easy transport and lifting (1 kg CFRP strips can replace 30 kg steel plates for equivalent strength); and avoids the need for support during curing of the adhesive as with steel plates. Prepreg can be simply rolled onto the repair;
4. reels of prepreg are endless, so joints are not necessary;
5. in contrast to steel plates, CFRP strips subjected to compressive stresses resist low-load debondment and increase flexural and shear strength; and
6. minimal disruption to the bridge function: overhead clearance is not reduced and installation requires less time and labour.

10.6.1   Strengthening with Non-tensioned CFRP Strips

Shear crack formation may lead to peeling of the composite reinforcing strip. Flexural cracks are spanned by the CFRP strip and do not influence the load capacity. Post-strengthening strips result in a finer distribution of cracks. Load capacity of post-strengthened beams has also been shown to be unaffected by 100 frost cycles from $+20\,°C$ to $-25\,°C$, indicating that the introduction of thermal stresses is not significant.

Peeling of the CFRP strips or interlaminar shear failure within the strips are potential failure modes. These are related to the cohesive strength of the adhesive: interfacial bond strengths between the CFRP strip and the adhesive and between the concrete and the adhesive.

10.6.2   Strengthening with Pre-tensioned Strips

It can be advantageous to provide pre-stressing to the flexural strengthening strips since the durability of the bridge structure can be improved and shear failure of the concrete in the tension zone can be avoided.

The pre-tensioning load needs to be carefully controlled. If too large, high shear stresses develop in the concrete adjacent to the CFRP strip in the end regions, where careful design and construction is required. Without end anchoring, the CFRP strips will debond under shear at relatively low values of pre-stress. For technical and economic reasons, a higher pre-stress is essential; therefore, end-anchoring systems have been developed.

In contrast to pure shear strengthening, the advanced composites that wrap around the flexural strips must definitely be pre-stressed. This will build up the maximum multi-axial stress in the concrete. In this way, failure at the two ends of the CFRP strips can be avoided. Two application technologies are suggested:

1. *Pultruded CFRP strips* require good preparation for good adhesion. The outermost matrix-rich layer needs to be removed to expose the fibres and ensure that the surface is clean. The surface skin of the concrete should be removed by sandblasting or related technique to expose the aggregates to ensure good bonding of the epoxy adhesive.
2. *Sheets (wet lay-up)* for post-strengthening are best suited for wrapping of columns with a rectangular cross-section. A minimum radius of curvature of ≈25 mm should be employed. These sheets are impregnated flexible fabrics and also require careful surface treatment for good bonding. Further details are given elsewhere [16,21].

10.6.3   Bridge Pillars and Columns Rehabilitation

Fibre-reinforced plastics composites are used to considerably increase strength and ductility without increasing stiffness. In seismic retrofit applications this technique can help to prevent the need to retrofit other parts of the structure. Sheets of FRP, such as prepreg, have been widely used for repairing and strengthening reinforced concrete members over the last 20 years. The advantages of high strength, low weight, and workability have promoted these applications:

1. Column strengthening: retrofit columns with premature termination of longitudinal concrete reinforcement.
2. Retrofitting of beam–column joints: bridge columns are most vulnerable to seismic activity and can be reinforced by jacketing the plastic hinge circumferentially.
3. Retrofitting of reinforced concrete beams: composite solutions are an attractive alternative to steel and other conventional techniques for seismic retrofitting or strengthening of bridges and buildings because of these advantages:
   1. high specific stiffness;
   2. high specific strength;
   3. their mechanical properties can be customized to the application;
   4. ease of handling and installation, avoiding the need for heavy equipment;
   5. high corrosion resistance – FRP composites can protect the inner reinforcement from rusting in harsh environments;
   6. resistance to extreme environmental conditions and temperatures;
   7. the durability of FRP ensures reduced maintenance costs; and
   8. they offer economic and viable solutions.

The construction industry has used traditional materials for hundreds of years, but the introduction of composite materials has reshaped the technologies employed. In particular, for retrofitting and rehabilitation, composite materials have proved superior for seismic retrofitting [22,23]. Historical structures are also benefiting from the application of composite materials in restoration and preservation. The preservation of old masonry using CFRP and confining cross-ties has been validated [24,25].

While seismic rehabilitation and protection of historic buildings is complex, the judicious use of FRP is a cost-effective technique for preserving our cultural heritage and architecture.

10.6.3.1   Concrete Confined with FRP Tubes

Under an axial compressive load, concrete expands laterally because of its positive Poisson's ratio of 0.278 (concrete based on ordinary Portland cement). By encasing a concrete column in a shell, such as a tube, the lateral expansion is resisted. Confinement of this nature modifies the stress–strain response of concrete and increases its compressive strength.

The improved performance of concrete encased in steel tubes is well recognized. FRP tubes for encasing concrete columns have the advantage of eliminating corrosion of the confining tube. Their low weight provides easy handleability.

The confining pressure of the tube puts the concrete into a triaxial stress state. The stiffness of the FRP tube contributes to the constraint, which prevents the shell from buckling inwards. The shell also protects the concrete surface from physical damage and environmental damage, such as carbonation and chloride penetration.

Concrete confined with FRP is a technically attractive system for piles, overhead highway signs, and other structures under compression. The confinement of concrete with a GFRP or CFRP tube has been modelled, with good predictions, and confirms that concrete confined with GFRP exhibits better ductility [22,23].

10.6.3.2   Seismic Performance of FRP-Retrofitted Lap-Spliced Columns

Considerable research has been directed at developing and applying FRP retrofit strategies to upgrade the seismic performance of deficient lap-spliced columns [22]. To postpone the onset of splitting of deficiently lap-spliced reinforcements and to reduce the severity of the subsequent deterioration, a hoop strain of 1000 $\mu\varepsilon$ or 0.1% is appropriate for the design of the composite jacket for circular and rectangular columns.

For FRP-retrofitted circular columns, FRP jackets designed for a jacket strain of 0.1–0.4% provide a minimum confinement pressure of 1–2.0 MPa within the lap-splice zone. Composite jackets for columns show a significant improvement in their cyclic performance.

On the other hand, the response of square-jacketed columns and those with a quasi-circular cross-section with continuous confinement had a very limited improvement in clamping on the lap-splice zone. Direct application of the FRP or a steel jacket to large rectangular reinforced concrete columns is ineffective [22]. To improve the confinement efficiency of direct CFRP jackets, a carbon-fibre steel retrofit method has been proposed by Chang et al. [26]. In this method, steel plates are attached before over-wrapping with FRP, which increases the confinement stress and energy dissipation capacity. The steel jacket may have either an elliptical or octagonal shape.

10.6.3.3 Seismic Performance of FRP-Retrofitted Columns with Shear Deficiency

In recent years, due to some of the excellent properties of fibre-reinforced polymers, retrofitting structures with FRP has drawn increased attention, especially for improving the shear strength and ductility of reinforced concrete columns. To avoid shear failure, horizontal FRP wrapping should limit the dilation of the column in the loading direction to a strain of $<0.4\%$.

It was observed that one CFRP layer was sufficient to increase the theoretical strength of the column, but insufficient to change the failure mode from shear to flexural. However, 2.5 and 4.5 layers were shown to change the failure mode to flexural and induce a ductile failure.

10.6.3.4 Seismic Performance of FRP-Retrofitted Columns with Flexural Deficiency

Continuous fibres have been used in a circumferential direction to confine the columns, since this can improve the inelastic deformation capacity of flexural plastic hinge regions. The thickness of the jacket should be designed accordingly [27].

10.7 Energy Generation: Wind Turbine Blades

The main requirements for wind turbine blades are:

1. resistance to extreme wind load and gravitational load, achieved with high-strength materials and design;
2. stable, optimal, aerodynamic blade shape and orientation and clearance between blade and tower, achieved with high-stiffness structure;
3. high fatigue resistance and reliability for >20 years and 10^8 cycles; and
4. low weight, to reduce the load on the tower.

Materials with high strength, fatigue resistance, and stiffness can meet these requirements. Composites are preferred as other materials, such as metals, alloys, or wood, do not meet these requirements fully [28].

10.7.1 Construction, Loads, and Requirements

The blades and nacelles of wind turbines can be manufactured from composite materials. A nacelle provides the mechanical components with weather protection. Thus, low weight, high strength, and corrosion resistance are essential requirements of the materials. Typically, they are made from glass-fibre composites. Turbine blades have more critical requirements, as detailed above, determining performance and lifetime. Early installations exhibited a failure rate of $\approx20\%$ within three years [29]. The early blades used hand lay-up manufacture, so to increase reliability and lifetime the manufacturer needed to employ more accurate fibre placement. Adaptation of aerospace understanding while maintaining economic processing was essential.

A blade consists of two faces (the low-pressure and high-pressure sides), joined and stiffened by either integral (shear) webs or a box beam (box spar with shell fairings). Figure 10.16 is a schematic of the construction of a turbine blade consisting of two shells and two shear webs [30].

Figure 10.17 shows a cross-section though a blade illustrating the loaded and adhesive regions of a blade. The tension–tension, compression–compression, and compression–tension stressed regions of the structure dictate the choice and orientation of the fibre reinforcement. A box beam or spar can provide the additional stiffness required for longer blades. In this case, the box beam or spar are adhered to the shells and choice of adhesives is critical to the blade life.

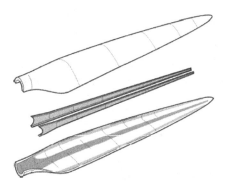

Figure 10.16 Wind turbine rotor blade manufacture by assembling and bonding two shells and two shear webs. The grey areas indicate the main load-bearing composites [30].

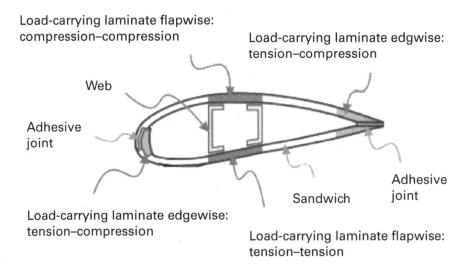

Figure 10.17 Cross-section through a wind turbine blade [30].

Wind turbine blades are subject to external loads resulting from:

- flap-wise bending;
- edge-wise bending;
- gravity and loading;
- inertia forces;
- pitch acceleration; and
- torsion.

Flap-wise loads mainly result from the wind pressure, while edge-wise loads arise from gravitational and torque loads. The main edge-wise bending moment occurs at the blade root. These bending loads place high longitudinal, tensile, and compressive stresses onto the material. The up-wind side of a blade is under a tensile stress, and the downwind side is put into compression. The flap-wise and edge-wise bending loads are responsible for the growth of damage through fatigue. Cyclic loads also result from variations in wind speed, turbulence, and air pressure around the tower. Table 10.5 summarizes the roles of the component parts of wind turbine blades in maintaining the blade shapes.

The blades are mainly manufactured with multi-axial fabrics: biaxial $\pm 45°$ laminates are generally used for the blade skins and the shear webs; triaxial materials of $\pm 45°/90°$ are used in the root area; unidirectional composites, with some biaxial plies, are employed for the spar caps. Figure 10.18 illustrates a typical fibre orientation employed in the shells, shear webs, and box section. Accurate fibre orientation has an essential role in providing good performance and durability of a blade.

The quality of the manufactured structure was improved by the use of vacuum infusion and prepreg-based processes. For example, the Vestas Company (Denmark) employs prepreg technology for the manufacture of blades.

The most widely employed processes is resin infusion, such as RTM and VARTM, while SCRIMP (Seemann composites resin infusion moulding process) is employed

Table 10.5. Functions of components for maintaining the shape of wind turbine blades [28]

Component	Function	Choice of materials
Blade shell	Maintenance of blade shape; resist wind and gravitational forces	Strong, lightweight composites
Unsupported areas of the shell	Resist buckling loads	Sandwich structures with lightweight cores and multidirectional fibre laminates
Integral web, spars, or box beam	Resist buckling of the shell and shear stresses arising from flap-wise bending	Biaxial lay-ups with fibres at $\pm 45°$
Adhesives for composite laminates with the web and the blade shell	Maintain out-of-plane strength and stiffness of the blade	Optimum adhesives for the chosen matrix

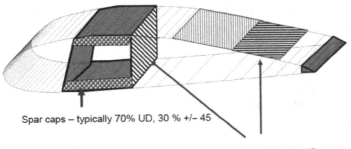

Spar caps – typically 70% UD, 30 % +/– 45

Skins and shear webs, +/– 45

Figure 10.18 Cross-section of a wind turbine blade showing a typical choice of fibre orientation in selected composite areas [31].

when non-crimp fabric reinforcements are utilized with specific fibre orientation. Resin infusion is a lower-cost process than employing prepreg technology, whereas the latter achieves structures with less variability in fibre orientation and volume fraction and hence control over mechanical properties. Prepreg enables a better choice of matrix resins but also lends itself to a higher degree of automation. The employment of 3D woven and glass, carbon, and hybrid fibre textile reinforcements is being examined and it has been demonstrated that spar caps with improved performance and lower weight can be made.

The unsupported areas of the blade shell are manufactured from sandwich composites, which ensure the shape stability of the blade shell. The sandwich structures provide much higher flexural stiffness to simple laminates, where the loads are mainly transverse to the fibres. Typically, polymer foams, balsa wood, or in some cases honeycomb structures such as those manufactured from Nomex fibre composites are used as sandwich core materials.

10.7.2   Wind Turbine Power

Wind power depends on wind speed. On excessively windy days blade rotation cut-off occurs at a speed of ≈80 km h^{-1} [28,32].

The power of a wind turbine is given by

$$P = \frac{1}{2}\eta\rho Av^3 = \frac{1}{2}\eta\rho\pi r^2 v^3, \tag{10.3}$$

where P is performance of the wind turbine; η is efficiency of the wind turbine ($\eta = \frac{E_{out}}{E_{in}}$); E_{out} is energy out; E_{in} is energy in; ρ is air density (1.2 kg m^{-3}); A is the swept area of wind turbine ($A = \pi L^2$); L is the length of the blade (in effect the radius (r) of the swept circle); and v is the wind velocity.

We see from eq. (10.3) that the power output from a turbine is proportional to the square of the length of the turbine blade:

$$P \alpha L^2. \tag{10.4}$$

Therefore, for high power generation we need to maximize the length of the turbine blades. To achieve a long blade that meets the requirements given above, high-modulus fibres, such as carbon, need to be used for blade construction. The higher specific stiffness of carbon fibres and their composites helps to offset the weight penalty associated with increasing blade length. The development of blade design to accommodate the use of carbon fibres is underway and represents a future opportunity for the industry.

10.8 Conclusions

The aim of these examples of composite applications is to demonstrate how the use of composite materials has developed over recent decades and to identify future uses. One of the important issues is the demand for carbon fibres for use in the civil engineering and automotive sectors. Without research and development into new fibre-spinning options, the demand created by these successes is limited by the available supply and the need for lower-cost precursor systems for carbon fibres.

10.9 Discussion Points

1. Consider each example in turn and identify the principle reasons for the use of composites in each application.
2. Utilize your conclusions to refresh your understanding of composite micromechanics and choice of fibres and resin matrix, as presented elsewhere.

References

1. D. H. Middleton, Case histories. In *Composite materials in aircraft structures*, ed. D. H. Middleton (Harlow: Longman, 1990), pp. 228–390.
2. www.classroom.materials.ac.uk/images/heli
3. Dowty Propellers, http://dowty.com/feature-stories/dowty-80-years/attachment/dowtypropellers-80year-anniversary-story-photo3.
4. Dowty Propellers, http://dowty.com/feature-stories/dowty-80-years/attachment/dowtypropellers-80year-anniversary-story-photo4
5. K. Mason, CFRP module saves weight on rocket design. *Composites World* **5** (2019), 44–47.
6. K. Friedrich and A. A. Almajid, Manufacturing aspects of advanced polymer composites for automotive applications. *Appl. Compos. Mater.* **20** (2013), 107–128.
7. D. H.-J. A. Lukaszewicz, Automotive composite structures for crashworthiness. In *Advanced composite materials for automotive applications: structural integrity and crashworthiness*, ed. A. Elmarakbi. (Chichester: Wiley, 2013), pp. 99–127

8. D. Hu, Y. Wang, L. Dang, and Q. Pan, Energy absorption characteristics of composite tubes with different fibres and matrix under axial quasi-static and impact crushing conditions. *J. Mech. Sci. Technol.* **20** (2018), 2587–2599.

9. D. Hull, A unified approach to progressive crushing of fibre-reinforced composite tubes. *Compos. Sci. Technol.* **40** (1991), 377–421.

10. D. Y. Hu, M. Luo, and J. L. Yang, Experimental study on crushing characteristics of brittle fibre/epoxy hybrid composite tubes. *Int. J. Crashworthiness* **15** (2010), 401–412.

11. J. M. Skinner, A critical analysis of the Aberfeldy footbridge, Scotland. In *Proceedings of bridge engineering 2 conference, 2009* (Bath: University of Bath, 2009).

12. J. T. Mottram and J. Henderson, ed., *Fibre-reinforced polymer bridges: guidance for designers* (London: CIRIA and Composites UK, 2018).

13. W. J. Harvey, A reinforced plastic footbridge, Aberfeldy, UK. *Struct. Eng. Int.* **4** (1993), 229–232.

14. T. Stratford, The condition of the Aberfeldy footbridge after 20 years in service. Paper presented at Structural Faults and Repair 2012, Edinburgh (2012).

15. J. Cadei and T. Stratford, The design, construction and in-service performance of the all-composite Aberfeldy footbridge. In *Advanced polymer composites for structural applications in construction*, ed. R. Shenoi, S. Moy, and L. Holloway (London: Thomas Telford, 2002), pp. 445–455.

16. T.A. Hoffard, L. J. Malvar, *Fiber-reinforced polymer composites in bridges: a state-of-the-art report*, Naval Facilities Engineering Service Center (NAVFAC) Technical Memorandum TM-2384-SHR (2005).

17. U. Meier, The development of composites and their utilisation in Switzerland. In *Advanced composites in bridge construction and repair*, ed. Y. Kim (Cambridge: Woodhead Publishing, 2014), pp 23–31.

18. U. Meier, Carbon fiber reinforced polymer cables: why? Why not? What if? *Arab. J. Sci. Eng.* **37** (2012), 399–411.

19. U. Meier, Carbon fiber reinforced polymers: modern materials in bridge engineering. *Struct. Eng. Int.* **2** (1992), 7–12.

20. U. Meier, Proposal for a carbon fibre reinforced composite bridge across the Strait of Gibraltar at its narrowest site. *Proc. Instn. Mech. Engrs. B2* 201 (1987), 73–78.

21. U. Meier, Composite materials in bridge repair. *Appl. Compos. Mater.* **7** (2000), 75–94.

22. M. F. M. Fahmy, Z. Wu, and G. Wu, Post-earthquake recoverability of existing RC bridge piers retrofitted with FRP composites. *Construct. Building Mater.* **24** (2010) 980–998.

23. Y.-K. Yeh and Y. L. Mo, Shear retrofit of hollow bridge piers with carbon fiber-reinforced polymer sheets. *J. Compos. Construct.* **9** (2005), 327–336.

24. S. S. Sivaraja, T. S. Thandavamoorthy, S. Vijayakumar, S. M. Aranganathan, and A. K. Dasarathy, Preservation of historical monumental structures using fibre reinforced polymer (FRP): case studies. *Procedia Engineering* **54** (2013) 472–479.

25. P. G. Asteris and V. Plevris, *Handbook of research on seismic assessment and rehabilitation of historic structures* (Hershey, PA: Engineering Science Reference, 2015).

26. K. C. Chang, L. L. Chung, B. J. Lee, et al. Seismic retrofit study of RC bridge columns. Technical Report, National Center for Research on Earthquake Engineering, Taiwan (2000).

27. P. Sarker, M. Begum, and S. Nasrin, Fiber reinforced polymers for structural retrofitting: a review. *J. Civil Eng.* **39** (2011), 49–57.

28. L. Mishnaevsky Jr., Composite materials in wind energy technology. In *Thermal to mechanical energy conversion: engines and requirements*, ed. O. N. Favorsky (Paris: UNESCO-EOLSS).
29. R. Richardson, New design tool improves manufacture of composite wind turbine blades. *Power* **154** (2010), 62–65.
30. L. Mishnaevsky Jr., K. Branner, H. N. Petersen, et al., Materials for wind turbine blades: an overview. *Materials* **10** (2017), 1285–1309.
31. P. Hogg, Wind turbine blade materials, SUPERGEN wind phase 1 final assembly. Presentation at University of Loughborough (2010).
32. R. Szabó and L. Szabó Composite materials for wind power turbine blades. In *Proceedings of EXPRES 2017* (2017), pp. 44–50.

Index

Printed in the United States
by Baker & Taylor Publisher Services